ANALYSIS WITH ULTRASMALL NUMBERS

TEXTBOOKS in MATHEMATICS

Series Editors: Al Boggess and Ken Rosen

PUBLISHED TITLES

ABSTRACT ALGEBRA: AN INQUIRY-BASED APPROACH
Jonathan K. Hodge, Steven Schlicker, and Ted Sundstrom

ABSTRACT ALGEBRA: AN INTERACTIVE APPROACH
William Paulsen

ADVANCED CALCULUS: THEORY AND PRACTICE
John Srdjan Petrovic

ADVANCED LINEAR ALGEBRA
Nicholas Loehr

ANALYSIS WITH ULTRASMALL NUMBERS
Karel Hrbacek, Olivier Lessmann, and Richard O'Donovan

APPLYING ANALYTICS: A PRACTICAL APPROACH
Evan S. Levine

COMPUTATIONS OF IMPROPER REIMANN INTEGRALS
Ioannis Roussos

CONVEX ANALYSIS
Steven G. Krantz

COUNTEREXAMPLES: FROM ELEMENTARY CALCULUS TO THE BEGINNINGS OF ANALYSIS
Andrei Bourchtein and Ludmila Bourchtein

DIFFERENTIAL EQUATIONS: THEORY, TECHNIQUE, AND PRACTICE, SECOND EDITION
Steven G. Krantz

DIFFERENTIAL EQUATIONS WITH MATLAB®: EXPLORATION, APPLICATIONS, AND THEORY
Mark A. McKibben and Micah D. Webster

ELEMENTARY NUMBER THEORY
James Kraft and Larry Washington

ELEMENTS OF ADVANCED MATHEMATICS, THIRD EDITION
Steven G. Krantz

EXPLORING LINEAR ALGEBRA: LABS AND PROJECTS WITH MATHEMATICA®
Crista Arangala

TEXTBOOKS in MATHEMATICS

ANALYSIS WITH ULTRASMALL NUMBERS

Karel Hrbacek
The City College of New York, USA

Olivier Lessmann
Collège Rousseau, Geneva, Switzerland

Richard O'Donovan
CEC André-Chavanne, Geneva, Switzerland

CRC Press
Taylor & Francis Group
Boca Raton London New York

CRC Press is an imprint of the
Taylor & Francis Group an **informa** business

A CHAPMAN & HALL BOOK

CRC Press
Taylor & Francis Group
6000 Broken Sound Parkway NW, Suite 300
Boca Raton, FL 33487-2742

© 2015 by Taylor & Francis Group, LLC
CRC Press is an imprint of Taylor & Francis Group, an Informa business

No claim to original U.S. Government works

Printed on acid-free paper
Version Date: 20141008

International Standard Book Number-13: 978-1-4987-0265-2 (Hardback)

Visit the Taylor & Francis Web site at
http://www.taylorandfrancis.com

and the CRC Press Web site at
http://www.crcpress.com

Contents

Preface

Infinitesimals then and now

The first calculus textbook, *Analyse des Infiniment Petits* by the Marquis de L'Hôpital, was published in 1696. As the title indicates, the presentation was based on "infinitely small" or "infinitesimal" quantities, introduced by Gottfried Wilhelm von Leibniz, one of the co-discoverers of calculus. For one hundred and fifty years Leibniz's method of infinitesimals served as the standard way of doing calculus, in preference to Isaac Newton's method of fluxes. It reached high sophistication in the hands of masters such as the Bernoulli brothers and Leonhard Euler. From its inception it was also criticized for the lack of firm foundations (as was Newton's method). Bishop Berkeley [2] famously pointed out the logical discrepancies that appear when dividing by nonzero quantities on the one hand, but then ignoring them in the results as though they were "ghosts of departed quantities" on the other hand.

The work of nineteenth century mathematicians, in particular of Augustin-Louis Cauchy and Karl Weierstrass, succeeded in giving a rigorous treatment of Newton's approach, culminating in the concept of limit defined by the now classical epsilon–delta method. As a result, infinitesimals disappeared from modern mathematical texts. The rigorous foundations provided by the epsilon–delta method enabled an unprecedented flowering of mathematical analysis. Nevertheless, physical scientists have been reluctant to give up on the simplicity and intuitive appeal of infinitesimals, which still persist in some form in contemporary scientific thinking.

A rigorous theory of infinitesimals consistent with the contemporary understanding of mathematical analysis was established in 1960 by Abraham Robinson. His book *Nonstandard Analysis* [24] provided paraphrases of many classical arguments, as well as numerous new results. At the research level, Robinson's methods have found significant applications in analysis, number theory, mathematical physics and other areas of pure and applied mathematics. The underlying framework of nonstandard analysis is model–theoretic, usually based on ultraproducts or superstructures, concepts unsuitable for elementary level exposition; see Goldblatt [4] or Vakil [28] for an excellent graduate level introduction to nonstandard analysis. Even at the research level, the need to invoke model theory is a bothersome distraction from the essential ideas.

In mathematical education the abandonment of infinitesimals had perhaps the greatest impact. The epsilon–delta definition of limit and the proofs based on this definition are just too complicated for an average student to master quickly, if ever. As a result, rigor has disappeared from many modern introductory calculus courses. They are usually taught in a way that leaves the basic concepts undefined and the fundamental theorems unproved. This "faith-based" approach runs counter to the conception of mathematics as a rigorous deductive science that one tries to convey to students in high school algebra and geometry. Many teachers (and some students) are justifiably bothered by this state of affairs.

Some attempts to teach elementary calculus using nonstandard analysis have been made; two nice calculus textbooks in this vein are Keisler [16] and Stroyan [27]. The model–theoretic prerequisites have been circumvented by an axiomatic treatment of an extension of the real number field called the *hyperreals*. Yet it seems fair to say that these attempts have not been as successful as the intuitive simplicity of the concept of infinitesimal would lead one to expect. The third author tried to teach elementary calculus using Keisler's book [16]; this experience and the pedagogical difficulties it uncovered are described in [17]. Besides the need to learn a new non-Archimedean number system while students still struggle to adequately understand the real numbers, there is the need to distinguish between internal and external objects and the potential of the latter to provide distracting, pathological examples. There is also the fact that the infinitesimal definitions of the basic concepts of calculus (derivative, limit, integral) apply only to standard objects. The epsilon–delta definition is still needed to make sense of, say, $f'(x)$ when either x or f is not standard.

Axiomatic nonstandard set theories have been proposed as a way to make nonstandard methods more accessible. Such theories were introduced in the mid-1970s independently by the first author [7, 8], Edward Nelson [18] and Petr Vopěnka [29]; we refer the interested reader to Kanovei and Reeken's comprehensive monograph [14]. Nelson's theory **IST** has found a significant following; see Robert [23] for a nice exposition. The axiomatic framework alleviates some of the pedagogical difficulties of the model–theoretic approach. In the simpler theories, like **IST** or its bounded variant **BST**, there are no external objects and no hyperreals. However, all these axiomatic approaches still have a significant "overhead" of logical formalism. Also, the fixed division of mathematical entities into "standard" and "internal" postulated by these theories means that the last difficulty mentioned above, to wit, that the infinitesimal definitions of the calculus concepts apply only to standard objects, remains in full force (see [19] and [10] for a fuller discussion of this point).

Following an idea of Guy Wallet, Yves Péraire in a series of papers beginning in 1989 ([21] is the most fundamental) developed an axiomatic nonstandard set theory **RIST**, where the notion of "standard" (and, consequently, also of "infinitesimal") is *relative*; every mathematical entity can be regarded as "standard" when viewed in the context of its own appropriate universe. The first author in [9] and [11] strengthened the axioms of Péraire's theory (axiomatic set theories **FRIST** and **GRIST**) and simplified its formalism.

About this book

The theory on which this book is based is a fragment of the bounded version of **RIST** (**RBST**; see the Appendix). It is a result of a long series of simplifications and modifications influenced by classroom experience over a period of ten years. Since the word "standard" in common usage, and even in nonstandard analysis, has an absolute connotation: "usual, ordinary, traditional, prevailing," we use "*observable*" for the relativized version of the concept. Every mathematical object can be regarded as observable relative to a suitable context. The fundamental Principle of Stability asserts, roughly speaking, that objects have exactly the same properties relative to any context where they are observable. In particular, relative to any context there are infinitesimal and infinitely large real numbers (we call them *ultrasmall* and *ultralarge* numbers, respectively, for reasons explained in the Introduction).

A major advantage of the relative approach is that the infinitesimal definitions (of derivative, limit and so on) apply uniformly to all functions and all of their arguments; thus there is no need for the epsilon–delta mechanism. One can completely eliminate it from elementary calculus if one so desires.

An important feature of our approach is the *contextual notation*: notions that depend on the context, such as "observable," "ultrasmall" and "ultralarge," are understood to be relative to the context of the theorem, definition or proof in which they are mentioned (unless explicitly stated otherwise). In conjunction with the Stability Principle, this convention minimizes the need to pay explicit attention to the context and greatly simplifies the presentation. The presentation is axiomatic, based on six principles. The Existence Principle and the Relative Observability Principle set up the basic structure of observability. The Closure Principle asserts, in effect, that objects definable from observable parameters are observable, and the Observable Neighbor Principle asserts that every real number that is not ultralarge has to be ultraclose to some observable real number. The last two important principles, Stability and Definition, are rarely appealed to explicitly; they provide the background justification for the contextual convention.

Of course, we do not expect students (or even trained mathematicians) to prove theorems formally from the axioms. Some intuitive representation of what the axioms are about is necessary. There are in fact two ways to view the axioms (of any nonstandard set theory) intuitively. In the *internal view*, advocated by Nelson, the numbers and sets of the theory are regarded as the usual sets and numbers we are all familiar with. In this view, no new objects are added to the usual mathematical universe; it is only the language that is being extended. The standardness (or, observability) predicate is a linguistic device that singles out some of the familiar objects for special attention. This idea is attractive to those who can reconcile their view of natural numbers with the existence of properties that do not satisfy the Principle of Mathematical Induction. Such properties can be expressed in the extended language; for example, "x is standard" is such a property: 1 is standard; if n is standard, then $n + 1$ is standard, but not all natural numbers are standard. We had originally presented the material in this book from the internal point of view (see [13]). It works quite well in the classroom, but it seems that most mathematicians find it incompatible with their ideas about natural numbers.

The alternative is the "standard view" proposed in [7]. In this view, which is adopted in this book, we identify the *standard* (observable in every context) sets with the familiar sets of traditional mathematics. But these sets are seen as also having a plethora of nonstandard, ideal elements, such as the infinitesimal and infinitely large elements of the set $\mathbb{R}$. See Section 1.1 for more details.

Admittedly, this picture still represents a change from the traditional view in which there are no infinitesimals in $\mathbb{R}$, but we think that it should be more easily acceptable. An important point is that the two views differ only philosophically. They are concerned with the intuitive interpretation; the actual mathematics is the same in either view.

The book develops the usual topics from calculus of one real variable. The presentation is based on ultrasmall numbers. It demonstrates that mathematics with ultrasmall numbers can be practiced in a style that is just as informal and natural as the traditional treatments, but with important advantages. Use of ultrasmall numbers is more intuitive and it disposes with the epsilon–delta machinery and with the associated bookkeeping. The proofs become simpler and more focused on the "combinatorial" heart of arguments. Fundamental results, such as the Extreme Value Theorem, can be fully proved from the axioms immediately, without the need to master notions of supremum or compactness. As a result, calculus can be presented as *mathematics*—with proofs—even at a student level where vague arguments about "approaching" have become the norm. Derivatives and definite integrals can be developed before limits, and independently of each other. The relative framework allows arguments involving two or more levels of observability simultaneously. (This is a feature not easily available in the Robinsonian or Nelsonian framework. It simplifies many proofs, especially where double limits are involved.) A rigorous theory of ultrasmall and ultralarge numbers also enables the construction of entirely new models of mathematical and physical phenomena.

Intended audience

In this book, perhaps for the first time, definitions and arguments involving infinitesimals are presented in a style that is both as informal and as rigorous as is customary in standard textbooks of introductory analysis. We eschew both the ultraproduct construction of the model–theoretic nonstandard analysis and the excessive formalism of the axiomatic approaches. This should make the book of interest to a wide audience of mathematically minded readers—mathematicians, teachers of mathematics at high school or college level, scientists and philosophers of mathematics—anybody looking for a simple but rigorous introduction to infinitesimal methods. Although some preliminary acquaintance with calculus would be helpful, the actual prerequisites do not go beyond high school algebra, geometry and trigonometry, making the book, especially Part I, accessible as an independent reading to ambitious beginning calculus students.

This is also the first time that an exposition of the relative framework for nonstandard analysis (allowing many levels of standardness) is given in a book format; until now, it has been available only in research papers. Thus perhaps even experts on nonstandard analysis will find here something of interest.

Our hope for the most significant impact of the book is in the teaching of introductory calculus at the high school or college level. We started this project in response to the high school syllabus of the canton of Geneva (Switzerland), where two of us teach, and which requires courses in calculus (as well as other mathematical subjects) to be taught in the standard mathematical fashion: definition, example, theorem and proof with a reasonable degree of rigor. This turned out to be impossible to do with the traditional epsilon–delta method. Our approach was developed explicitly to satisfy this requirement. It has been used in two Geneva high schools for the last ten years by up to as many teachers, and repeatedly and extensively modified in response to the classroom experience. It has been successful in remedying the situation: It provides simpler definitions for the basic concepts, allowing students to form a good intuition and actually prove things by themselves. Moreover, this approach does not require any additional "black boxes" once the initial axioms have been presented. Many theorems can be proved simply, without resorting to difficult concepts like compactness or completeness. The track record of former students is very encouraging. Those of our students who had to take a course in analysis during their first year at the university all passed the exam. They report no particular difficulties with switching to the standard epsilon–delta method at the university level, having had to work rigorously in analysis before. This contrasts with students exposed to the informal standard method, who encounter rigor in analysis for the first time at the university level. A report on an earlier stage of this project has been published in [20].

For teachers of mathematics who wish to present calculus at an introductory college level, or even high school, with at least some proofs, the text can serve as a reference and a sourcebook of ideas for such a course. This should be of particular interest in countries where proofs are part of the syllabus from the onset, such as Switzerland, France and others. At the introductory level one would aim to cover only some of the material in Part I. In particular, the technical aspects of the Closure and Stability Principles in Chapter 1 can be de-emphasized and/or introduced gradually, as needed in the subsequent chapters. A student handout that illustrates how the ideas from the book can be used at an elementary level is available on the website www.ultrasmall.org.

The format of our book differs from textbooks for traditional Calc 101 courses mainly in that we clearly have to start by convincing the teachers of such courses that ours is a worthwhile approach. They first have to master the techniques themselves, and for this purpose we wrote the book at a slightly higher level, including explanations and material beyond what would be presented to the beginning students. The book is intended to inspire teachers to supplement the usual Calc 101 and 102 material or to fashion their own courses on its basis.

The book is structured so that it could be used as a textbook for a course at a more advanced level, comparable to the (U.S.) first advanced calculus course. In this case, one would probably want to cover most of Parts I and II. This would be especially appropriate for courses directed towards physics or engineering majors, as arguments involving infinitesimals are common in the practice of those fields. We think that there are advantages to teaching with ultrasmall numbers even in a course oriented towards mathematics majors. It seems that many students, even at this level, find it difficult to understand, say, the distinction between pointwise and uniform convergence of a sequence of functions, based on the epsilon–delta definitions of these concepts; an initial approach via infinitesimals might be more intuitive. We recognize that students in a course of this nature have to learn the traditional epsilon–delta methods, and this book makes it possible to get used to them gradually, while maintaining full rigor from the start. The transition to traditional methods is motivated in Section 4.7 (on numerical integration), Chapter 10 (topology of the real line), and explicitly worked out in Section 5.2. We focus on those topics that best illustrate the variety of infinitesimal methods and de-emphasize those where algebraic or computational aspects predominate. (Yet, for the sake of providing a complete course, we also include some theorems whose proofs are not specific to our approach, some routine computational examples and many exercises.) The book could also serve as a text for a seminar or independent study with an emphasis on nonstandard methods.

There are 80 numbered exercises scattered throughout the text. They are an important part of the learning experience and the reader is encouraged to attempt all of them. In many cases, the results are used later in the text. They all have worked out solutions starting on page 241. Additional exercises (without answers) are placed at the end of each chapter (170 in all), ranging from the routine to the more challenging.

Chapter-by-chapter summary

Part I includes material that—probably with omission of some of the more difficult proofs—could be covered in an elementary calculus course. In an advanced calculus course one would want to include all the proofs.

Chapter 1 provides some intuition about how to interpret the nontraditional concept of observability on which our approach is based. It formulates the basic principles that govern observability and defines the key concepts: ultrasmall and ultralarge numbers and observable neighbors.

Chapter 2 studies continuity and limits. In particular, simple proofs of the Intermediate Value Theorem and the Extreme Value Theorem are given; they do not rely on the notion of supremum or topological properties such as compactness. Uniform continuity is also introduced, and the theory of exponentiation with real exponents is developed.

Chapter 3 develops elementary differential calculus and Chapter 4, integration of continuous functions. All relevant theorems are fully proved.

Part II contains material that would not usually be found in a first calculus course, but that should be included in advanced calculus.

Sections 5.1 and 5.2 in Chapter 5 discuss the notion of supremum, completeness of the real numbers, mathematical induction, and the epsilon–delta method. With the exception of induction, this material is almost never used in the rest of the book and can be omitted or postponed. Section 5.3 establishes a useful equivalent version of the definition of limit.

Chapter 6 proves various versions of L'Hôpital's Rule, introduces higher derivatives, and defines the Taylor polynomial.

Chapter 7 develops the usual material on sequences and series in our framework. Uniform convergence of sequences of functions is studied in Section 7.4.

The last three chapters of Part II are independent of each other. Chapter 8 begins with some elementary material on differential equations, and then follows with a nonstandard proof of the Peano theorem about the existence of solutions of first order differential equations. The proof of the uniqueness theorem assuming the Lipschitz condition is also given. Chapter 9 develops the theory of the Riemann integral. Chapter 10 illustrates the nonstandard treatment of topological concepts, such as open, closed, dense and compact sets, in the simple setting of sets of real numbers.

The Appendix, intended for mathematically more sophisticated readers, gives a formal outline of the foundations on which our approach rests. After a brief review of logical notation and the role of axioms and proofs, we state formally the axioms of the nonstandard set theory **RBST** and deduce from them the principles used in the text. We then discuss consistency of **RBST** and its extensions and provide a guide to the history and literature of the subject.

Preface for Students

Calculus was developed by Isaac Newton (1642–1727) and Gottfried Wilhelm von Leibniz (1646–1716) in the last third of the seventeenth century as a general method for the study of changing quantities (functions). It has found extensive applications in every field of science concerned with change: physics, chemistry, geology, ecology, economics; in engineering, finance and many other areas. Newton and Leibniz discovered calculus independently and approached it from different viewpoints. In order to understand the difference, let us look at a simple example of an important problem of calculus.

We consider a point-like object P moving in a straight line. The position of P at time t is determined by the distance $s(t)$ of P from a fixed origin O.

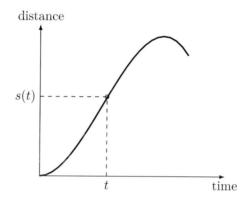

A fundamental assumption of mechanics is that the moving object has, at each time t, a definite *instantaneous velocity* $v(t)$, and one of its basic problems is to determine this instantaneous velocity, assuming that the distance function is known.

We begin by observing that the *average velocity in an interval*, say from t to $t + \Delta t$ where $\Delta t > 0$, can be obtained by a straightforward algebraic computation.

If $s(t)$ is the distance of the object from the origin at time t, $s(t+\Delta t)$ is its distance from the origin at time $t + \Delta t$, hence, during the time interval from t to $t + \Delta t$ the object has travelled the net distance Δs equal to $s(t + \Delta t) - s(t)$, with the average velocity

$$\frac{\Delta s}{\Delta t} = \frac{s(t + \Delta t) - s(t)}{\Delta t}. \tag{1}$$

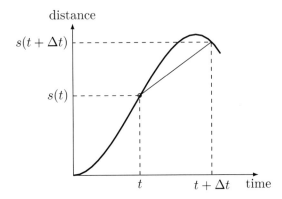

As an instant has no measurable duration, one might think that the instantaneous velocity $v(t)$ at time t could be obtained from equation (1) by setting $\Delta t = 0$. However, this idea does not work because the resulting expression $0/0$ is mathematically meaningless. It does not follow that there is *no way* of obtaining $v(t)$ from equation (1); however, to do so we have to employ some reasoning, in addition to algebra.

Let us consider a specific example: a small ball in free fall. It has been determined experimentally by Galileo Galilei (1564–1642) that the distance of the falling ball from the point of release is $s(t) = ct^2$, where the constant c has approximate numerical value 5 (if time is measured in seconds and distance in meters). For an object moving according to $s(t) = 5t^2$ we have

$$\Delta s = 5(t + \Delta t)^2 - 5t^2 = 10t(\Delta t) + 5(\Delta t)^2$$

and the average velocity is

$$\frac{\Delta s}{\Delta t} = \frac{10t(\Delta t) + 5(\Delta t)^2}{\Delta t} = 10t + 5\Delta t. \tag{2}$$

Can the instantaneous velocity at time t be obtained from this formula? Intuitively, the instantaneous velocity is approximated by the average velocity when Δt is very small, and it has to depend only on the time t, not on the arbitrary choice of Δt we use to compute the average velocity. The expression on the right side of equation (2) is a sum of two terms: the term $10t$ that depends only on t, and the term $5\Delta t$ that depends on Δt; moreover, if Δt is very small, this second term is also very small. We conclude that the first term $10t$ represents the instantaneous velocity $v(t)$ at time t, and the second term $5\Delta t$ represents the difference between $v(t)$ and the average velocity in the interval $[t, t + \Delta t]$. The challenge is to convert this reasoning into rigorous mathematics.

We pause to consider again an idea that does not work: letting $\Delta t = 0$. True, if one does that in the expression on the right of equation (2), one obtains $10t$, which is indeed the instantaneous velocity at time t. However, note that setting $\Delta t = 0$ does not make sense in the expression in the middle of equation (2); it works on the right because the factor Δt cancelled out. This is something that happens only in simple special cases. It cannot be used to obtain $v(t)$ if the formula for $s(t)$ is more complicated (try it for $s(t) = \sin(t)$), and it certainly does not give us a general definition of instantaneous velocity for an arbitrary $s(t)$, in the absence of any specific formula. In order to obtain such a general definition, we have to engage in some reasoning about the formula for $\Delta s/\Delta t$.

Looking at the right side of equation (2) we see that the term $5\Delta t$ gets smaller and smaller as Δt does. For example, letting successively $\Delta t = 0.1, 0.01, 0.001, 0.000001$ gives the values $0.5, 0.05, 0.005$ and 0.000005 for $5\Delta t$. In other words, as Δt *approaches* 0, $5\Delta t$ also *approaches* 0 and the average velocity $\Delta s/\Delta t$ *approaches* the instantaneous velocity $v(t) = 10t$. This is an informal, intuitive argument that goes back to Newton. But the informality also involves a lack of clarity that, in more complicated situations, may lead to confusion or even to a contradiction. One needs a precise definition of what is meant by "getting smaller and smaller" or "approaching." Mathematicians spent a great deal of effort trying to put arguments about approaching on a firm foundation. It culminated in the introduction of the notion of limit by Augustin-Louis Cauchy (1789–1857) and the rigorous definition of limit given by Karl Weierstrass (1815–1897) (see Section 5.2). His epsilon–delta definition of limit forms the cornerstone of almost all contemporary texts on calculus or mathematical analysis. However, in the process some of the simplicity and intuitiveness of the original idea has been lost. The epsilon–delta machinery feels artificial, is notoriously hard to learn and, when used in proofs, often requires bookkeeping whose details are irrelevant and distract from issues at hand.

An alternative way to reason, originating with Leibniz, goes as follows. Let us take the duration Δt of an instant to be *infinitesimal*, that is, smaller than every positive real number, but yet not zero. Then the average velocity (equation (2)) differs from $10t$ by the infinitesimal amount $5\Delta t$, and we can identify $v(t)$ with the "real part" $10t$ of $\Delta s/\Delta t$, and discard $5\Delta t$ as an "infinitesimal error" caused by the fact that our algebraic formula still computes only the average velocity, albeit over an "infinitely short" interval. Leibniz's approach via infinitesimals was the standard way to do calculus until the mid-nineteenth century. Mathematicians today still use the notations df/dx for derivatives and $\int f(x)\,dx$ for integrals that Leibniz invented. However, the task to make rigorous sense of infinitesimals turned out to be even more difficult than the formalization of Newton's approach, and in fact, mathematicians in the nineteenth century largely gave up on infinitesimals in favor of the epsilon–delta method. It was only in 1960 that Abraham Robinson (1918–1974) succeeded in developing a rigorous mathematical theory of infinitesimals that could be used to realize both Leibniz's original ideas and many new ones. Robinson called his theory *nonstandard analysis*, in order to contrast it with the epsilon–delta approach that was "standard" by then. Robinson's original presentation relied on techniques from model theory, an advanced branch of mathematical logic, but developments in the subsequent decades produced frameworks for nonstandard analysis that are more elementary. This book takes advantage of these recent developments to present calculus with infinitesimals in a style that is both as informal and as rigorous as is traditional in textbooks of introductory analysis, but without any appeal to model theory or the excessive formalism of some of the axiomatic approaches.

We briefly highlight the key ideas on which our approach is based. They are elaborated in detail in Chapter 1 and developed and used throughout the text.

We assume that the set of real numbers $\mathbb{R}$ contains, in addition to the familiar, describable, "observable" real numbers like 1, $\sqrt{2}$ and π, also many ideal, unobservable numbers. Some of these are smaller (in absolute value) than all observable real numbers, but yet not 0. They can be used to represent the duration of an instant and to compute instantaneous rate of change, as indicated above. They are the infinitesimals, or, as we prefer to call them for reasons explained in the Introduction, *ultrasmall numbers*. There are also many unobservable numbers that are not infinitesimal; for example, the reciprocals of ultrasmall numbers are ultralarge. The key point is that the unobservable numbers follow the same rules of arithmetic, and altogether have the same properties, as the familiar real numbers.

With some practice, the reader should be able to develop a helpful intuitive picture of our view of the real line. But, in order not to rely on intuition alone, we proceed axiomatically. This is the time-tested approach to mathematics dating to Euclid's treatise on geometry. In Chapter 1 we state a few precise principles that our intuitive picture of the real line satisfies. We then carefully derive all our assertions from these principles. With the help of mathematical logic it has been established that all results obtained in this way have to agree with the claims of traditional mathematics (see the Appendix for details).

Why study analysis with ultrasmall numbers? First and most important, the approach makes analysis more intuitive, simpler and easier to learn. The intuitive meaning of limits, derivatives and integrals becomes more transparent. One learns a whole new set of useful tools, unavailable to the traditionally trained. And one learns the traditional epsilon–delta ideas too, but this happens gradually, almost as an afterthought, in the context where they are really indispensable (estimation in Sections 4.7 and 5.2) or advantageous (elementary topology of the real line in Chapter 10).

Who should read this book? The minimal prerequisite is a familiarity with pre-calculus material: intervals of real numbers, the coordinate system, functions and some trigonometry. For the first-time student of calculus the book offers an easier and more intuitive approach while also providing rigor often lacking in traditional introductory courses. The student handout, available at www.ultrasmall.org, provides additional explanations, motivation and exercises. The book can be used together with a traditional calculus textbook, or independently. Readers who already studied calculus, but without emphasis on proofs, will find here a rigorous development in which all assertions are proved from a small number of intuitive axioms in a way close to the classical ideas of Leibniz. The arguments are easier to understand than the traditional epsilon–delta proofs. Again, at this level the book can be used either by itself or in conjunction with a traditional advanced calculus or elementary analysis textbook. Finally, for those who already mastered the epsilon–delta theory of calculus the book provides an easy introduction to nonstandard analysis and its methods. These methods have found many applications in various areas of mathematics and science, and will enrich the reader's mathematical toolkit.

The book covers the usual syllabus of a first course in analysis. The reader should start with Chapter 1 and Section 2.1. It is here that the groundwork is laid and the most basic ideas as to how it is applied to the study of functions appear. Chapters 2 through 7 treat the standard topics of calculus: continuity, limits, derivatives, integrals and infinite sequences and series. One can of course omit certain parts (for example, the construction of exponential and logarithmic functions in Section 2.4) if the results are known or accepted without proof. We emphasize the nonstandard methods; readers who are not familiar with the epsilon–delta techniques and wish to learn them at this stage should pay extra attention to Sections 4.7, 5.2, and Chapter 10, and supplement them with material from traditional textbooks. The last three chapters are independent of each other. The Appendix presents the foundations in a more formal way; it is written at a somewhat more advanced level.

Acknowledgments

Of course, none of our work would be possible without Abraham Robinson and the subsequent development of nonstandard analysis. We have drawn freely on the literature of nonstandard analysis; while our framework is different, the "combinatorial kernel" of most proofs traces back to arguments found in Robinson [24] (and sometimes perhaps Euler and Leibniz).

A handbook on a classical subject such as elementary analysis is not likely to contain original mathematical results, and we claim none. The contents and organization of the material follow the usual syllabus of advanced calculus in one variable, such as Ross [25] or Bartle and Sherbert [1].

Some specific acknowledgments are due. We are happy to acknowledge our debt to Yves Péraire. He originated the relative approach to nonstandard analysis, and many ideas elaborated in this book have their root in his writings. We are grateful to him for his sympathetic reception of this project and valuable comments and suggestions. Our treatment of applications of the integral in Chapter 4.5 is modeled on Keisler's Infinite Sum Theorem [16, 15]. We note also that Evgeni Gordon [5, 6] developed an approach to relative standardness that is different from that of Péraire.

A number of people commented on various stages of this work, in particular on the preliminary version published as an article in the *American Mathematical Monthly* [13]. We are very grateful to them for all their constructive criticism.

Authors

KAREL HRBACEK earned an RNDr. in mathematical logic under Petr Vopěnka at Charles University in Prague. After stays at the University of California at Berkeley (1968–69) and the Rockefeller University (1969–71), he joined the Department of Mathematics at the City College of New York in 1971. He became a full professor in 1983 and retired as professor emeritus in 2008. He published over thirty papers in set theory, model theory and theory of computation and his 1979 expository paper *Nonstandard set theory* was awarded the Lester Ford prize by the MAA. Hrbacek's textbook *Introduction to Set Theory*, written jointly with Thomas Jech, is now in its third edition and still widely used. His continuing research interests are in the foundations of nonstandard analysis and set theory. He is an editor for the *Journal of Logic and Analysis*.

OLIVIER LESSMANN received his diploma from the Swiss Institute of Technology in 1991 with a specialization in analysis and earned his PhD from Carnegie Mellon University in 1998 in the area of model theory. He was a research assistant professor at the University of Illinois-Chicago and a researcher at Oxford University (UK). He published over twenty papers in model theory in both logic and mainstream mathematics journals, such as the *Journal of the AMS*. Always very interested in mathematics education, he won a couple of teaching awards and earned a teaching degree in 2006 in Switzerland. He currently teaches for the bilingual program in the Geneva secondary school system (Switzerland).

RICHARD O'DONOVAN was a carpenter for ten years, then a musical instrument maker for another ten years. He earned his MA from the University of Geneva in 1998 and his teaching degree in 2000. He earned his PhD from the Blaise-Pascal University (France) in 2011 in the area of nonstandard analysis and alternative set theory under Yves Péraire. He published several articles on the links between pedagogy and the use of nonstardard analysis in high school since 2000. He currently teaches for the bilingual program in the Geneva secondary school system (Switzerland).

Part I

Elementary Analysis

1

Basic Concepts

1.1 Introduction

The fundamental problem of calculus is to define and study *instanta-neous rate of change*, that is, the rate of change of some variable quantity at a given instant. By an *instant* we understand intuitively a duration of time shorter than any measurable time interval, but yet not zero. A "number" that describes the duration of an instant thus has to be smaller than any positive real number, but bigger than zero. Gottfried Wilhelm von Leibniz, who pioneered the use of such numbers in the seventeenth century, called them "infinitesimals." In his view, infinitesimals have all the properties of the usual real numbers. For example, they obey the basic laws of arithmetic, such as the commutative and associative laws for addition and multiplication and the distributive law.

The challenge that faced Leibniz and his followers was to make their intuitions about infinitesimals sufficiently clear and rigorous to avoid errors and misunderstandings. The founders of calculus were not able to meet this challenge adequately and, at least in part for this reason, the method of infinitesimals was gradually abandoned by mathematicians. A fully rigorous treatment of infinitesimals suitable for the needs of calculus was given only in the mid-twentieth century by Abraham Robinson.

Robinson showed how to construct an extension of the real num-ber system $\mathbb{R}$ to a larger system of numbers (so-called *hyperreals*) that contains, among others, also infinitesimals. In this he followed the long-established precedent of extending a familiar number system to a larger, more comprehensive one. For example, Richard Dedekind (1831–1916) in the 1870's showed how to construct real numbers as cuts in the set of rational numbers $\mathbb{Q}$. It is also well known that complex numbers can be constructed as ordered pairs of real numbers, and other examples of similar extensions abound.

These constructions have some common features. They tend to be rather complicated (Dedekind's construction is usually not taught even in advanced calculus courses). There is no need to know them in order to develop proficiency in mathematics with the objects that were so constructed. For example, when working with real numbers we rely on our intuitive understanding, the mental representation of real numbers as points on the number line, and on axioms that list the essential properties of $\mathbb{R}$ (various laws of arithmetic, the Completeness Axiom, and so on—see Section 5.1). It is not particularly helpful to know Dedekind's construction, and there is never any practical need to refer to it.

This book uses a similar approach. A construction of an appropriate extension of $\mathbb{R}$ in the style of Robinson is quite complicated. There is no need to study it unless one is concerned about the consistency of our approach. We base our presentation instead on the intuitive picture outlined below and in Section 1.5, and, rigorously, on the axioms formulated in Chapter 1.

The difficulty with infinitesimals lies in the need to reconcile two seemingly contradictory ideas. On the one hand, infinitesimals cannot *be* the "usual" real numbers. On the other hand, they *have all the usual properties* of real numbers, and so we would like to treat them as such. To reconcile these two conflicting ideas, we adopt a somewhat different view of sets than is customary. **In this book, we consistently adopt the point of view that the usual, standard sets can contain, besides their usual, standard elements, also ideal, fictitious elements with the same properties as the standard ones.**

For a picturesque example, let us consider the standard set of all mammals. The usual, standard view of this set is that it has elements such as lions, horses, bats, whales and kangaroos. In our view it has also ideal elements, such as unicorns and yetis. These fictitious mammals share all the properties of standard, "real" mammals. They are warm-blooded, females lactate after giving birth, and so on.

Turning to mathematics, the standard set of natural numbers $\mathbb{N}$ has standard elements such as 0, 1, 2, 17, 324 and so on. In our view, it has also nonstandard, ideal elements. Let $N \in \mathbb{N}$ be such a nonstandard element. What can we say about N? Well, certainly $0 < N$, because N is assumed to have all the properties of natural numbers and there are no natural numbers less than 0; also $N \neq 0$ because N is not standard. Similarly, $1 < N$, because the only natural number less than 1 is 0, and $N \neq 0, 1$. By the same argument it follows that $2 < N$, $3 < N$ and, in general, $n < N$ for any standard n. We still call N a "natural number," because, in our view, it is an element of the standard set of natural numbers $\mathbb{N}$. But it is an "infinitely large" natural number, in the sense that it is bigger than all standard natural numbers. Nevertheless, N has all the usual properties of natural numbers. For example, like all natural numbers, it has an immediate successor $N + 1$ and (since $N \neq 0$) an immediate predecessor $N - 1$. The number $2N$ is even and $2N + 1$ is odd. Of the two numbers N and $N+1$, one has to be even and the other odd (which of the two alternatives actually occurs depends on which particular nonstandard N is under consideration).

More interesting for our purposes, the reciprocal $1/N \in \mathbb{R}$ is not zero (because "$1/x \in \mathbb{R}$ *and* $1/x \neq 0$" is a property that all standard real numbers have, hence the ideal number N has it as well), but from $n < N$ it follows that $1/N < 1/n$ for all standard n. Since for every standard real number $r > 0$ there is some standard n such that $1/n < r$, it follows that $1/N < r$ for all standard real $r > 0$. The number $\varepsilon = 1/N$ is thus infinitesimal in the sense the concept was understood by Leibniz.

We have to elaborate on the claim that the new, ideal elements "have the same properties as the standard ones." What exactly is that supposed to mean? In our view, the universe of mathematical objects is a much richer place than is the standard view, full of ideal elements of all sorts. But the presence of the ideal elements in the standard sets does not change the properties of these sets. **Every fact (be it axiom or theorem) of traditional mathematics remains true.** Thus the arithmetic operations $+$, $-$ and $\times$ are defined for all real numbers, whether these are standard or not, and satisfy the usual axioms. Division is defined whenever the denominator is not 0; in particular, it is defined for infinitesimal denominators. Similar remarks apply to other functions and operations of traditional mathematics: $\sqrt{x}$, $\sin(x)$, $\log(x)$ and so on. For every real number r there is a natural number n such that $n \leq r < n+1$ (of course, if r is "infinitely large," then n is also "infinitely large"). Every nonempty set of natural numbers has a least element. Every continuous function defined on a closed bounded interval attains its maximum there. These are just a few facts of traditional mathematics; they all remain valid in our view. They justify the use of the familiar notation for the traditional mathematical concepts, in spite of the change of viewpoint.

The second aspect of the claim is that there are no ideal elements with genuinely new properties. **If there is an object with some property, then there is also a standard object with this property.** We call this statement the Closure Principle; one of its consequences is that standard operations performed on standard objects yield the usual, standard results. It has to be noted that the Closure Principle applies only to properties that can be described in the language of traditional mathematics. For example, there exist infinitesimal real numbers, but there are no standard infinitesimal real numbers. We elaborate on this matter in Section 1.4.

Adding ideal natural numbers to the set $\mathbb{N}$ has consequences outside the domain of natural numbers. One example given already is the existence of infinitesimals. Here is another: In standard mathematics, a set is finite if it can be enumerated by natural numbers up to some $n \in \mathbb{N}$; say $\{a_0, a_1, \ldots, a_{n-1}\}$. In our view, N is also a natural number, albeit an ideal one, so a set that can be enumerated by it, such as $\{a_0, a_1, \ldots, a_{N-1}\}$, is also finite, albeit in an ideal sense. In particular, the set $\{0, 1, \ldots, N-1\}$ having N elements is finite. It is customary to call natural numbers like N "infinitely large," but this would be very confusing in our context. It is mainly for this reason that we abandon the traditional terminology "infinitely large" and "infinitesimal" in favor of "ultralarge" and "ultrasmall," respectively.

Let us consider the set $\{0, 1, 2, 3, \ldots\}$. One has to be careful about the interpretation of "..." ("*and so on*"). In our view it indicates a run through all natural numbers, standard or not, so this set is just $\mathbb{N}$, the set of all natural numbers. It is of course an infinite set. But readers conditioned by years of traditional mathematical training may be tempted to take $\{0, 1, 2, 3, \ldots\}$ to be the collection of only the standard natural numbers, that is, what can be described as $\{n \in \mathbb{N} : n \text{ is standard}\}$. We stress that this is *not* our view. For us, every standard infinite set contains both standard and nonstandard elements, intermingled together and without the possibility of sharply separating the ones from the others. The set $\mathbb{N}$ is the usual standard infinite set of natural numbers, only we view it as having, besides the standard elements, also some ideal, ultralarge elements that the usual viewpoint disregards. We never consider "bare" collections that separate the standard elements of a set from the nonstandard ones (except in the Appendix). The collection $\{n \in \mathbb{N} : n \text{ is standard}\}$ is **not a set** (either standard or nonstandard) in our view, and it is not used in the book. One of our axioms makes it clear which properties can be used to define sets.

We do not mean to say that one could not admit such "bare" collections into the theory. It can be done consistently, as long as one does not confuse them with sets (they can be called "classes" or "external sets"). But doing this involves mixing two very different points of view on the same objects: on the one hand, our view that the set $\mathbb{N}$, say, contains also ideal elements, and on the other hand, the "standard view" that it does not. The two views are compatible, but at the cost of substantial complications. For example, one would have to have two names for "the same" concept from the two points of view; say $\mathbb{N}$ for the set of natural numbers from our point of view, and $^\circ\mathbb{N}$ for the external set of the standard natural numbers only. This is essentially how things are handled in Robinson's model–theoretic approach. Clearly it involves a great increase in the complexity of the framework. More details can be found in the Appendix. The combined viewpoint has some advantages in more advanced mathematics, but it is not necessary for the development of calculus. We urge the readers of this book to try to adopt our point of view. It is the price (a small one, in our opinion) one pays for having a truly elementary account of infinitesimal calculus.

In the rest of this chapter we develop our point of view systematically and more formally. As noted above, we use "ultrasmall" and "ultralarge" in place of the established terminology "infinitesimal" and "infinitely large." These concepts are defined in terms of the fundamental distinction between "observable" and "non-observable" objects. Observability is a primitive concept whose properties are specified by our axioms. Intuitively one should think of "observable" as synonymous with "standard," for the time being. An explanation of the distinction and of the full meaning of observability is given in Section 1.5. The consistency of our axiomatic system is discussed in the Appendix.

1.2 Observability

Every mathematical book has to start with some concepts that are *primitive*, not defined in terms of simpler notions, and take some basic properties of these primitive concepts for granted. As is traditional in books on analysis, we assume familiarity with sets, natural numbers 0,1,2,..., the set of all natural numbers $\mathbb{N}$, the set of all integers $\mathbb{Z}$, the set of all rational numbers $\mathbb{Q}$, the set of all real numbers $\mathbb{R}$, and the usual arithmetic operations $+, -, \times, /$ and ordering $\leq$ on $\mathbb{R}$, but this is not meant to be an exhaustive list. These concepts are not defined in this book; we take them as primitive and we take it for granted that the reader is acquainted with elementary properties of these notions. As explained in the Introduction, all such results remain valid in our extended view of the mathematical universe, and we use them without comment.

Our book differs from traditional analysis textbooks by introducing an additional primitive concept: observability. For now, one should intuitively identify the observable objects with the standard objects of traditional mathematics and view unobservable objects as ideal, fictitious elements of standard sets. This is not the whole story, but we postpone the full explanation of our understanding of observability until Section 1.5. In any case,

$$p \text{ is observable}$$

is a primitive property that has no counterpart in traditional mathematics. Here p can be any mathematical object: a number, function, set, operation, geometric figure, and so on. Like other primitive concepts, observability has no explicit definition in terms of more fundamental concepts. Its meaning is specified implicitly by the axioms that are formulated in this chapter. All our reasoning about observability is based on these axioms.

We begin by stating two of our key definitions.

Definition 1.

> *(1) A real number is **ultrasmall** if it is nonzero and its absolute value is less than any observable positive real number.*
>
> *(2) A real number is **ultralarge** if its absolute value is greater than any observable positive real number.*

More formally, $\varepsilon \in \mathbb{R}$ is ultrasmall if $\varepsilon \neq 0$ and $|\varepsilon| < r$ for all $r > 0$, r observable. Similarly, $M \in \mathbb{R}$ is ultralarge if $|M| > r$ for all $r > 0$, r observable.

Ultralarge numbers are somewhere over there.

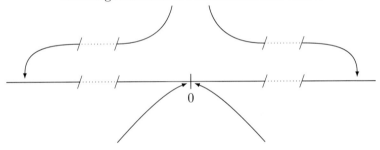

Ultrasmall numbers are somewhere here.

Intuitively, ultrasmall numbers cluster about the origin; they are smaller in absolute value than $1/n$, for every observable natural number n. Ultralarge numbers are very far from the origin; farther than any observable natural number n.

The assumption that ultrasmall numbers exist is what makes the infinitesimal approach to calculus possible. Our first principle merely records this assumption formally.

Existence Principle

There exist ultrasmall real numbers.

Exercise 1 (Answer page 241)

> (1) If x is such that $0 < |x| < |\varepsilon|$ and ε is ultrasmall, then x is ultrasmall.

(2) If x is such that $|M| < |x|$ and M is ultralarge, then x is ultralarge.

1.3   First Principles

In this section we develop systematic rules for computing with ultra-small and ultralarge numbers. Before starting on this project, we need some principle that would connect the notion of observability with the traditional mathematical concepts. For now, we postulate only a very special case of such a principle (see Section 1.4 for a more general version). We stress once more that the arithmetic operations $+$, $-$ and $\times$ are defined for all real numbers, whether observable or not, and have the usual properties. Division is defined as long as the denominator is not 0.

Closure Principle for Elementary Arithmetic Operations

The real number 1 is observable. If the real numbers x and y are observable, then $x \pm y$, $x \cdot y$, and x/y (if $y \neq 0$) are observable.

Assuming that we identify observable numbers with the standard numbers of traditional mathematics, the intuitive validity of the Closure Principle is obvious: Arithmetic operations applied to standard real numbers yield standard results. It is of course equally obvious that 2, 3 and 17 are observable, but these facts are not included in the statement of the Closure Principle. Therefore they should be proved, but this is very easy.

Exercise 2 (Answer page 241)

(1) Apply the Closure Principle and conclude that the number 0 is observable.

(2) Similarly, conclude that 2 is observable.

(3) Prove that $1/2$, 4 and 17 are observable.

Caution: From the Closure Principle we may deduce that if n is observable, then $n + 1$ is also observable. But it would be erroneous to appeal to the Principle of Mathematical Induction and conclude that *all* natural numbers are observable! Induction is a property of sets of natural numbers. In the classical setting, every statement about natural numbers defines a set; hence the Principle of Induction is valid for all statements of traditional mathematics. In our system induction remains valid for the same statements, that is, for those that do not refer to observability. See Section 1.6 for an overview of induction in our approach. The statement "*n is observable*" is not a traditional statement, so it may not be used in induction nor to define a set. As discussed in the Introduction, there is no set containing all and only numbers which are observable.

Theorem 1. *If ε is ultrasmall, then ε is not observable. If M is ultralarge, then M is not observable.*

Proof. If ε were observable, then $|\varepsilon|$ would also be observable, by the Closure Principle (because $|\varepsilon|$ is either ε or $-\varepsilon = 0-\varepsilon$). We could take $r = |\varepsilon|$ and obtain a contradiction $|\varepsilon| < |\varepsilon|$. Similarly, if M were ultralarge and observable, we could let $r = |M|$ and obtain a contradiction $|M| > |M|$. □

Thus, for any real number x, exactly one of the three alternatives occurs:

- $x = 0$ or x is ultrasmall.

- $r_1 < |x| \leq r_2$ for some observable $r_1, r_2 > 0$.

- x is ultralarge.

Observable numbers are neither ultrasmall nor ultralarge, but there are also many unobservable numbers that are neither ultrasmall nor ultralarge. For example, $1 + \varepsilon$ is not observable if ε is ultrasmall (why?), but $1/2 < 1 + \varepsilon < 3/2$ (because $-1/2 < \varepsilon < 1/2$), so $1 + \varepsilon$ is neither ultrasmall nor ultralarge. It differs from 1 by an ultrasmall amount ε. Around each observable number a there is a cluster of unobservable numbers $a + \varepsilon$, for ε ultrasmall, that differ from a only by an ultrasmall amount.

Rule 1. *Let x, y, h, k be real numbers.*

(1) If x, y are not ultralarge, then $x \pm y$ and $x \cdot y$ are not ultralarge.

(2) If h, k are ultrasmall and x is not ultralarge, then $h \pm k$ and $x \cdot h$ are ultrasmall or zero.

(3) h is ultrasmall if and only if $\frac{1}{h}$ is ultralarge.

Proof.   (1) If x, y are not ultralarge, then $|x| \leq r$ and $|y| \leq s$ for some observable $r, s > 0$. It follows that $|x \pm y| \leq |x| + |y| \leq r + s$ and $|x \cdot y| = |x| \cdot |y| \leq r \cdot s$, where $r + s$, $r \cdot s$ are observable, by the Closure Principle.

(2) Let $r > 0$ be observable. Then, by the Closure Principle, $2 = 1+1$ is observable, and $\frac{r}{2}$ is also observable. Hence, $|h| < \frac{r}{2}$ and $|k| < \frac{r}{2}$, and therefore $|h \pm k| \leq |h| + |k| < \frac{r}{2} + \frac{r}{2} = r$.

Let $|x| \leq r_0$, where $r_0 > 0$ is observable. For every observable $r > 0$, $\frac{r}{r_0} > 0$ is also observable (Closure Principle again). Hence $|h| < \frac{r}{r_0}$, and so $|x \cdot h| = |x| \cdot |h| < r_0 \cdot \frac{r}{r_0} = r$. The case where the result is zero occurs if $h = -k$ (for $h + k$) or if $x = 0$ (for $x \cdot h$).

(3) Assume h is ultrasmall. Then, for every observable $r > 0$, $|h| < \frac{1}{r}$, and so $\left|\frac{1}{h}\right| > r$. The converse is similar. $\qquad \square$

Theorem 2. *There exist ultralarge natural numbers.*

Proof. By the Existence Principle, there is an ultrasmall real number h; we can take $h > 0$. The real number $x = \frac{1}{h}$ is then ultralarge (Rule 1 (3)). The natural number n for which $n \leq x < n + 1$ is ultralarge. (If not, $n \leq r$ for some observable r, so $x < n + 1 \leq r + 1$, where $r + 1$ is observable by Closure, and we have a contradiction.) $\qquad \square$

The readers who feel that the existence of "huge" natural numbers is more intuitive than the existence of ultrasmall numbers can replace the Existence Principle with Theorem 2. The proof that there are then ultrasmall numbers is a simple exercise.

Definition 2. *We say that a and b are **ultraclose**, or that a and b are **neighbors**, written*

$$a \simeq b,$$

if $a - b$ is ultrasmall or 0.

We reformulate some of the results of the previous rule using this new terminology.

Rule 2. *Let a, b, x, h be real numbers.*

(1) *If $a, b \simeq 0$, then $a \pm b \simeq 0$ and $a \cdot b \simeq 0$.*

(2) *If x is not ultralarge and $h \simeq 0$, then $x \cdot h \simeq 0$.*

(3) *If h is ultrasmall and $x \not\simeq 0$, then $\frac{x}{h}$ is ultralarge.*

Proof. Only item (3) requires some argument. Since x is neither ultrasmall nor 0, there is an observable $r_0 > 0$ such that $|x| \geq r_0$. We know that $\frac{1}{h}$ is ultralarge, hence, for any observable $r > 0$, $\frac{1}{|h|} > \frac{r}{r_0}$ and $\frac{|x|}{|h|} > r_0 \cdot \frac{r}{r_0} = r$. $\qquad \square$

Rule 3. *Let a, b be real numbers. If $a \simeq b$ and a and b are observable, then $a = b$.*

Proof. The assumptions imply that $a - b \simeq 0$ (by Definition 2) and $a - b$ is observable (by Closure). Therefore $a - b$ is not ultrasmall (by Theorem 1); hence $a - b = 0$ and $a = b$. $\qquad\qquad\qquad$ □

Exercise 3 (Answer page 241)
Let a, b, x, y be real numbers such that $a \simeq x$, $b \simeq y$ and a and b are observable, and let $h \simeq 0$.

(1) Show that if $a \neq 0$, then $a + h \not\simeq 0$.

(2) Show that $a < b$ implies $x < y$.

(3) Show that $x \leq y$ implies $a \leq b$.

(4) Show that converses of (2) and (3) do not hold.

We now show that $\simeq$ has the properties of an equivalence relation.

Rule 4. *Let a, b, c be real numbers. Then*

(1) $a \simeq a$.

(2) If $a \simeq b$, then $b \simeq a$.

(3) If $a \simeq b$ and $b \simeq c$, then $a \simeq c$.

Proof. As $a - a = 0$, it is immediate that $a \simeq a$. If $a - b$ is ultrasmall or 0, then so is $b - a$, so $a \simeq b$ implies $b \simeq a$. The third point follows from Rule 2. Assume $a = b + \varepsilon$ and $b = c + \delta$ with ε, δ ultrasmall or zero. Then $a = c + \varepsilon + \delta$, and by Rule 2(1), $\varepsilon + \delta \simeq 0$, hence $a \simeq c$. $\qquad$ □

Rule 5. *Let a, b, x, y be real numbers.*

(1) If $x \simeq a$ and $y \simeq b$, then

$$x \pm y \simeq a \pm b.$$

(2) If x and y are not ultralarge and if $x \simeq a$ and $y \simeq b$, then

$$x \cdot y \simeq a \cdot b.$$

(3) If $x \simeq a$, $y \simeq b$, x is not ultralarge and $y \not\simeq 0$, then

$$\frac{x}{y} \simeq \frac{a}{b}.$$

Proof. We can write $a = x + \varepsilon$ with $\varepsilon \simeq 0$ and $b = y + \delta$ with $\delta \simeq 0$.

(1)
$$a \pm b = (x + \varepsilon) \pm (y + \delta) = x \pm y + \underbrace{(\varepsilon \pm \delta)}_{\simeq 0},$$

hence $a \pm b \simeq x \pm y$.

(2)
$$a \cdot b = x \cdot y + \underbrace{x \cdot \delta}_{\simeq 0} + \underbrace{y \cdot \varepsilon}_{\simeq 0} + \underbrace{\varepsilon \cdot \delta}_{\simeq 0}$$

by Rule 2, hence $a \cdot b \simeq x \cdot y$.

(3) Assume first that y is ultralarge. Then b is also ultralarge, and the reciprocals $\frac{1}{y}, \frac{1}{b}$ are ultrasmall, by Rule 1(3). As x and a are not ultralarge, we have $\frac{x}{y} \simeq 0 \simeq \frac{a}{b}$, by Rule 2(2).

Assume next that y is not ultralarge. We show that the difference $\frac{a}{b} - \frac{x}{y}$ is ultraclose to zero.

$$\frac{a}{b} - \frac{x}{y} = \frac{a \cdot y - b \cdot x}{b \cdot y} = \frac{1}{b \cdot y} \cdot (y \cdot \varepsilon - x \cdot \delta).$$

As x and y are not ultralarge and ε and δ are ultraclose to zero, we have $(y \cdot \varepsilon - x \cdot \delta) \simeq 0$. By Rule 2, it suffices to show that $\frac{1}{b \cdot y}$ is not ultralarge. The hypotheses imply that there is an observable $r > 0$ such that $|y| > r$ and $|b| > r$. Then $\frac{1}{|b \cdot y|} < \frac{1}{r^2}$, with the latter being observable, by Closure. Hence $\frac{1}{b \cdot y}$ is not ultralarge, so $\frac{a}{b} \simeq \frac{x}{y}$.

$\square$

Exercise 4 (Answer page 241)

(1) Give an example of x and y such that $x \simeq y$ but $x^2 \not\simeq y^2$.

(2) Give an example of x and y such that $x \simeq y$ but $\frac{1}{x} \not\simeq \frac{1}{y}$.

Exercise 5 (Answer page 241)
Is it possible to have ultrasmall ε and δ such that ε/δ is

(1) neither ultralarge nor ultrasmall?

(2) ultralarge?

(3) ultrasmall?

Exercise 6 (Answer page 242)
If $x \cdot y$ is not ultralarge and y is neither ultralarge nor $\simeq 0$ and if $x \simeq a$ and $y \simeq b$, then $x \cdot y \simeq a \cdot b$.

Exercise 7 (Answer page 242)
In the following, assume that ε, δ are positive ultrasmall numbers and H, K positive ultralarge numbers. Determine whether the given expression yields an ultrasmall number, an ultralarge number, or a number which is neither ultrasmall nor ultralarge.

(1) $1 + \dfrac{1}{\varepsilon}$

(2) $\dfrac{\sqrt{\delta}}{\delta}$

(3) $\sqrt{H+1} - \sqrt{H-1}$

(4) $\dfrac{H + K}{H \cdot K}$

(5) $\dfrac{2 + \varepsilon}{5 + \delta} - \dfrac{2}{5}$

(6) $\dfrac{\sqrt{1+\varepsilon} - 2}{\sqrt{1+\delta}}$

Exercise 8 (Answer page 243)
Prove that if h is ultrasmall, then $\sqrt{1 + h} \simeq 1$.

Exercise 9 (Answer page 243)
Prove that if N is an ultralarge positive integer, then $\sqrt[N]{N} \simeq 1$.

Exercise 10 (Answer page 243)
For $x, y \in \mathbb{R}$ define: $x \sim y$ if $x - y$ is not ultralarge. Prove Rules 3 and 4 with $\sim$ in place of $\simeq$. Give an example of $x, y \in \mathbb{R}$ such that $x \simeq y$ but not $\frac{1}{x} \sim \frac{1}{y}$.

The following principle characterizes the completeness of the real number system in terms of observability. It can be deduced from more fundamental principles (see the Appendix), but this version is sufficient for the purposes of developing calculus.

Observable Neighbor Principle

If a real number is not ultralarge, then there is an observable real number that is ultraclose to it.

Theorem 3. *If a real number x is not ultralarge, then there is a unique observable real number r ultraclose to x.*

Proof. The existence is given by the Observable Neighbor Principle. For uniqueness, let observable r_1 and r_2 be such that $x \simeq r_1$ and $x \simeq r_2$. This implies that $r_1 \simeq r_2$, hence $r_1 = r_2$, by Rule 3. □

A consequence of the Observable Neighbor Principle and Theorem 3 is that a real number x is not ultralarge if and only if it can be written as $x = r + \varepsilon$ where r is observable and $\varepsilon \simeq 0$. This uniquely determined r is said to be the **observable neighbor** of x.

In general, if we have $x \simeq y$, we say that x and y are *neighbors*. If one of the two is observable, then it is *the observable* neighbor of the other number. Intuitively, about each observable a there is a cluster of its neighbors, all of which are ultraclose to a.

It is worth noting that the observable neighbor of $x \in \mathbb{Q}$ need not be in $\mathbb{Q}$. Consider $\sqrt{2}$, which is observable. Let x be a rational number whose difference with $\sqrt{2}$ is ultrasmall (for example, let x be given by the first N digits of $\sqrt{2}$, for ultralarge $N \in \mathbb{N}$). Then $\sqrt{2}$ is the observable neighbor of x, but it is not rational.

Intervals of real numbers are written in the usual way: $[a, b]$ stands for all real values between a and b including a and b. The interval $[a, b)$ stands for all real values between a and b, including a but not b. Similarly for $(a, b]$ and (a, b), where the square bracket means that the endpoint is included and the parenthesis means that the endpoint is not included. An interval of the form (a, ∞) stands for all real numbers greater than a and $(-\infty, a)$ for all real numbers less than a. Notice that $+\infty$ and $-\infty$ are not real numbers but indicate that an interval has no upper or no lower bound.

Exercise 11 (Answer page 244)
Show that if a, b are observable and $x \in [a, b]$, then the observable neighbor of x exists and is in $[a, b]$. Does the statement remain true if $[a, b]$ is replaced by (a, b)?

1.4   Closure

The Closure Principle in Section 1.3 applies only to the four basic arithmetic operations. Before postulating the Closure Principle in full generality we need to clarify some features of mathematical statements, in particular the distinction between free and bound variables.

Some mathematical statements are about particular objects, whether primitive or previously defined. Such statements are either true or false. Two examples are "$\sqrt{2} < 1$" (a false statement) and "$\sin^2(\pi) + \cos^2(\pi) = 1$" (a true statement). Other statements are general; they contain *parameters* (also called *free variables*), usually letters like $x, n, A, f, \ldots$. Such statements become true or false only after particular objects (*values*) have been assigned to the parameters; typically, they are true for some values of the parameters and false for others. Some examples of general statements are "$x < 1$" (true for $x = 0, \frac{1}{2}, -3, \ldots$ and false for $x = 1, \frac{3}{2}, \sqrt{2}, \ldots$), "$\sin^2(x) + \cos^2(x) = 1$" (true for all $x \in \mathbb{R}$), "$x^2 < y$" (true for example for $x = 2$, $y = 5$, and false for example for $x = 1$, $y = 0$), and "$f(0) < \frac{1}{2}$" (true if f is the function defined by $f(x) = x^2$ and false if we take $f(x) = \cos(x)$).

An important point is that *bound variables* (also called *dummy variables*), those preceded by expressions "for some," "there exists," "for every," "for all" (so-called *quantifiers*), are *not* parameters. In order to determine truth or falsity of a statement containing a bound variable, we do not assign a particular value to that variable; rather, we consider *all* values the variable can have, and determine whether the statement is true for some, or all of them, as appropriate. Thus the statement "*For all $x \in \mathbb{R}$, $x^2 \geq 0$*" has no parameters (it is in fact true). The variable x is not assigned any particular value; one could say, equivalently, "*For all $z \in \mathbb{R}$, $z^2 \geq 0$*." The statement "*There exists $k \in \mathbb{N}$ such that $k < n$*" has the parameter n (but not k); it is false for $n = 0$ and true for $n = 1, 2, \ldots$.

The Closure Principle below applies to statements of traditional mathematics. This means mathematical statements that do not refer to the notion of observability, either directly or indirectly. To be more specific, we call the notions "*observable,*" "*ultrasmall,*" "*ultralarge,*" "*ultraclose*" ($\simeq$) and "*observable neighbor*" **relative concepts**. (The reason for this terminology is explained in Section 1.5.) For the purposes of this book, statements of traditional mathematics are statements in which no relative concepts are mentioned.

Closure Principle, Existential Version

Given a statement of traditional mathematics that has parameters p, $p_1, \ldots, p_k$:

If $p_1, \ldots, p_k$ are observable and there exists some object p for which the statement is true, then there exists some observable object p for which the statement is true.

Example. Consider the statement

"x is a real number, $x > 0$, and $x \cdot x = x$."

In this statement there is only the parameter x; there are no additional parameters. The Existential Closure Principle applied to this statement asserts that if there exists some x for which the statement is true, then there exists some observable x for which the statement is true. The only number x that satisfies this statement is 1. Therefore 1 is observable. (In Section 1.3 this is postulated explicitly.)

We describe a general form of this type of argument. A *definition* gives a name to a unique object. Thus, given a statement with a parameter x, about which we know two things:

(1) there is an x such that the statement is true;

(2) this x is unique; that is, if the statement is true both for x_1 and for x_2, then $x_1 = x_2$;

we can give a name to this object, say we call it C. Then C is "the unique object such that the statement holds for it." More formally, "for every x, $x = C$ if and only if the statement is true for x."

It is obvious that we can use the same reasoning as in the above example and exercises to conclude:

Any mathematical object defined (without additional parameters) as a unique object by a statement of traditional mathematics is observable.

Here are some further examples of uniquely defined objects.

Numerical constants: $1, 2, 3, 196883, -5, \frac{2}{3}, -\frac{17}{324}, e$ and π. All of these numbers are observable.

Sets: The set $\mathbb{N}$ of all natural numbers, the sets $\mathbb{Z}, \mathbb{Q}, \mathbb{R}, \mathbb{N} \times \mathbb{N}, \mathbb{R}^3$, and the closed interval $[1, 3] = \{x \in \mathbb{R} : 1 \leq x \leq 3\}$. All of these sets are observable.

Caution: This does *not* mean that all *elements* of these sets are also observable!

The *function* $f : x \mapsto x^2$, as well as the functions $\sin$, $\tan$, $\exp$, $\log$, and all other functions that can be defined without parameters, are observable.

An important special case of the Closure Principle is:

The value of an observable function at an observable argument is observable.

Hence 10^{10}, $\sqrt{5}$, $\sin(\pi/7)$ and $\log 35$ are observable.

More generally, the statement used in a definition may depend on one or more additional parameters $x_1, \ldots, x_n$; the name given to the unique object should then indicate the parameters on which it depends; thus $C(x_1, \ldots, x_n)$ could be used. C can be viewed as an *operation*, defined for those values of $x_1, \ldots, x_n$ for which such unique object exists.

The above argument applies to concepts that depend on parameters:

Any mathematical object uniquely defined from parameters $x_1, \ldots, x_n$ by a statement of traditional mathematics is observable provided $x_1, \ldots, x_n$ are observable.

In particular, if C is an operation defined in traditional mathematics and $x_1, \ldots, x_n$ are observable, then $C(x_1, \ldots, x_n)$ is observable.

Example.

- If the real numbers x and y are observable, then $x + y$, $x - y$, $x \cdot y$, and x/y (if $y \neq 0$) are observable.

- Let N be a positive integer. The numbers $-N$, $\frac{1}{N}$, $\frac{N}{2N}$, $3 + \frac{2}{N}$, $\sqrt{N}$, $\sqrt[N]{N}$ are all observable whenever N is observable. We note that $\frac{N}{2N} = \frac{1}{2}$, and hence it is observable, even when N is not.

- The absolute value of a, $|a|$, is observable whenever a is observable.

- Let A and B be sets. Their union $A \cup B$ is observable whenever A and B are.

- For any $a, b \in \mathbb{R}$,

$$(a, b] = \{x \in \mathbb{R} : a < x \leq b\}$$

is observable whenever a and b are observable.

- The set $\{x \in \mathbb{N} \mid x < k\}$ is observable whenever the natural number k is observable.

- Let a, b and c be fixed real numbers. The function $f : \mathbb{R} \to \mathbb{R}$ defined by

$$f : x \mapsto ax^2 + bx + c$$

is observable whenever a, b, c are observable.

  If x_0 is also observable, then the value $f(x_0)$ is observable. If x_0 is not observable, then $f(x_0)$ may or may not be observable.

- Let r_1 be the smaller of the two roots of the equation $x^2 - (N+1)x + N = 0$, where N is a positive integer. Then r_1 is uniquely defined from N, and hence r_1 is observable whenever N is observable. However, we can determine by factoring that the two roots are 1 and N, and so obtain a stronger result that $r_1 = 1$ is observable even if N is not.

- If $\frac{1}{N}$ is observable, then N is also observable. We can see it as follows: Let $h = \frac{1}{N}$. We assume that h is observable, hence also $N = \frac{1}{h}$ is observable.

Exercise 12 (Answer page 244)
Prove that if $3 + \frac{2}{N}$ is observable, then N is observable.
Similarly for $\sqrt{N}$, $\sqrt[N]{3}$, $\{n \in \mathbb{N} : n \leq N\}$.

We use the Closure Principle to derive some further properties of the concepts introduced in Section 1.3.

Theorem 4. *Let n be an integer; if n is not observable, then n is ultralarge.*

Proof. Assume that n is not ultralarge. By the Observable Neighbor Principle, there is an observable r such that $n \simeq r$. But n is the unique integer in the interval $[r - 0.5,\ r + 0.5)$, hence n is observable by Closure, contradicting our assumption. $\qquad \square$

Corollary. *If $k, n \in \mathbb{N}$, $k \leq n$, and n is observable, then k is observable.*

Another important corollary is the following observation.

Theorem 5. *If A is an observable finite set, then each element of A is observable.*

Proof. To say that A is finite means that there is a sequence $\langle a_1, \ldots, a_n \rangle$, $n \in \mathbb{N}$, such that $A = \{a_1, \ldots, a_n\}$. This is a statement with parameter A. By Closure, there is an observable sequence with this property. The number n is uniquely determined by the sequence (it is the largest element of its domain); hence it is also observable. By the above Corollary, every $i \leq n$ is observable. Therefore a_i, the unique value of the sequence at i, is observable. $\qquad \square$

Theorem 5 should be contrasted with the behavior of infinite sets, such as the set $\mathbb{N}$ of all natural numbers. $\mathbb{N}$ is observable, but there are elements of $\mathbb{N}$ that are not observable (Theorem 2).

Rule 6. *Let m be a positive integer and $x, x_1, \ldots, x_m, a_1, \ldots, a_m$ be real numbers.*

(1) If $x_1, \ldots, x_m$ are not ultralarge and m is observable, then

$$\sum_{i=1}^{m} x_i \ \text{and} \ \prod_{i=1}^{m} x_i$$

are not ultralarge.

(2) If $x_i \simeq a_i$ for $i = 1, \ldots, m$ and m is observable, then

$$\sum_{i=1}^{m} x_i \simeq \sum_{i=1}^{m} a_i.$$

(3) If $x \simeq 1$ and m is observable, then $x^m \simeq 1$.

(4) If $x_i \simeq a_i$ for $i = 1, \ldots, m$, each a_i is observable, and m is observable, then

$$\prod_{i=1}^{m} x_i \simeq \prod_{i=1}^{m} a_i.$$

Proof.   (1) First note that for every x which is not ultralarge there is some observable positive integer n such that $|x| < n$. [If $|x| \leq b$ where b is an observable real number, take n to be the least positive integer greater than b; n is observable by Closure.] Hence the *least* positive integer N such that $|x| < N$ is also observable, by the Corollary to Theorem 4 (as $N \leq n$).

Let N_i be the least positive integer such that $|x_i| < N_i$ and let $N = \max\{N_1, \ldots, N_m\}$. Then N is observable and

$$\left| \sum_{i=1}^{m} x_i \right| \leq \sum_{i=1}^{m} |x_i| \leq N \cdot m \qquad \text{and}$$

$$\left| \prod_{i=1}^{m} x_i \right| = \prod_{i=1}^{m} |x_i| \leq N^m,$$

where $N \cdot m$ and N^m are observable.

(2) We assume that each $x_i = a_i + \varepsilon_i$ where $\varepsilon_i \simeq 0$. Then

$$\sum_{i=1}^{m} x_i = \sum_{i=1}^{m} a_i + \sum_{i=1}^{m} \varepsilon_i.$$

Let $\varepsilon = \max\{|\varepsilon_1|, \ldots, |\varepsilon_m|\}$ and note that $\varepsilon \simeq 0$. We have

$$\left| \sum_{i=1}^{m} \varepsilon_i \right| \leq \sum_{i=1}^{m} |\varepsilon_i| \leq \sum_{i=1}^{m} \varepsilon = \varepsilon \cdot m \simeq 0,$$

because m is not ultralarge.

(3) Write $x = 1 + \varepsilon$ with $\varepsilon \simeq 0$. By the Binomial Theorem,

$$x^m = (1+\varepsilon)^m = 1 + \binom{m}{1} \cdot \varepsilon + \binom{m}{2} \cdot \varepsilon^2 + \ldots + \binom{m}{m} \cdot \varepsilon^m \simeq 1,$$

because the binomial coefficients are not ultralarge since $\binom{m}{k} \leq m^k \leq m^m$. Thus each term except the first is ultrasmall or 0, and their sum is ultrasmall or 0 by (2).

(4) We first assume that $a_i \neq 0$ for all i. We let $\xi_i = x_i/a_i$; then $\xi_i \simeq 1$ by Rule 5(3), and it suffices to prove that

$$\prod_{i=1}^{m} \xi_i \simeq 1$$

and then multiply both sides by the non-ultralarge number $\prod_{i=1}^{m} a_i$ by (1) and Rule 5(2).

Write $\xi_i = 1 + \varepsilon_i$ with $\varepsilon_i \simeq 0$ and let $\varepsilon = \max\{|\varepsilon_1|, \ldots, |\varepsilon_m|\}$. Then

$$(1 - \varepsilon) \leq 1 - |\varepsilon_i| \leq \xi_i \leq 1 + |\varepsilon_i| \leq (1 + \varepsilon)$$

holds for each i, so

$$(1 - \varepsilon)^m \leq \prod_{i=1}^{m} \xi_i \leq (1 + \varepsilon)^m.$$

As both $(1 - \varepsilon)^m \simeq 1$ and $(1 + \varepsilon)^m \simeq 1$ by (3), the claim $\prod_{i=1}^{m} \xi_i \simeq 1$ follows.

Now assume that some $a_i = 0$; without loss of generality $a_1 = 0$. Then $\prod_{i=1}^{m} x_i = x_1 \cdot \prod_{i=2}^{m} x_i \simeq 0 = \prod_{i=1}^{m} a_i$, because $x_1 \simeq a_1 = 0$ and $\prod_{i=2}^{m} x_i$ is not ultralarge. $\qquad \square$

Exercise 13 (Answer page 244)
Show that (1)–(4) need not hold if m is not observable.

The following example is characteristic of how the Closure Principle can be used in proofs.

Let f be a function defined on an interval I and bounded from above.

There is $M \in \mathbb{R}$ such that $f(x) \leq M$, for all $x \in I$.

This statement has parameters f and I; assume that they are observable. By the Closure Principle, there exists an observable M such that $f(x) \leq M$, for all $x \in I$. Thus, if an observable function f is bounded above on an observable interval I, then it has an observable upper bound.

Exercise 14 (Answer page 244)
Let f be an observable function defined on an observable open interval I. Assume that $f(x)$ is positive ultralarge, for some $x \in I$. Show that f is unbounded above; that is, for each $M \in \mathbb{R}$ there is $x \in I$ such that $f(x) \geq M$.

Exercise 15 (Answer page 245)
Let f be an observable function defined on an observable interval I. Show that if there exists a $c \in I$ such that $f(c) = 0$, then it is possible to find such a $c \in I$ which is observable.

Exercise 16 (Answer page 245)
Let f be an observable function. Show that if there exist M and L such that $f(x) = L$ for all $x \geq M$, then it is possible to choose observable M, L with this property; in particular $f(x) = L$ for all ultralarge positive x.

We conclude this section with a contrapositive version of the Closure Principle, equivalent to the existential version, but sometimes more convenient to use.

Closure Principle, Universal Version

Given a statement of traditional mathematics that has parameters among $p, p_1, \ldots, p_k$: If the statement is true for all observable p, then the statement is true for all p.

Proof. We proceed by contradiction, and assume that the statement is true for all observable p, but not for all p. Then there exists some p for which the negation of the statement is true. By the existential version of Closure applied to the negation of the original statement, there is some observable p for which the negation of the statement is true. In other words, the original statement is not true for all observable p, contradicting the assumption.   □

Exercise 17 (Answer page 245)
Deduce the existential version of the Closure Principle from the universal version.

1.5   Relativization and Stability

Our goal in the main body of the book is to use ultrasmall numbers to develop differential and integral calculus. But there is still an important issue that has to be addressed first. Let us consider the definition of the instantaneous rate of change in some more detail. For a function f and a point x, the average rate of change in an ultrasmall interval $[x, x + h]$ is defined by the ratio

$$\frac{f(x + h) - f(x)}{h}.$$

The instantaneous rate of change (also called the *derivative*) of f at x is the observable part of this ratio. As an example, let $f(x) = x^2$. We get

$$\frac{f(x+h) - f(x)}{h} = \frac{(x+h)^2 - x^2}{h} = \frac{2x \cdot h + h^2}{h} = 2x + h.$$

If x is observable, then $2x + h$ is ultraclose to the observable number $2x$ and we conclude that the derivative of $f(x) = x^2$ at x is $2x$.

But what if x is not observable? One would like to conclude that the derivative of $f(x) = x^2$ is $2x$ for all x, not just the observable ones. Indeed it has to be so, because derivatives can be defined by methods of traditional mathematics (albeit in a more complicated way), and we assert that all results of traditional mathematics are valid for all x, whether standard or not. However, the simpler, nonstandard definition given above does not apply to all x (as yet). It does not give correct results when x is not observable. For example, let $x = h$. The correct value for the derivative of $f(x) = x^2$ at $x = h$ is $2x = 2h$. However, a calculation gives

$$\frac{f(x+h) - f(x)}{h} = \frac{(h+h)^2 - h^2}{h} = \frac{(2h)^2 - h^2}{h} = 3h$$

and the observable part of $3h$ is 0, not $2h$.

It is easy to see where the problem is: The ultrasmall number h that we took to represent the duration of an instant is negligible when compared to an observable value of x, but it is *not* negligible when compared to $x = h$. To make the calculation work correctly for $x = h$, we need to use an instant whose duration is negligible relative to h, that is, *ultrasmall relative to h*. The same issue arises when one tries to compute the derivative of $f(x) = ax^2$ for non–observable values of a, and in the nonstandard approach to other fundamental concepts of calculus, such as continuity, limits and integrals. Our present framework would allow us to define these concepts for standard functions at standard points, but not in general.

We resolve this issue by making observability a *relative concept*. That is, we assume that the universe of mathematical objects (including both the standard and the ideal ones) is stratified into *levels of observability*. The standard objects are always observable. If, say, h is ultrasmall (relative to the standard objects), then it is not observable relative to the standard objects; but the standard objects, as well as h itself and other objects uniquely definable from h (such as $2h$, $h^3/2$, $1/h$) are observable relative to h. However, there are also numbers that are not observable relative to h. Among them are numbers ultrasmall relative to h. Such "second-order" ultrasmall numbers can then be used in the definition of the derivative at h. There are also numbers unobservable relative to these, and so on. The guiding principle is that *all levels of observability should have the same properties*. In particular, each level satisfies the Closure Principle, and so it is closed under traditional mathematical operations. This is a strong, uniform version of the principle that can be traced to Leibniz, namely, that the ideal elements have the same properties as the standard ones.

An analogy with physics may be helpful in visualizing relative observability. The literature of physics is full of references to phenomena at various *scales*: the macroscopic scale, the microscopic scale, the atomic scale, large scale versus small scale and so on. The quantities at the macroscopic scale are those observable with a naked eye (they are "always observable"). Optical technology (microscopes and telescopes) enables us to observe objects otherwise invisible, such as bacteria and faint stars, but all objects that are observable with a naked eye also remain visible at this level of technology. Compared to macroscopic quantities, such as a diameter of a soccer ball, quantities at the microscopic scale, such as the diameter of a bacterium, are "ultrasmall"; more precisely, they are so small as to be negligible in any considerations of macroscopic phenomena. A higher level of technology (electron microscopes, radio telescopes) allows for observation of additional objects, such as molecules and quasars. Diameter of a molecule is negligible compared to microscopic quantities, such as the diameter of a bacterium. Yet higher levels of technology (particle accelerators) enable even finer observations (subatomic particles).

The approach taken in this book is a very idealized version of this point of view. The standard objects are those that are always observable. Every ideal, nonstandard object is observable at some level, although not at the level of standard objects. For every object p (standard or not) there exist nonzero real numbers smaller (in absolute value) than all positive real numbers that are observable at the level where p is observable; they are ultrasmall relative to that level and not observable at that level. The reciprocals of the ultrasmall numbers are larger (in absolute value) than every real number observable at that level; they are ultralarge relative to that level.

We now proceed to describe this intuition axiomatically.

Observability is a primitive relation of two arguments:

q is observable relative to p.

For variety, we sometimes rephrase it informally as "*q is observable when p is observable*" or "*q is as observable as p.*"

We state precisely the axioms on which our reasoning about observability is based.

We begin with three elementary properties; they postulate that observability is a total pre-ordering.

Relative Observability Principle
For all p, q and r:

   (1) *p is observable relative to p.*

   (2) *If p is observable relative to q and q is observable relative to r, then p is observable relative to r.*

   (3) *If p is not observable relative to q, then q is observable relative to p.*

We say that q **is observable relative to** p_1 **and** p_2 if q is observable relative to p_1 or q is observable relative to p_2.

More generally, given a list $p_1, \ldots, p_k$, we say that

$$q \text{ \textbf{is observable relative to} } p_1, \ldots, p_k$$

if q is observable relative to some (at least one) p_i, $i = 1, \ldots, k$, and we refer to the list $p_1, \ldots, p_k$ as the **context**. The term "list" always means an explicitly given finite collection. The empty collection is also allowed; by definition, objects are observable relative to the empty context if they are observable relative to every context. We call them **standard** (and identify them intuitively with the objects of traditional mathematics).

Exercise 18 (Answer page 245)
If x is observable relative to $q_1, \ldots, q_\ell$, then x is observable relative to $p, q_1, \ldots, q_\ell$. If x is observable relative to $p, q_1, \ldots, q_\ell$ and p is observable relative to $q_1, \ldots, q_\ell$, then x is observable relative to $q_1, \ldots, q_\ell$.

In accordance with the idea that all levels of observability should have the same properties, we re-interpret the definitions and axioms given so far as applicable to every level.

Definitions 1 and 2 and the definition of observable neighbor apply to any context. The Existence, Closure and Observable Neighbor Principles are valid relative to any context.

Example.

   (1) The number 1 is standard (observable relative to every context); therefore every x observable relative to 1 is standard.

(2) If the real numbers x and y are observable relative to $p_1, \ldots, p_k$, then $x \pm y$, $x \cdot y$, and x/y (if $y \neq 0$) are observable relative to $p_1, \ldots, p_k$.

(3) A real number is ultrasmall relative to $p_1, \ldots, p_k$ if it is nonzero and its absolute value is less than any positive real number observable relative to $p_1, \ldots, p_k$. Similarly for ultralarge and ultraclose numbers.

(4) The Observable Neighbor Principle asserts, in detail:

Given any context $p_1, \ldots, p_k$: If a real number x is not ultralarge relative to $p_1, \ldots, p_k$, then there is a real number r observable relative to $p_1, \ldots, p_k$ that is ultraclose to x relative to $p_1, \ldots, p_k$.

As before, this real number r is unique; it is called the observable neighbor of x relative to $p_1, \ldots, p_k$.

(5) The Closure Principle asserts:

Given a statement of traditional mathematics with parameters $p, p_1, \ldots, p_k$:

If $p_1, \ldots, p_k$ are observable relative to $q_1, \ldots, q_\ell$ and there exists some object p for which the statement is true, then there exists some object p observable relative to $q_1, \ldots, q_\ell$ for which the statement is true.

Therefore, **all the results obtained in the previous sections are valid relative to any given context.**

Example.

(1) For any $p_1, \ldots, p_k$: If x, y are not ultralarge relative to $p_1, \ldots, p_k$, then $x \pm y$ and $x \cdot y$ are not ultralarge relative to $p_1, \ldots, p_k$. [Rule 1 (1).]

(2) For every $p_1, \ldots, p_k$ there exist natural numbers ultralarge relative to $p_1, \ldots, p_k$. [Theorem 2.]

Let ε be ultrasmall relative to 1; then there exists an ultrasmall number, say δ, which is ultrasmall relative to ε (hence also ultrasmall relative to 1).

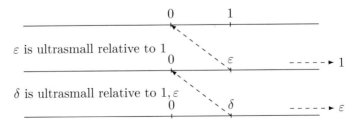

Exercise 19 (Answer page 245)

Let p be observable relative to $q_1, \ldots, q_\ell$. Show that x is ultrasmall relative to $p, q_1, \ldots, q_\ell$ if and only x is ultrasmall relative to $q_1, \ldots, q_\ell$. Similarly for ultralarge, ultraclose and observable neighbor.

It may seem that the need to pay attention to the context of observability could become a major headache, but this is not the case. We next introduce a key convention that takes care of the context automatically, and often eliminates the need to pay it explicit attention. It is used throughout the rest of the book, from this section on (but *not* in Sections 1.2–1.4, where relative concepts are defined and their properties proved relative to *any* context, as explained above).

Relative concepts (observability, ultrasmall, ultralarge, ultraclose and observable neighbor) usually occur not on their own, but in definitions, theorems and proofs. Theorems are statements, and hence may have parameters; we call them the **context of the theorem**. Similarly in a definition, some new concept is defined by a statement involving previously defined concepts, and that statement may have parameters; we call them the **context of the definition**. Unless explicitly stated otherwise, we take the **context of a proof** to be the context of the theorem being proved.

The following convention greatly reduces notational burden and simplifies the presentation.

Convention about contexts

In a theorem, definition, or proof, whenever a relative concept is used without explicit specification of its context, it is understood to be relative to the context of that theorem, definition, or proof.

For example, in Section 2.1 we define continuity of a function f at a by the defining statement "*For all x, if $x \simeq a$, then $f(x) \simeq f(a)$.*" The parameters of this statement are f, a; this is the context of the definition. According to the convention about contexts, this statement is to be understood as "*For all x, if $x \sim a$ relative to f, a, then $f(x) \simeq f(a)$, relative to f, a.*"

Definition 3. *(1) A statement is **internal** if the context of every relative concept that occurs in it is given by the parameters of the statement. We refer to the parameters of the internal statement as its* context.

*(2) An **internal concept** is a concept defined by an internal statement.*

(3) Previously defined internal concepts can be used in subsequent internal statements.

In particular, all statements in the language of traditional mathematics are internal, because no references to relative concepts occur in them. The statement *"For all x, if x $\simeq$ a relative to f, a, then f(x) $\simeq$ f(a), relative to f, a"* is internal. Therefore, the concept *"f is continuous at a,"* which is defined by it, is an internal concept. Hence the statement *"f is continuous at a for every a in its domain"* is also internal.

The statements *"y is ultrasmall relative to x"* and *"There exists x such that y is ultrasmall relative to x"* are not internal, but the statement *"There exists y such that y is ultrasmall relative to x"* is internal (and true for all x).

Obviously, all concepts defined according to our convention about contexts are automatically internal. They include the fundamental notions of calculus: continuity (see Definition 4), differentiability (Definition 20) and others.

It gets better! Relative concepts, that is, concepts dependent on the context, are not used in traditional mathematics. Hence, in order to be of interest to traditional mathematicians, the results obtained in this book have to be independent of the context. The internal statements have precisely this property. This is the content of the Stability Principle.

Stability Principle

An internal statement is equivalent to the statement obtained from it by extending its context by additional parameters.

We give some examples.

- *"For every x, x $\simeq$ a implies 2x $\simeq$ 2a"* is a theorem (see Rule 5). Recall that by our convention $\simeq$ is to be understood relative to a, the context of the theorem. By Stability, *"For every x, x $\simeq$ a implies 2x $\simeq$ 2a, where $\simeq$ is understood to be relative to a and $q_1, \ldots, q_\ell$"* is also true (for any $q_1, \ldots, q_\ell$). Assume now that a is observable relative to $q_1, \ldots, q_\ell$. By Exercise 19, $\simeq$ relative to $a, q_1, \ldots, q_\ell$ is equivalent to $\simeq$ relative to $q_1, \ldots, q_\ell$. Hence the statement *"For every x, x $\simeq$ a implies 2x $\simeq$ 2a"* is true when $\simeq$ is understood to be relative to $q_1, \ldots, q_\ell$. That is, it is true in any context where a is observable.

Arguably, this example is not very impressive, because the conclusion can be obtained directly from Rule 5, but it verifies the validity of Stability in this case. However, in general Stability provides information that is not obtainable otherwise.

- Consider once again the statement *"For all x, if x $\simeq$ a, then f(x) $\simeq$ f(a)."* By our convention about contexts, $\simeq$ is to be taken relative to the context of the statement, that is, f, a. By Stability, the statement is true (in its context f, a) if and only if it is true in the context $f, a, q_1, \ldots, q_\ell$, for any $q_1, \ldots, q_\ell$. As in the previous example, if f, a are observable relative to $q_1, \ldots, q_\ell$, then $\simeq$ holds relative to $f, a, q_1, \ldots, q_\ell$ if and only if it holds relative to $q_1, \ldots, q_\ell$. It follows that the statement is true in *some* context where the parameters f and a are observable, if and only if it is true in *every* context where the parameters f and a are observable.

The last example is of great importance, and applies generally. By our convention, if a theorem does not specify the context of the relative concepts used in it, then we understand this context to be that of its parameters. By Stability and Exercises 18, 19, the theorem is then true in every context where the parameters are observable. Conversely, if the theorem is true in some context where its parameters are observable, then it is true also in the context specified by the parameters. Similar remarks apply to definitions. In summary:

When giving definitions or stating theorems and their proofs according to our conventions, the precise specification of the context is unimportant. The only requirement is that the parameters of the definition or theorem be observable relative to it.

Caution: Proofs can introduce auxiliary unobservable objects, so not every statement in a proof need be internal.

It is still a good idea to be aware of the context, if only in order to avoid the error of treating some auxiliary unobservable object, introduced in the course of a proof, as observable. We often point out at the beginning of a proof what parameters its context has to include. Usually, they are the parameters of the theorem being proved. But it is not necessary to pay excessive attention to this matter; any context where all the parameters are observable will do.

We conclude this section by stating the final version of the Closure Principle.

Closure Principle, Existential Version

Given an internal statement with parameters $p, p_1, \ldots, p_k$:
If $p_1, \ldots, p_k$ are observable and there exists some object p for which the statement is true, then there exists some observable object p for which the statement is true.

1.6 Sets and Induction

The last principle we need deals with the way one usually defines sets and functions. If $\mathcal{P}(x)$ describes some property of x and A is a given set, then there exists a unique set X such that, for all x, $x \in X$ if and only if $x \in A$ and $\mathcal{P}(x)$; we denote it

$$\{x \in A : \mathcal{P}(x)\}.$$

Similarly, when defining a function, one has to specify a set A (the domain of the function) and a rule (statement, formula) $\mathcal{P}(x,y)$ that assigns to each $x \in A$ a unique value y; then we can define a function f with domain A by

$$f(x) = y \quad \text{if and only if} \quad \mathcal{P}(x,y) \text{ holds.}$$

In elementary textbooks it is customary to say that the function f *is* the rule, but this idea is open to some objections. In particular, different rules can assign the same value to each $x \in A$, and thus describe the same function. For example, $f(x) = 1$ and $f(x) = \sin^2(x) + \cos^2(x)$ are different rules, but they describe the same function with domain $\mathbb{R}$. For this and other reasons, it is more correct to say that a function is a set:

$$f = \{\langle x, y \rangle : x \in A \text{ and } \mathcal{P}(x,y)\}.$$

In other words, we identify functions with their graphs.

One of the parameters of such a definition is A. The **defining statements** $\mathcal{P}(x)$ or $\mathcal{P}(x,y)$ can involve additional parameters. The variables x and y are also parameters of the defining statements, but they do not count as parameters of the **definition**, because they become bound in it; for example, $\{x \in A : \mathcal{P}(x)\}$ is the set of *all* $x \in A$ such that $\mathcal{P}(x)$; the variable x is bound. One could just as well describe this set as $\{z \in A : \mathcal{P}(z)\}$.

Definition Principle

Internal defining statements can be used to define sets and functions. These sets and functions are observable whenever all the parameters of their definition are observable.

It follows from the Definition Principle that the Principle of Mathematical Induction applies to internal statements (see Section 5.1 for details). On the other hand, statements that are not internal (*external statements*) need not define sets and the Induction Principle may fail for such statements. External statements allow us to single out the observable elements of A from among all the elements. But, as discussed in the Introduction, no (infinite) set can contain observable elements only. However, it is perfectly legitimate to make external statements and to use them in proofs, as long as one avoids collecting all objects that satisfy such a statement into a set.

Example.

(1) Consider the statement "n is observable relative to p," for a fixed p. There is no set S such that $n \in S$ if and only if $n \in \mathbb{N}$ and n is observable relative to p. In other words, the "collection" $\{n \in \mathbb{N} : n \text{ is observable relative to } p\}$ is not a set.

Proof: Assume $S = \{n \in \mathbb{N} : n \text{ is observable relative to } p\}$ is a set. Then

(a) $0 \in S$, because $0 \in \mathbb{N}$ and 0 is observable relative to p.

(b) If $n \in S$, then $n \in \mathbb{N}$ and n is observable relative to p, so $n+1 \in \mathbb{N}$ and $n+1$ is observable relative to p (the latter follows from the Closure Principle); hence $n + 1 \in S$.

By the Principle of Mathematical Induction applied to the statement (of traditional mathematics) "$n \in S$" we conclude that $\mathbb{N} = S$, that is, all $n \in \mathbb{N}$ are observable relative to p. This is a contradiction with Theorem 2.

(2) Consider

$$x \mapsto \begin{cases} 1 & \text{if } x \text{ is observable relative to } p; \\ 0 & \text{otherwise.} \end{cases}$$

The defining statement is not internal and does not define a function. If such a function g existed, then $\{x \in \mathbb{R} : x \in \mathbb{N} \text{ and } g(x) = 1\} = \{x \in \mathbb{N} : x \text{ is observable relative to } p\}$ would be a set, contradicting (1).

Exercise 20 (Answer page 245)
Which of the following statements define functions? For those that do, when is the function observable?

(1) $x \mapsto x^2$, $x \in \mathbb{R}$.

(2) Let a be a positive number ultralarge relative to p;
$x \mapsto \frac{1}{x}$, $x \in (0, a]$.

(3) Let b be ultralarge relative to p;

$$x \mapsto \frac{b}{2b}x, \ x \in \mathbb{R}.$$

(4) $x \mapsto \begin{cases} 2x & \text{if } x \text{ is ultrasmall relative to } p; \\ 0 & \text{otherwise.} \end{cases}$

1.7   Summary

For easy reference, we list below all the axioms that deal with observability.

> ### Relative Observability Principle
> *For all p, q and r:*
>
> (1) *p is observable relative to p.*
>
> (2) *If p is observable relative to q and q is observable relative to r, then p is observable relative to r.*
>
> (3) *If p is not observable relative to q, then q is observable relative to p.*

This principle allows us to fix the context.

> ### Stability Principle
> *An internal statement is equivalent to the statement obtained from it by extending its context by additional parameters.*

All of following principles are relative to a given context:

> ### Existence Principle
> *There exist ultrasmall real numbers.*

> ### Closure Principle   *Given an internal statement with parameters $p, p_1, \ldots, p_k$: If $p_1, \ldots, p_k$ are observable and there exists some object p for which the statement is true, then there exists some observable object p for which the statement is true.*

> ### Observable Neighbor Principle
> *If a real number is not ultralarge, then there is an observable real number that is ultraclose to it.*

Definition Principle

Internal defining statements can be used to define sets and functions. These sets and functions are observable whenever all the parameters of their definition are observable.

The Closure Principle is actually a consequence of the Stability Principle. The Observable Neighbor Principle and the Definition Principle can both be deduced from the so-called Standardization Principle. We leave these matters for the Appendix.

1.8   Additional Exercises

Exercise 1.1
The number ε is ultrasmall if and only if $\varepsilon \neq 0$ and $|\varepsilon| \leq r$ for all observable $r > 0$. Similarly, $M \in \mathbb{R}$ is ultralarge if and only if $|M| \geq r$ for all observable $r > 0$.

Exercise 1.2
Prove that if $h > 0$ is ultrasmall, then $\sqrt{h}$ is ultrasmall.

Exercise 1.3
If $n \in \mathbb{N}$, $n > 0$, n is observable and $h > 0$ is ultrasmall, then $\sqrt[n]{h}$ is ultrasmall.

Exercise 1.4
Assume that ε is positive and ultrasmall and H is positive and ultralarge. Determine whether the given expression yields an ultrasmall number, an ultralarge number, or a number which is neither ultrasmall nor ultralarge.

(1) $\dfrac{\sqrt{\varepsilon}}{\varepsilon + 1}$

(2) $\dfrac{H - 1}{H^2 - 1}$

(3) $\dfrac{\varepsilon^2 - 5\varepsilon + 1}{5\varepsilon^2 + 4}$

(4) $\dfrac{\sqrt{1 + \varepsilon} - 1}{\varepsilon}$

(5) $\dfrac{1}{\varepsilon} \cdot \left(\dfrac{1}{2 + \varepsilon} - \dfrac{1}{2} \right)$

(6) $\dfrac{H^2 - 5H + 1}{5H^2 + 4}$

Exercise 1.5

(1) Apply the Closure Principle to the statement "x is a real number and $x + x = x$" and conclude that the number 0 is observable.

(2) Similarly, use the statement "x is a real number, $x > 0$ and $x \cdot x = x + x$" to conclude that 2 is observable.

(3) Consider the statement "x is a real number, $x > 0$ and $x^2 = 2$" and conclude that $\sqrt{2}$ is observable. (We can reach the same conclusion even more directly, by considering the statement "$x = \sqrt{2}$.")

Exercise 1.6
For each of the following mathematical objects, determine when it is observable.

(1) Numbers $a + b$, $3xy$, e^x.

(2) The function $f : \mathbb{R} \to \mathbb{R}$ defined by $f(x) = e^x$.

(3) The function $g : \mathbb{R} \times \mathbb{R} \to \mathbb{R}$ defined by $g(x, y) = x + y$.

(4) The set $C = \{\langle x, y \rangle \in \mathbb{R} \times \mathbb{R} : x^2 + y^2 = a^2\}$.

Exercise 1.7
Show that the assertion in the preceding exercise need not hold when n is not observable.
Hint: Use Exercise 9.

Exercise 1.8
If x is ultrasmall [respectively, ultralarge] relative to p_1, p_2, then x is ultrasmall [respectively, ultralarge] relative to p_1.
If $x \simeq y$ relative to $p, p_1, \ldots, p_k$, then $x \simeq y$ relative to $p_1, \ldots, p_k$.

Exercise 1.9
Show that (relative to a fixed context):

(1) $\{x \in \mathbb{R} : x \text{ is not ultralarge}\}$ is not a set.

(2) $\{h \in \mathbb{R} : h \text{ is ultrasmall}\}$ is not a set.

(3) For any $x \in \mathbb{R}$, $\{y \in \mathbb{R} : y \simeq x\}$ is not a set.

Exercise 1.10
Use the Principle of Mathematical Induction to prove that, for all $n \geq 1$,

$$1^2 + 2^2 + \ldots + n^2 = \frac{n(n+1)(2n+1)}{6}.$$

Exercise 1.11
Let $x_1 = \sqrt{2}$, $x_{n+1} = \sqrt{2 + x_n}$. Use the Principle of Mathematical Induction to prove that $x_n \leq 2$, for all $n \geq 1$.

Exercise 1.12
Let $a > 0$; prove by induction that $(1 + a)^n \geq 1 + na$, for all $n \in \mathbb{N}$.

Exercise 1.13
Prove by induction: If $0 < \varepsilon < 1$, then $0 < \varepsilon^{n+1} < \varepsilon^n$ holds for all $n \in \mathbb{N}$. Conclude that if ε is ultrasmall, then ε^n is ultrasmall, for all $n > 0$. Similarly, if H is ultralarge, then H^n is ultralarge for all $n > 0$.

2

Continuity and Limits

2.1 Continuity

Intuitively, a function is continuous if an ultrasmall change of the argument produces an ultrasmall change (or no change) of the value of the function.

In order to discuss continuity of a function f at a point a, we require f to be defined at least on an open interval containing a.

Definition 4. *Let f be a function defined on an open interval containing a. We say that f is **continuous at** a if*

$$f(x) \simeq f(a), \qquad whenever \ x \simeq a.$$

The above definition is about f and a. The expression "whenever $x \simeq a$" is a paraphrase of "for all $x \simeq a$," hence x is a bound variable. The parameters therefore are f and a, and according to our context convention, the symbol $\simeq$ is to be understood relative to any context where f and a are observable.

It follows from the Closure Principle that f is defined on some *observable* open interval containing a. Such interval contains all $x \simeq a$, so $f(x)$ is defined for all $x \simeq a$.

Upon letting $h = x - a$ we see that f is continuous at a if and only if

$$f(a + h) \sim f(a), \qquad whenever \ h \simeq 0.$$

Example. We examine the question of continuity of a few functions.

(1) The function f defined by $f : x \mapsto x^3$ is continuous at a for all $a \in \mathbb{R}$.

Proof. The function f is standard, and a is a parameter; hence $\simeq$ is to be understood relative to a context where a is observable.

Let $x \simeq a$; then $f(x) = x^3 \simeq a^3 = f(a)$, by Rule 5. □

(2) The function f defined by $f : x \mapsto \sqrt{x}$ is continuous at all $a > 0$.

Proof. The parameter is a. We write $f(x) = \sqrt{x} = \sqrt{a} \cdot \sqrt{\frac{x}{a}}$. For $x \simeq a$, $\frac{x}{a} \simeq 1$ by Rule 5, and then $\sqrt{\frac{x}{a}} \simeq 1$ (Exercise 8). Hence $f(x) \simeq \sqrt{a} = f(a)$ ($\sqrt{a}$ is not ultralarge). $\qquad\qquad\square$

(3) The function f defined by $f : x \mapsto x^m$, where m is a positive integer, is continuous at a for all $a \in \mathbb{R}$.

Proof. The parameters are a and m. Let $x \simeq a$; then $f(x) = x^m \simeq a^m = f(a)$, by Rule 6. $\qquad\qquad\square$

(4) The function f defined by $f : \theta \mapsto \sin(\theta)$ is continuous at θ for all $\theta \in \mathbb{R}$.

Proof. We give a geometric proof. The sine function is standard. Let θ be given and let δ be ultrasmall (relative to θ). Consider the point B on the unit circle given by angle θ and the point C given by the angle $\theta + \delta$. The chord BC, being a straight line, is shorter than the arc δ from B to C. Hence the length of the chord $\overline{BC}$ is ultrasmall.

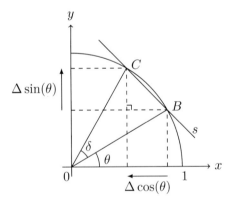

The variation along the x-axis is $\Delta \cos(\theta) = \cos(\theta+\delta) - \cos(\theta)$ and the variation along the y-axis is $\Delta \sin(\theta) = \sin(\theta + \delta) - \sin(\theta)$. By the Pythagorean Theorem,

$$(\Delta \cos(\theta))^2 + (\Delta \sin(\theta))^2 = \overline{BC}^2 \simeq 0.$$

Hence, $(\Delta \sin(\theta))^2 \simeq 0$ and $(\Delta \cos(\theta))^2 \simeq 0$. But we showed in Exercise 1.2 that this implies $\Delta \sin(\theta) \simeq 0$ and $\Delta \cos(\theta) \simeq 0$. Hence sine and cosine are continuous at all $\theta \in \mathbb{R}$. $\qquad \square$

(5) The Heaviside function H is given by

$$H(x) = \begin{cases} 1 & \text{if } x \geq 0; \\ 0 & \text{if } x < 0. \end{cases}$$

Then H is not continuous at 0.

Proof. Let h be ultrasmall; if $h > 0$, then $H(h) = 1 = H(0)$, but if $h < 0$, then $H(h) = 0 \not\simeq 1 = H(0)$. $\qquad \square$

(6) The function f given by

$$f(x) = \begin{cases} \sin\left(\frac{1}{x}\right) & \text{if } x \neq 0; \\ 0 & \text{if } x = 0 \end{cases}$$

is not continuous at 0.

Proof. Let $N \in \mathbb{N}$ be ultralarge. Then $h = \frac{1}{2\pi N + \pi/2}$ is ultrasmall and $f(h) = \sin\left(2\pi N + \frac{\pi}{2}\right) = 1$ is not ultraclose to $0 = f(0)$. $\qquad \square$

(7) The function g given by

$$f(x) = \begin{cases} x \cdot \sin\left(\frac{1}{x}\right) & \text{if } x \neq 0; \\ 0 & \text{if } x = 0 \end{cases}$$

is continuous at all $a \in \mathbb{R}$.

Proof. Let h be ultrasmall (relative to a). For $a = 0$, $|f(a + h)| = |f(h)| = \left|h \cdot \sin\left(\frac{1}{h}\right)\right| \leq |h|$, so $f(a + h) \simeq 0 = f(a)$. For $a \neq 0$ we have $a + h \simeq a$, $\frac{1}{a+h} \simeq \frac{1}{a}$ (Rule 5) and $\sin\left(\frac{1}{a+h}\right) \simeq \sin\left(\frac{1}{a}\right)$ (by (3)), hence $f(a + h) = (a + h) \cdot \sin\left(\frac{1}{a+h}\right) \simeq a \cdot \sin\left(\frac{1}{a}\right) = f(a)$. $\qquad \square$

(8) The Dirichlet function, defined by

$$f(x) = \begin{cases} 0 & \text{if } x \text{ is rational}; \\ 1 & \text{if } x \text{ is irrational}, \end{cases}$$

is not continuous at a for any $a \in \mathbb{R}$.

Proof. Fix $a \in \mathbb{R}$ and let $h > 0$ be ultrasmall (relative to a). We recall that every interval contains both rational and irrational numbers. Take $a', a'' \in [a, a+h]$ where a' is rational and a'' is irrational; then $a' \simeq a$, $a'' \simeq a$, but $f(a') = 0$ and $f(a'') = 1$, so one of these values is not ultraclose to $f(a)$. $\square$

The next theorem states that continuity is preserved under the basic operations.

Theorem 6. *Let f and g be functions. If f and g are continuous at a, then*

    (1) $f \pm g$ is continuous at a.

    (2) $f \cdot g$ is continuous at a.

    (3) $\frac{f}{g}$ is continuous at a, provided $g(a) \neq 0$.

Proof. Assume that f and g are continuous at a. Let $x \simeq a$. Then by continuity we have $f(x) \simeq f(a)$ and $g(x) \simeq g(a)$. By Closure, $f(a)$ and $g(a)$ are observable, so none of the numbers $f(x)$, $g(x)$, $f(a)$, $g(a)$ are ultralarge. By Rule 5 we have $f(x) \pm g(x) \simeq f(a) \pm g(a)$, $f(x) \cdot g(x) \simeq f(a) \cdot g(a)$. For the quotient, notice in addition that if $g(a) \neq 0$, then $g(x) \not\simeq 0$; hence $\frac{f(x)}{g(x)} \simeq \frac{f(a)}{g(a)}$. We conclude that $f \pm g, f \cdot g$ and $\frac{f}{g}$ are continuous at a $\square$

Note that the context of this proof is specified by f, g and a. In most cases, it is clear what the parameters are and we do not spell them out explicitly.

The *composition* of functions f and g is the function $f \circ g$, defined by

$$f \circ g : x \mapsto f(g(x)).$$

Theorem 7. *If g is continuous at a and f is continuous at $g(a)$, then $f \circ g$ is continuous at a.*

Proof. Let $x \simeq a$. By continuity of g at a, we have $g(x) \simeq g(a)$. By continuity of f at $g(a)$ (by Closure, $g(a)$ is observable), we have $f(g(x)) \simeq f(g(a))$. Hence $(f \circ g)(x) \simeq (f \circ g)(a)$. $\square$

If a is a left or right endpoint of an interval on which the function is defined, it makes sense to consider continuity on the right or on the left.

Definition 5. *Let f be a function and $a \in \mathbb{R}$.*

(1) *Suppose that f is defined on an interval of the form $(b, a]$, with $b < a$. We say that f is* **continuous on the left at** *a if*

$$f(x) \simeq f(a), \qquad \text{whenever } x \simeq a \text{ and } x < a.$$

(2) *Suppose that f is defined on an interval of the form $[a, b)$, with $a < b$. We say that f is* **continuous on the right at** *a if*

$$f(x) \simeq f(a), \qquad \text{whenever } x \simeq a \text{ and } x > a.$$

Example. The function $f : [0, \infty) \to \mathbb{R}$ defined by $f(x) = \sqrt{x}$ is continuous on the right at 0. Let $x > 0$ be such that $x \simeq 0$. Then $\sqrt{x} \simeq 0$, by Exercise 1.2.

Definition 6. *A function f defined on an interval I is* **continuous on** *I if it is continuous at every a in I, with the understanding that if a is the left (respectively, right) endpoint of I, then only continuity on the right (respectively, on the left) is required.*

A direct consequence of previous theorems is that, if f and g are functions continuous on an interval I, then the sum, difference, and product are also continuous on I. The quotient is continuous on the interval I, provided the function in the denominator is nowhere equal to zero on the interval. In summary, rational functions are continuous wherever they are defined. If g is continuous on I and f is continuous on an interval containing $g(I) = \{g(x) : x \in I\}$, then $f \circ g$ is also continuous on I.

Example. These theorems can be used to show the continuity of more complicated functions, such as the function

$$f \ : \ x \mapsto \sin\left(\frac{\sqrt{x^2 + 1}}{|x|}\right) \qquad \text{on its domain } \mathbb{R} \setminus \{0\}.$$

This is done in stages. The constant function $x \mapsto 1$ is continuous at each $a \in \mathbb{R}$, so is $x \mapsto x^2$, and therefore also the sum $x \mapsto x^2 + 1$. Now $x \mapsto \sqrt{x}$ is continuous at positive a, so the composition $x \mapsto \sqrt{x^2 + 1}$ is continuous at each $a \in \mathbb{R}$. It is easy to verify that $x \mapsto |x|$ is continuous everywhere, and $|x| = 0$ only at $x = 0$, hence $x \mapsto \frac{\sqrt{x^2+1}}{|x|}$ is continuous at each $a \in \mathbb{R} \setminus \{0\}$. Finally, since the sine function is continuous everywhere, we indeed have the claim.

2.2   Properties of Continuous Functions

This section begins with two important theorems about continuous functions on a closed interval. The method of proof is worth particular attention. It is used in other arguments, and it is typical of one way ultrasmall numbers are employed in analysis. The general idea is to "approximate" the closed interval $[a, b]$ by a finite set of points $\{x_0, x_1, \ldots, x_N\}$ of $[a, b]$, each ultraclose to the next.

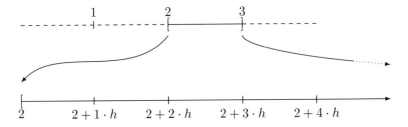

When faced with a task of determining some feature of a function f defined on $[a, b]$ that seemingly requires an examination of the values $f(x)$ for all (infinitely many) $x \in [a, b]$, we instead determine an "approximation" to the feature of interest by examining $f(x_0), f(x_1), \ldots, f(x_N)$—a much easier task—and then use this approximation to obtain the desired feature of f. More specifically, in the two theorems below one wants to produce an element $c \in [a, b]$ with a particular property. The parameters are a, b and f. We choose a positive ultralarge integer N; hence $h = \frac{b-a}{N}$ is ultrasmall. We consider $N + 1$ points $x_i = a + i \cdot h$, for $i = 0, \ldots, N$. We then find among these points, a point x_j which is the best approximation to the one we are looking for. As x_j is bounded by a and b, its observable neighbor c exists and c is in $[a, b]$ (because the interval $[a, b]$ is closed). Continuity of f at c is then used to show that c has the desired property.

Theorem 8 (Intermediate Value). *Let f be a continuous function on $[a, b]$ and let d be between $f(a)$ and $f(b)$. Then there is $c \in [a, b]$ such that $f(c) = d$.*

This theorem is about f, a, b and d; these are the parameters of the theorem.

Proof. Without loss of generality, we may assume that $f(a) < f(b)$ (otherwise consider $-f$ and $-d$). Let N be an ultralarge positive integer. Let $h = \frac{b-a}{N}$ and notice that h is ultrasmall. Consider $x_i = a + i \cdot h$, for

$i = 0, \ldots, N$. Then $a = x_0$ and $x_N = b$. Choose the least index j such that $f(x_{j+1}) \geq d$.

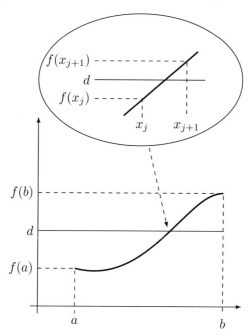

By the choice of j we have

$$f(x_j) < d \leq f(x_{j+1}).$$

Let c be the observable neighbor of x_j (it exists because $x_j \in [a, b]$, and $c \in [a, b]$ by Exercise 11). Then $x_j \simeq c$, and $c \simeq x_{j+1}$ because $x_j \simeq x_{j+1}$. By continuity of f at c we have

$$d > f(x_j) \simeq f(c) \quad \text{and} \quad f(c) \simeq f(x_{j+1}) \geq d.$$

Necessarily $d \simeq f(c)$, but since both numbers are observable, we conclude that $f(c) = d$. $\qquad \square$

Definition 7. *A function attains its **maximum** (respectively **minimum**) on an interval I if there is a $c \in I$ such that for any $x \in I$ we have $f(c) \geq f(x)$ (respectively $f(c) \leq f(x)$).*

Theorem 9 (Extreme Value). *Let f be a continuous function on $[a, b]$. Then f attains its maximum and minimum on $[a, b]$.*

Proof. Without loss of generality, we consider the case of a maximum (for the minimum, replace f by $-f$). Let N be an ultralarge positive integer and let $h = \frac{b-a}{N}$. Consider $x_i = a + i \cdot h$, for $i = 0, \ldots, N$. Choose j such that $f(x_j) \geq f(x_i)$, for all $i = 0, \ldots, N$; we can do this because the set $\{f(x_0), f(x_1), \ldots, f(x_N)\}$ is finite (albeit of ultralarge size).

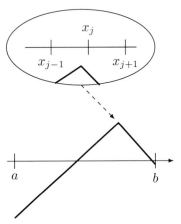

Let c be the observable neighbor of x_j (it exists because $x_j \in [a, b]$, and $c \in [a, b]$ by Exercise 11). By continuity of f at c we have $f(x_j) \simeq f(c)$, and by Closure $f(c)$ is observable.

Let $x \in [a, b]$ be observable. There is an i such that $x_i \leq x \leq x_{i+1}$. Hence $x_i \simeq x$ and $f(x_i) \simeq f(x)$, because f is continuous at x and x is observable. By definition of x_i and c we have

$$f(x) \simeq f(x_i) \leq f(x_j) \simeq f(c).$$

As $f(x)$ and $f(c)$ are both observable, this implies that $f(x) \leq f(c)$ (see Exercise 3).

We have established that $f(x) \leq f(c)$ holds for all observable x, where "observable" refers to any context containing the parameters f, a and b. But every x is observable relative to some such context, for example f, a, b, x. Hence $f(x) \leq f(c)$ is true for all $x \in [a, b]$, and f attains its maximum at c. (Alternatively, one can appeal to the Universal Closure Principle.) $\qquad\square$

The proof shows that the maximum of the function is achieved for some observable c. This is, in fact, a consequence of the Closure Principle: If there is a c such that $f(c)$ is maximum, then some such c must be observable. In particular, the maximum value $f(c)$ is observable.

Together, the previous theorems imply that the image of a closed bounded interval under a continuous function is a closed bounded interval (or a singleton set). Let m and M be, respectively, the minimum and the maximum value of f on $[a, b]$. Then $f([a, b]) = [m, M]$ if $m < M$, and $f([a, b]) = \{m\}$ if $m = M$. The second case occurs precisely when the function f is constant on $[a, b]$, with value m. We prove a more general result.

Theorem 10. *Let I be an interval. If f is continuous on I, then $f(I)$ is either an interval or a singleton set.*

Proof. We give the proof for the case $I = (a, \infty)$; other cases are handled by appropriate combinations of the two techniques used in this case.

If f is a constant function with value m, then $f(I) = \{m\}$ is clear. Henceforth we assume that f is not constant. The context is specified by f and a. We fix $a' > a$, $a' \simeq a$, and b' positive ultralarge so that f is not constant on $[a', b']$. The function f is continuous on $[a', b']$, and hence it attains there a minimum value m' and a maximum value M'; moreover, $f([a', b']) = [m', M']$.

If m' is not ultralarge, we let m be the observable neighbor of m'; otherwise we set $m = -\infty$; similarly, if M' is not ultralarge, we let M be the observable neighbor of M'; otherwise we set $M = \infty$.

For every observable $x \in I$, $a' < x < b'$, so $m' \leq f(x) \leq M'$, and, as $f(x)$ is observable, $m \leq f(x) \leq M$. By Universal Closure, for every $x \in I$, $m \leq f(x) \leq M$.

Conversely, if $m < y < M$ and y is observable, then $m' \leq y \leq M'$, and so there is $x \in I$ such that $y = f(x)$. By Universal Closure again, for every y such that $m < y < M$ there is $x \in I$ such that $y = f(x)$.

Putting these two results together we see that $f(I)$ is one of the intervals with endpoints m and M. The endpoints are included according to whether m or M belongs to the range of f. □

Definition 8.

(1) *A function f is **one-to-one** on I if $x_1 \neq x_2$ implies $f(x_1) \neq f(x_2)$, for all $x_1, x_2 \in I$.*

(2) *A function f is **strictly increasing** on I if $x_1 < x_2$ implies $f(x_1) < f(x_2)$, for all $x_1, x_2 \in I$.*

(3) *A function f is **strictly decreasing** on I if $x_1 < x_2$ implies $f(x_1) > f(x_2)$, for all $x_1, x_2 \in I$.*

Theorem 11. *If f is continuous and one-to-one on an open interval I, then $f(I)$ is an open interval.*

Proof. Let $I = (a, b)$ and assume to the contrary that $f(I) = [c, d)$, for example. Because f is one-to-one, there is a unique x_c with $a < x_c < b$ such that $c = f(x_c)$. For each $x \in (a, b)$ we have $f(x_c) = c \leq f(x)$. Take $x', x'' \in (a, b)$ such that $x' < x_c < x''$, and let s be such that

$c < s < \min\{f(x'), f(x'')\}$. By the Intermediate Value Theorem there are r', r'' such that $x' < r' < x_c$, $x_c < r'' < x''$, and $f(r') = s = f(r'')$, which contradicts the assumption that f is one-to-one. □

Exercise 21 (Answer page 246)
If f is continuous and one-to-one on a closed interval $I = [a, b]$, then either f is strictly increasing and $f(I) = [f(a), f(b)]$, or f is strictly decreasing and $f(I) = [f(b), f(a)]$.

Exercise 22 (Answer page 246)
Prove that if f is continuous and one-to-one on I, then f is either strictly increasing or strictly decreasing on I, for any interval I.

If $f : x \mapsto f(x)$ is a one-to-one function that maps an interval I onto an interval J, then $f^{-1} : J \to I$, the **inverse function** of f, is defined for all $y \in J$ by

$$f^{-1}(y) = x \quad \text{if and only if} \quad f(x) = y.$$

It is thus the unique function such that

$$f(f^{-1}(y)) = y \quad \text{for all } y \in J \quad \text{and} \quad f^{-1}(f(x)) = x \quad \text{for all } x \in I.$$

Exercise 23 (Answer page 246)
If $f : I \to J$ is strictly increasing [respectively, strictly decreasing] on I, then $f^{-1} : J \to I$ is strictly increasing [respectively, strictly decreasing] on J.

We now show that the inverse of a continuous function is continuous. As the inverse of f is defined from the parameter f, it is as observable as f.

Theorem 12. *Let $f : I \to J$ be a continuous one-to-one function. Then $f^{-1} : J \to I$ is continuous.*

Proof. We assume that f is strictly increasing (the case when f is strictly decreasing is similar). Let $d \in J$; we assume that d is not an endpoint of J (the cases when it is one of the endpoints are similar). The context of the proof is specified by f, I, J and d. Let $y \simeq d$. We must show that

$$f^{-1}(y) \simeq f^{-1}(d).$$

Fix observable $a, b \in I$ such that $a < f^{-1}(d) < b$ and note that $f(a) < d < f(b)$ and $f(a), f(b)$ are observable. Hence also $f(a) < y < f(b)$

and $a < f^{-1}(y) < b$, so $f^{-1}(y)$ is not ultralarge and has an observable neighbor c. Since $c \simeq f^{-1}(y)$ and f is continuous at c, we have $f(c) \simeq f(f^{-1}(y))$, that is, $f(c) \simeq y \simeq d$. By Closure, $f(c)$ is observable, so $f(c) = d$ and $c = f^{-1}(d)$. We conclude that $f^{-1}(y) \simeq f^{-1}(d)$. $\square$

Continuity at a point is a statement about the function and the point, hence the context for continuity is determined by the function and the point. We now define uniform continuity, which is a statement about the function and an interval.

Definition 9. *Let f be a function and I an interval. We say that f is **uniformly continuous on I** if*

$$f(x) \simeq f(y), \quad \text{whenever } x \simeq y, \quad x, y \in I.$$

*A function is **uniformly continuous** if it is uniformly continuous on its domain.*

We stress that in the previous definition the symbol $\simeq$ is to be taken relative to f and I, independently of the particular x or y.

Example. We examine uniform continuity of a few functions.

(1) The function f defined by $f : x \mapsto x^2$, for $x \in (0,1)$, is uniformly continuous.

Proof. Let $x, y \in (0,1)$ be such that $x \simeq y$. Then x and y are not ultralarge, so $x^2 \simeq y^2$, that is, $f(x) \simeq f(y)$. Hence f is uniformly continuous. $\square$

(2) The function g defined by $g : x \mapsto \frac{1}{x}$, for $x \in (0,1]$, is continuous, but not uniformly continuous.

Proof. Since g is a rational function defined for all $x \in (0,1]$, it is continuous. Let $h > 0$ be ultrasmall. Then $h \simeq 2h$, but

$$g(h) - g(2h) = \frac{1}{h} - \frac{1}{2h} = \frac{1}{2h} \quad \text{is ultralarge.}$$

$\square$

This example shows that the reciprocal of a uniformly continuous function need not be uniformly continuous.

(3) The function h defined by $h : x \mapsto \frac{1}{x}$, for $x \in [1,\infty)$, is uniformly continuous.

Proof. Let ε be ultrasmall. Then

$$h(x) - h(x + \varepsilon) = \frac{1}{x} - \frac{1}{x + \varepsilon} = \frac{\varepsilon}{x(x + \varepsilon)} \simeq 0,$$

because $x \geq 1$, and so $x(x + \varepsilon) \not\simeq 0$. $\square$

(4) The function k given by $k : x \mapsto \sin\left(\frac{1}{x}\right)$, for $x \in (0, 1]$, is continuous, but not uniformly continuous.

Proof. The function k is continuous because it is the composition of two continuous functions. As in Example 6, let $N \in \mathbb{N}$ be ultralarge. Define $h_1 = \frac{1}{2\pi N}$ and $h_2 = \frac{1}{2\pi N + \pi/2}$. Then $h_1 \simeq h_2$, but $k(h_1) = 0 \not\simeq 1 = k(h_2)$. $\square$

Uniform continuity of f on I implies continuity of f on I. Indeed, if f is uniformly continuous on I, then for all $x, y \in I$, $x \simeq y$ implies $f(x) \simeq f(y)$ for any context where f, I are observable. Hence, given any fixed $x \in I$ and any context where f, I and also x are observable, for all $y \in I$, $x \simeq y$ implies $f(x) \simeq f(y)$; that is, f is continuous at x.

Theorem 13. *If f is continuous on $[a, b]$, then f is uniformly continuous on $[a, b]$.*

Proof. Let $x, y \in [a, b]$ with $x \simeq y$. Let c be the observable neighbor of x (x is not ultralarge); then $c \in [a, b]$. But $x \simeq c$ and also $y \simeq c$ (because $x \simeq y$). Since f is continuous at c and c is observable, we have $f(x) \simeq f(c)$ and $f(y) \simeq f(c)$. This implies that $f(x) \simeq f(y)$. $\square$

The argument used above does not work if the interval is not closed and bounded. For example, take $f : x \mapsto \frac{1}{x}$ on the interval $(0, 1]$. For $x > 0$ and $x \simeq 0$, the observable neighbor of x is 0, which is not in $(0, 1]$. As shown above, the function $x \mapsto \frac{1}{x}$ is not uniformly continuous on $(0, 1]$.

2.3   Limits

The existence of a limit of a function at a point a is a property that depends on the behavior of the function around the point a. The function must therefore be defined in a neighborhood of a, even though it need not be defined at a itself.

A **deleted** (open) **neighborhood** of a is a set of the form $(b,c) \setminus \{a\}$ with $b < a < c$. If an observable function f is defined in a deleted neighborhood of a, then it is defined in some observable deleted neighborhood of a, by Closure.

Intuitively, the limit of a function f at a is the "value" that f ought to have at a in order to be continuous. In more detail, we are looking for a number L such that the function

$$x \mapsto \begin{cases} f(x) & \text{if } x \neq a; \\ L & \text{if } x = a \end{cases}$$

is continuous at a. Moreover, we want this function to be as observable as f, which implies that L has to be as observable as f and a. From these considerations and the definition of continuity we get the following.

Definition 10. *Let f be a function defined on a deleted neighborhood of a. We say that f **has a limit at** a if there exists an observable real number L such that*

$$f(x) \simeq L \quad \text{whenever } x \simeq a, \text{ with } x \neq a.$$

*This number L is called a **limit of f at** a.*

Equivalently, a **limit of f at** a is an observable real number L such that

$$f(a+h) \simeq L, \quad \text{for all ultrasmall } h.$$

The existence of a limit, as defined above, is a property of f and a, so, according to our convention, observability and the $\simeq$ symbol are to be interpreted relative to some (any) context where f and a are observable.

Example. Consider the limit of

$$f : x \mapsto \frac{2x^2 - 7x + 3}{x - 3} \quad \text{at } x = 3.$$

The function is defined on a deleted neighborhood $\mathbb{R} \setminus \{3\}$ of 3, and is standard. Let h be ultrasmall. Then

$$f(3+h) = \frac{2(3+h)^2 - 7(3+h) + 3}{(3+h) - 3} = \frac{5h + 2h^2}{h} = 5 + 2\,h \simeq 5.$$

As 5 is observable and is ultraclose to $f(3+h)$, it is the limit.

Theorem 14. *If f has a limit at a, then this limit is unique.*

Proof. The limit is the observable neighbor of $f(x)$ (for any $x \simeq a$, $x \neq a$), and we show in Theorem 3 that this observable neighbor is unique. $\qquad \square$

We write
$$\lim_{x \to a} f(x) = L$$
if L is the limit of f at a.

The following theorem addresses a rather subtle issue. The statement that expresses the existence of the limit of f at a is internal, with parameters f, a. Therefore, the *existence* of a limit of f at a does not depend on the context, as long as f and a are observable. However, the defining statement for $\lim_{x \to a} f(x) = L$:

   L is observable and $f(x) \simeq L$ whenever $x \simeq a$, $x \neq a$, relative to f and a

is not internal, because its parameters are f, a and L. But in fact, this statement is equivalent to "$f(x) \simeq L$ whenever $x \simeq a$, $x \neq a$, relative to f, a, L," which is internal.

Theorem 15. *The following statements are equivalent:*

   (1) L is observable relative to f, a, and $f(x) \simeq L$ whenever $x \simeq a$, $x \neq a$, relative to f, a.

   (2) $f(x) \simeq L$ whenever $x \simeq a$, $x \neq a$, relative to f, a, L.

Proof. Assume (1). Since L is observable relative to f, a, ultracloseness relative to f, a is equivalent to ultracloseness relative to f, a, L (Exercise 19), and this immediately implies (2).

   Now assume (2). It suffices to show that L is observable relative to f, a.

   If there exists M observable relative to f, a, and such that $f(x) \simeq M$ whenever $x \simeq a$, $x \neq a$, relative to f, a, then for this (unique) M also $f(x) \simeq M$ whenever $x \simeq a$, $x \neq a$, relative to f, a, L, by Stability. From (2) and the uniqueness of the observable neighbor we conclude that $L = M$, so L is observable relative to f, a.

   If there is no such M, then, by Stability, there is no M observable relative to f, a, L, and such that $f(x) \simeq M$ whenever $x \simeq a$, $x \neq a$, relative to f, a, L. This is a contradiction with (2) (take $M = L$).   □

   In particular, the *value* of $\lim_{x \to a} f(x)$ does not depend on the context, as long as f and a are observable. Just as for continuity, when working with limits we can safely use any context where the parameters are observable. It is thus possible to compute limits of functions obtained by arithmetic operations in the same way as for continuity (see Theorem 6). We show here as an example that

$$\lim_{x \to a} (f(x) + g(x)) = \lim_{x \to a} f(x) + \lim_{x \to a} g(x),$$

provided the limits on the right side exist. Suppose that f and g are functions defined on a deleted neighborhood of a and that $\lim_{x \to a} f(x) = L_f$ and $\lim_{x \to a} g(x) = L_g$. We work in a context where f, g and a are observable. Then L_f, L_g and $L_f + L_g$ are observable. We have $f(x) \simeq L_f$ and $g(x) \simeq L_g$ whenever $x \simeq a$ $(x \neq a)$. By Rule 5 we deduce

$$f(x) + g(x) \simeq L_f + L_g,$$

which proves our claim.

Theorem 16. *Suppose* $\lim_{x \to a} f(x)$ *and* $\lim_{x \to a} g(x)$ *exist. Then*

(1) $\lim_{x \to a} (f(x) \pm g(x)) = \lim_{x \to a} f(x) \pm \lim_{x \to a} g(x).$

(2) $\lim_{x \to a} (f(x) \cdot g(x)) = \lim_{x \to a} f(x) \cdot \lim_{x \to a} g(x).$

(3) *If* $\lim_{x \to a} g(x) \neq 0,$ *then* $\lim_{x \to a} \left(\dfrac{f(x)}{g(x)} \right) = \dfrac{\lim_{x \to a} f(x)}{\lim_{x \to a} g(x)}.$

(4) *For all* $\lambda \in \mathbb{R},$ $\lim_{x \to a} \lambda = \lambda.$

Exercise 24 (Answer page 246)
Prove the rest of Theorem 16.

The next theorem summarizes the relationship between continuity and limits; its proof is immediate from the definitions.

Theorem 17. *Let f be defined on an open interval containing a. Then f is continuous at a if and only if*

$$\lim_{x \to a} f(x) = f(a).$$

Definition 11.

(1) *Let f be a function defined on an interval (a, b), for some $b > a$. A **right-hand limit of f at a** is an observable real number L such that*

$$f(x) \simeq L \quad \text{whenever } x \simeq a, \text{ with } x > a.$$

(2) *Let f be a function defined on an interval (c, a), for some $c < a$. A **left-hand limit of f at a** is an observable real number L such that*

$$f(x) \simeq L \quad \text{whenever } x \simeq a, \text{ with } x < a.$$

As before, if f has a one-sided limit at a, then this limit is unique. We write

$$\lim_{x \to a^+} f(x) = L$$

if L is the right-hand limit of f at a. Similarly, we write

$$\lim_{x \to a^-} f(x) = L$$

if L is the left-hand limit of f at a.

It is immediate that $\lim_{x \to a} f(x) = L$ if and only if

$$\lim_{x \to a^-} f(x) = L = \lim_{x \to a^+} f(x).$$

Theorem 16 holds for one-sided limits. We now extend the definition of limit to the case where f is unbounded near a.

Notation: Relative to a context, we use the abbreviation $M \simeq +\infty$ to denote that M is positive and ultralarge, and $M \simeq -\infty$ to denote that M is negative and ultralarge. The "+" sign can be omitted.

Definition 12. *Let f be a function defined in a deleted neighborhood of a. We write*

$$\lim_{x \to a} f(x) = \infty$$

if $f(x) \simeq +\infty$ whenever $x \simeq a$, $x \neq a$. Similarly, we write

$$\lim_{x \to a} f(x) = -\infty$$

if $f(x) \simeq -\infty$ whenever $x \simeq a$, $x \neq a$.

We can similarly adapt these definitions to left-hand and right-hand limits.

Caution: In this book, to say that a "limit exists" means that the limit is a real number. Although we use the notation $\lim_{x \to a} f(x) = \infty$ and $\lim_{x \to a} f(x) = -\infty$ in the situations described by Definition 12, the symbols ∞ and $-\infty$ are not real numbers and the limits do not exist in these situations.

Exercise 25 (Answer page 247)
Show: If $f(x)$ is ultralarge whenever $x \simeq a$, $x \neq a$, relative to the context f, a, then $f(x)$ is ultralarge whenever $x \simeq a$, $x \neq a$, relative to any extended context; and conversely, if $f(x)$ is ultralarge whenever $x \simeq a$, $x \neq a$, relative to an extended context, then $f(x)$ is ultralarge whenever $x \simeq a$, $x \neq a$ relative to the original context. Hence any context where the parameters f and a are observable can be used in Definition 12.

Definition 13. *We say that f has a* **vertical asymptote** *at $x = a$ if*

$$\lim_{x \to a^-} f(x) = \pm\infty \quad or \quad \lim_{x \to a^+} f(x) = \pm\infty.$$

Unravelling the definition: f has a vertical asymptote at $x = a$ if

$$x \simeq a \quad \text{and} \quad x < a \quad (\text{or } x > a) \quad \text{implies} \quad f(x) \simeq \pm\infty.$$

Example. The function $f : x \mapsto \frac{1}{x}$ defined on $\mathbb{R} \setminus \{0\}$ has a vertical asymptote at $x = 0$: Let $h \simeq 0$ and $h > 0$, then $f(h) = \frac{1}{h}$ is positive ultralarge, and therefore $\lim_{x \to 0^+} f(x) = +\infty$. Similarly, $\lim_{x \to 0^-} f(x) = -\infty$.

Theorem 16 can be extended to limits with $\pm\infty$, but cases such as $\infty - \infty$, $0 \cdot \infty$ or $\frac{\infty}{\infty}$ are indeterminate.

If f is defined on an interval of the form $[b, \infty)$ or $(-\infty, b]$ (note that b can always be chosen observable relative to f), then we can consider its asymptotic behavior for ultralarge values of the argument.

Definition 14. *Let f be a function defined on an interval of the form $[b, \infty)$ (respectively $(-\infty, b]$). We say that f has a* **horizontal asymptote** *$y = L$ at ∞ (respectively $-\infty$) if L is observable and*

$$x \simeq \infty \quad (\text{respectively } x \simeq -\infty) \quad \text{implies} \quad f(x) \simeq L;$$

we then write

$$\lim_{x \to \infty} f(x) = L \quad (\text{respectively} \lim_{x \to -\infty} f(x) = L).$$

Example. Consider

$$\lim_{x \to +\infty} \frac{2x^2 - 3x + 1}{x^2 + 1}.$$

The standard function f above is defined on $\mathbb{R}$. Let x be positive ultralarge. Then

$$f(x) = \frac{2x^2 - 3x + 1}{x^2 + 1} = \frac{x^2(2 - \frac{3}{x} + \frac{1}{x^2})}{x^2(1 + \frac{1}{x^2})} = \frac{2 - \frac{3}{x} + \frac{1}{x^2}}{1 + \frac{1}{x^2}} \simeq 2.$$

Clearly, the same result also holds for x negative ultralarge, so f has a horizontal asymptote $y = 2$ at $\pm\infty$.

We can define similarly

$$\lim_{x \to \pm\infty} f(x) = \pm\infty$$

if $f(x)$ is ultralarge (positive or negative) for ultralarge x.

We conclude this section with a discussion of oblique asymptotes.

Definition 15. *The function f has an **oblique asymptote** $y = ax + b$ at $+\infty$ (respectively $-\infty$) if*

$$\lim_{x \to +\infty} (f(x) - (ax + b)) = 0 \quad (\text{respectively } \lim_{x \to -\infty} (f(x) - (ax + b)) = 0).$$

Using Stability one can prove that a, b have to be observable whenever f is observable. This also follows from Theorem 18.

Unravelling the definition: f has an oblique asymptote $y = ax + b$ at $+\infty$ (respectively $-\infty$) if a and b are observable and

$$x \simeq +\infty \quad (\text{respectively } x \simeq -\infty) \quad \text{implies} \quad f(x) \simeq ax + b.$$

Example.

$$f : x \mapsto \frac{x^3 + 2x^2 + x - 1}{x^2 + 1}$$

is defined for all values in $\mathbb{R}$. Using polynomial division we obtain

$$f(x) = x + 2 - \frac{3}{x^2 + 1}.$$

Let x be ultralarge relative to f. Then

$$f(x) - (x + 2) = \frac{-3}{x^2 + 1} \simeq 0,$$

because $x^2 + 1$ is ultralarge. We conclude that f has an oblique asymptote $y = x + 2$ at $\pm\infty$. Notice that $-3/(x^2 + 1) < 0$ whether x is positive or negative ultralarge, so the graph of f is below the asymptote at $\pm\infty$.

In the previous example, a and b seem to be unique. This is true in general; it is an immediate consequence of the following theorem, which also gives a method for finding the parameters of the asymptotic straight line, even when the function is not a rational function.

Theorem 18. *Let f be a function. Then f has an oblique asymptote $y = ax + b$ at $+\infty$ if and only if*

$$\lim_{x \to +\infty} \frac{f(x)}{x} = a \quad and \quad \lim_{x \to +\infty} (f(x) - ax) = b.$$

(The same holds at $-\infty$.)

Proof. First we assume that f has an oblique asymptote $y = ax + b$ at $+\infty$. We work in a context where f, a, b are observable. Let x be positive ultralarge. Since $f(x) - (ax + b) \simeq 0$, we have $f(x) - ax \simeq b$. Furthermore, $\frac{f(x) - ax}{x} \simeq \frac{b}{x}$, so $\frac{f(x)}{x} - a \simeq \frac{b}{x} \simeq 0$, hence $\frac{f(x)}{x} \simeq a$. We conclude that $\lim_{x \to +\infty}(f(x) - ax) = b$ and $\lim_{x \to +\infty} \frac{f(x)}{x} = a$.

For the converse, suppose that

$$\lim_{x \to +\infty} \frac{f(x)}{x} = a \quad \text{and} \quad \lim_{x \to +\infty} (f(x) - ax) = b.$$

Let f be observable; then a and b are also observable. If $x > 0$ is ultralarge, then $f(x) - ax \simeq b$. We immediately deduce (Rule 4) that $f(x) - (ax + b) \simeq 0$, so $\lim_{x \to +\infty} (f(x) - (ax + b)) = 0$ by definition. Thus f has an oblique asymptote $x \mapsto ax + b$. $\qquad\square$

Example. Consider $f : x \mapsto \sqrt{x^2 + 1}$ defined on $\mathbb{R}$. Let x be a positive ultralarge number. Then

$$\frac{f(x)}{x} = \frac{\sqrt{x^2 + 1}}{x} = \sqrt{\frac{x^2 + 1}{x^2}} = \sqrt{1 + \frac{1}{x^2}} \simeq 1.$$

Subsequently,

$$f(x) - x = \sqrt{x^2 + 1} - x = \frac{(\sqrt{x^2 + 1} - x)(\sqrt{x^2 + 1} + x)}{\sqrt{x^2 + 1} + x} = \frac{1}{\sqrt{x^2 + 1} + x}$$

and this last term is ultrasmall. Hence f has an oblique asymptote $y = x$ at $+\infty$. If x is a negative ultralarge number, we show similarly that $f(x)/x \simeq -1$. Then $f(x) - (-x) \simeq 0$. Hence f has an oblique asymptote $y = -x$ at $-\infty$.

━━━━━━━

2.4   Exponential and Logarithmic Functions

Fix a real number $a > 0$. The goal of this section is to define

$$a^b, \quad \text{for } b \in \mathbb{R}.$$

We quickly remind the reader how this is done for a rational exponent b and then proceed to do it for a real, not necessarily rational, exponent.

First, we define $a^1 = a$ and $a^0 = 1$. For n a **positive integer**, we define a^n by

$$a^n = \underbrace{a \cdot a \cdots \cdot a}_{n \text{ times}}.$$

With this definition, we have the following fundamental facts (see the exercise below): Let $a > 0$ and n be a positive integer ultralarge relative to a.

- If $a > 1$, then $a^n \simeq +\infty$.

- If $0 < a < 1$, then $a^n \simeq 0$.

Exercise 26 (Answer page 247)
The **binomial formula** is the formula

$$(a + b)^n = a^n + \binom{n}{1} a^{n-1} b + \binom{n}{2} a^{n-2} b^2 + \cdots + \binom{n}{n-1} ab^{n-1} + b^n,$$

where $\binom{n}{k} = \frac{n!}{k!(n-k)!}$, for $0 \le k \le n$. This formula generalizes the familiar
identities

$$(a + b)^2 = a^2 + 2ab + b^2 \quad \text{and} \quad (a + b)^3 = a^3 + 3a^2 b + 3ab^2 + b^3.$$

(1) Use the binomial formula to show that $(1 + b)^n \ge 1 + nb$, for
$b > 0$ and n a positive integer.

(2) Deduce from (1) that $a^n \simeq +\infty$ if $a > 1$ and n is a positive
integer ultralarge relative to a.

(3) Use $c = 1/a$ and (2) to deduce that $a^n \simeq 0$ if $0 < a < 1$ and
n is a positive integer ultralarge relative to a.

The function $x \mapsto x^n$ is strictly increasing on $(0, \infty)$, for $n \in \mathbb{N}$,
$n > 0$. Also $\lim_{x \to 0^+} x^n = 0$ and $\lim_{x \to \infty} x^n = \infty$. Further, by the rules
on products of continuous functions we have that $x \mapsto x^n$ is continuous.
(See Example (3) on page 38 or Example (1) on page 146.) It follows from
the Intermediate Value Theorem that $x \mapsto x^n$ is a continuous one-to-one
correspondence between $(0, \infty)$ and $(0, \infty)$.

For m a **negative integer**, we let

$$a^m = \frac{1}{a^{-m}}.$$

Clearly, if m is a negative integer, then the function $x \mapsto x^m$ is strictly
decreasing, satisfies $\lim_{x \to 0^+} x^m = \infty$ and $\lim_{x \to \infty} x^m = 0$, and maps
$(0, \infty)$ continuously onto $(0, \infty)$.

With these definitions we have, for $a > 0$ and $c, b \in \mathbb{Z}$:

$$a^{b+c} = a^b \cdot a^c, \quad a^{b-c} = \frac{a^b}{a^c}, \quad \text{and} \quad (a^b)^c = a^{b \cdot c}. \tag{2.1}$$

We also have the properties (for $a, b > 0$ and $c \in \mathbb{Z}$):

$$(a \cdot b)^c = a^c \cdot b^c \quad \text{and} \quad \left(\frac{a}{b}\right)^c = \frac{a^c}{b^c}. \tag{2.2}$$

Let n be a positive integer. We call the n-**th root**, written $x \mapsto \sqrt[n]{x}$,
the function which is the inverse of $x \mapsto x^n$ on $(0, \infty)$. The n-th root is

a continuous function on $(0, \infty)$ since it is the inverse of a continuous function. It is strictly increasing since $x \mapsto x^n$ is strictly increasing. We define

$$a^{1/n} = \sqrt[n]{a}.$$

We next extend this to more general **rational exponents**. For $q = \frac{m}{n}$, with $m \in \mathbb{Z}$ and $n \in \mathbb{N}$, $n \neq 0$, we let

$$a^q = a^{m/n} = \sqrt[n]{a^m}.$$

One checks easily that $a^{m/n} = a^{m'/n'}$ when $m/n = m'/n'$ (raise both sides to the power $n \cdot n'$). As a composition of two continuous functions, the function $x \mapsto x^q$ is continuous on $(0, \infty)$.

These definitions allow us to establish properties (2.1) and (2.2) for rational, rather than just integer, exponents. We have also the following monotonicity properties: For any $a, a_1, a_2 \in \mathbb{R}$ and any $c, c_1, c_2 \in \mathbb{Q}$,

$$0 < a_1 < a_2 \quad \text{and} \quad c > 0 \quad \text{implies} \quad a_1^c < a_2^c$$

and

$$1 < a \quad \text{and} \quad 0 < c_1 < c_2 \quad \text{implies} \quad a^{c_1} < a^{c_2}.$$

Using the monotonicity properties above we deduce that for $a > 1$ and $q \in \mathbb{Q}$

- $a^q \simeq 0$ if $q \simeq -\infty$.

- $a^q \simeq 1$ if $q \simeq 0$.

- $a^q \simeq +\infty$ if $q \simeq +\infty$.

Exercise 27 (Answer page 247)
Prove the above three properties using the fact that $a^n \simeq +\infty$, if n is a positive ultralarge integer.

We finally extend the definition of exponentiation to irrational exponents. Recall that $\mathbb{Q}$ is **dense** in $\mathbb{R}$: Whenever $a < b$, there is $c \in \mathbb{Q}$ such that $a < c < b$. In particular, given any context where $b \in \mathbb{R}$ is observable, we can find $b' \in \mathbb{Q}$ such that $b' \simeq b$. We call such a b' a **rational neighbor** of b.

Definition 16. *Let $a > 0$ and $b \in \mathbb{R}$ be given. We let*

$$a^b$$

be the observable neighbor of $a^{b'}$, where b' is a rational neighbor of b.

The parameters of this definition are a and b. We show that the number a^b is well-defined. Some rational neighbor b' of b exists, as observed before the definition. Moreover, the number $a^{b'}$ is not ultralarge. Let m be the integer part of b, that is, the unique integer such that $m \leq b < m + 1$; then m is observable and $m - 1 < b' < m + 1$, so $a^{b'}$ is between a^{m-1} and a^{m+1}, which are observable. This implies that the observable neighbor of $a^{b'}$ can always be found. To see that this observable neighbor is independent of the choice of b', suppose that b' and b'' are two rational neighbors of b. Then

$$a^{b'} = a^{b''+(b'-b'')} = a^{b''} \cdot a^{b'-b''} \simeq a^{b''} \cdot 1 = a^{b''},$$

since $b' - b'' \simeq 0$ is rational, so $a^{b'-b''} \simeq 1$. Thus, $a^{b'}$ and $a^{b''}$ have the same observable neighbor. Finally, the value of a^b does not depend on the context. If L is observable and $L \simeq a^{b'}$ for all rational $b' \simeq b$, relative to the context a, b, then $L \simeq a^{b'}$ for all rational $b' \simeq b$, relative to a, b, L (Exercise 19). This last statement is internal, and it follows by Stability that $L \simeq a^{b'}$ for all rational $b' \simeq b$, relative to any extended context.

The next two theorems extend the properties (2.1) and (2.2) of exponentiation to real, not necessarily rational, exponents.

Theorem 19. *Let $a > 0$ and $b, c \in \mathbb{R}$.*

    (1) $a^{b+c} = a^b \cdot a^c$.

    (2) $a^c \neq 0$ and $a^{b-c} = a^b / a^c$.

    (3) $(a^b)^c = a^{bc}$.

Proof. The parameters are a, b and c.

    (1) Let b' be a rational neighbor of b and c' be a rational neighbor of c. Then $b' + c'$ is a rational neighbor of $b + c$, so

$$a^{b+c} \simeq a^{b'+c'} = a^{b'} \cdot a^{c'} \simeq a^b \cdot a^c.$$

Since the left-hand side and the right hand side are observable, they must be equal.

    (2) is immediate from (1): $1 = a^0 = a^{c+(-c)} = a^c \cdot a^{-c}$, so $a^c \neq 0$ and $a^{-c} = 1/a^c$. Another application of (1) yields $a^{b-c} = a^{b+(-c)} = a^b \cdot a^{-c} = a^b / a^c$.

    (3) Let c' be a rational neighbor of c. Consider also the context where c' is an extra observable, and write $\overset{+}{\simeq}$ when working relative to a, b, c and also c'. Choose $b' \overset{+}{\simeq} b$ with $b' \in \mathbb{Q}$; then $a^b \overset{+}{\simeq} a^{b'}$. Since the function $x \mapsto x^{c'}$ is continuous and observable relative to a, b, c, c', we have

$$(a^b)^{c'} \overset{+}{\simeq} (a^{b'})^{c'}.$$

Using the definition and the properties of exponentiation for rationals (and Exercise 1.8) we obtain

$$(a^b)^c \simeq (a^b)^{c'} \simeq (a^{b'})^{c'} = a^{b' \cdot c'} \simeq a^{b \cdot c},$$

since $b' \cdot c' \in \mathbb{Q}$ and $b' \cdot c' \simeq b \cdot c$. Since the left-hand side and the right-hand side are observable, we have equality. $\square$

The proof of (3) above is worth special attention. It is the first example of an argument where two levels of observability are employed simultaneously. The ability to do this is particular to our relative framework and is often useful, especially in proofs that involve "double limits."

Theorem 20. *Let $a, b > 0$ and $c \in \mathbb{R}$.*

(1) $(a \cdot b)^c = a^c \cdot b^c$.

(2) $\left(\dfrac{a}{b}\right)^c = \dfrac{a^c}{b^c}$.

Proof. (1) Let c' be a rational neighbor of c relative to a, b, c. Then $a^c \simeq a^{c'}$, $b^c \simeq b^{c'}$ and $(a \cdot b)^c \simeq (a \cdot b)^{c'}$ (by Closure, $a \cdot b$ is observable). Then

$$(a \cdot b)^c \simeq (a \cdot b)^{c'} = a^{c'} \cdot b^{c'} \simeq a^c \cdot b^c.$$

Since the left-hand side and the right-hand side are observable, they must be equal.

(2) is similar. $\square$

Exercise 28 (Answer page 247)
Show, using the density of the rationals, that for all real a and c

$$1 < a \quad \text{and} \quad 0 < c \quad \text{implies} \quad 1 < a^c.$$

Deduce the following monotonicity properties.

(1) Let $a_1, a_2, c \in \mathbb{R}$. If $0 < a_1 < a_2$ and $0 < c$, then $a_1^c < a_2^c$.

(2) Let $a, c_1, c_2 \in \mathbb{R}$. If $1 < a$ and $0 < c_1 < c_2$, then $a^{c_1} < a^{c_2}$.

Definition 17. *Let $a > 0$. The **base a exponential** is the function from $\mathbb{R}$ to $(0, \infty)$ defined by*

$$\exp_a : x \mapsto a^x.$$

According to the remarks following Definition 16, the defining statement of $y = a^x$ is internal, so the function $\exp_a$ is well-defined and as observable as a, by the Definition Principle.

By the previous exercise, the function $\exp_a$ is strictly increasing when $a > 1$, and

$$\lim_{x \to -\infty} a^x = 0 \quad \text{and} \quad \lim_{x \to +\infty} a^x = +\infty.$$

The first limit shows that there is a horizontal asymptote $y = 0$ at $-\infty$.

Since $a^x = \left(\frac{1}{a}\right)^{-x}$, it follows that for $0 < a < 1$ the function $\exp_a$ is strictly decreasing and the horizontal asymptote $y = 0$ is at $+\infty$.

Theorem 21. *Let $a > 0$; then $\exp_a$ is continuous on its domain.*

Proof. Let $b \in \mathbb{R}$ be given; we show that $\exp_a$ is continuous at b. The parameters are a and b. Let $x \simeq b$; we need to establish that $a^x \simeq a^b$. Extend the context by x; we write $\overset{+}{\simeq}$ when working relative to a, b, x. Let $x' \overset{+}{\simeq} x$ and $b' \overset{+}{\simeq} b$, with $x', b' \in \mathbb{Q}$. Then by definition

$$a^b \overset{+}{\simeq} a^{b'} \quad \text{and} \quad a^x \overset{+}{\simeq} a^{x'}.$$

But $b' \simeq x'$, so $a^{b'} = a^{x'+(b'-x')} \simeq a^{x'}$, since $a^{b'-x'} \simeq 1$. This shows $a^b \simeq a^x$, as desired. □

By continuity, $\exp_a$ is a one-to-one correspondence between $\mathbb{R}$ and $(0, \infty)$.

Theorem 22. *Let $b \in \mathbb{R}$. The function $x \mapsto x^b$ is continuous at each $x > 0$.*

By the Definition Principle, the function $y = x^b$ is well-defined and as observable as b.

Proof. We first claim that, relative to a given context, if $x \simeq 1$ and $b' \in \mathbb{Q}$ is not ultralarge, then $x^{b'} \simeq 1$.

We prove this for $b' \geq 0$ and $x \simeq 1$ such that $x > 1$ (the other cases are proved similarly). The claim is true for observable $b' \in \mathbb{Q}$, since $x \mapsto x^{b'}$ is continuous at 1. Now suppose that b' is not ultralarge. Then there is an observable c' rational such that $0 \leq b' \leq c'$, and we have $1 = x^0 \leq x^{b'} \leq x^{c'} \simeq 1$. This implies that $x^{b'} \simeq 1$ and proves the claim.

We now deduce the theorem. Let $a > 0$ and let $x \simeq a$; we must show that $x^b \simeq a^b$. We extend the context to a, b, x and write $\overset{+}{\simeq}$ when working relative to a, b, x. Let $b' \overset{+}{\simeq} b$ with $b' \in \mathbb{Q}$. By definition, we have $x^b \overset{+}{\simeq} x^{b'}$ and $a^b \overset{+}{\simeq} a^{b'}$. To see that $x^{b'} \simeq a^{b'}$, notice that $\frac{x^{b'}}{a^{b'}} = \left(\frac{x}{a}\right)^{b'} \simeq 1$, since b' is not ultralarge and $\frac{x}{a} \simeq 1$. □

Definition 18. *Let $a > 1$. We define the **base** a **logarithm**, written*

$$\log_a : (0, \infty) \to \mathbb{R}$$

to be the inverse of the base a exponential.

The function $\log_a$ is observable when a is observable. We have $\log_a(1) = 0$, $\log_a(a) = 1$, and more generally

$$a^{\log_a(x)} = x \quad \text{(for } x > 0\text{)} \quad \text{and} \quad \log_a(a^x) = x \quad \text{(for } x \in \mathbb{R}\text{)}.$$

The next theorem is immediate from the properties of inverse functions established in Section 2.2.

Theorem 23. *Let $a > 1$. The base a logarithm is continuous on its domain and satisfies*

$$\lim_{x \to 0^+} \log_a(x) = -\infty \quad \text{and} \quad \lim_{x \to \infty} \log_a(x) = +\infty.$$

The first limit shows that there is a vertical asymptote at $x = 0$.

We deduce the following theorem from the properties of the base a exponential.

Theorem 24. *Let $a > 1$.*

(1) Let $b, c > 0$. We have $\log_a(b \cdot c) = \log_a(b) + \log_a(c)$.

(2) Let $b, c > 0$. We have $\log_a(b/c) = \log_a(b) - \log_a(c)$.

(3) Let $b > 0$ and $c \in \mathbb{R}$; then $\log_a(b^c) = c \cdot \log_a(b)$.

Proof. We prove (1) and leave the rest as exercise. Let $x = \log_a(b)$ and $y = \log_a(c)$. Then $b = a^x$ and $c = a^y$. This gives

$$\log_a(b \cdot c) = \log_a(a^x \cdot a^y) = \log_a(a^{x+y}) = x + y = \log_a(b) + \log_a(c).$$

$\square$

2.5 Additional Exercises

Exercise 2.1

(1) Show that $f : x \mapsto |x|$ is continuous at $x = 0$, at $x = 1$, at $x = -1$ and at x in general.

(2) Show that $f : x \mapsto \begin{cases} x^2 & \text{if } x \geq 0 \\ x^3 & \text{if } x < 0 \end{cases}$ is continuous at $x = 0$ and at x in general.

(3) Show that $f : x \mapsto \begin{cases} x^2 & \text{if } x \geq -1 \\ x^3 & \text{if } x < -1 \end{cases}$ is not continuous at $x = -1$ but it is continuous for all other values of x.

Exercise 2.2
Use the definition of continuity to show that the following functions are continuous on the given intervals.

(1) $f : x \mapsto \frac{1}{3}x + \sqrt{2}$ on $\mathbb{R}$.

(2) $f : x \mapsto x^2 - 3x - 1$ on $\mathbb{R}$.

(3) $f : x \mapsto \dfrac{x + 2}{x - 1}$ on $(1, +\infty)$.

Exercise 2.3
Prove that if $f, g : \mathbb{R} \to \mathbb{R}$ are continuous and $f(x) = g(x)$ for all $x \in \mathbb{Q}$, then $f(x) = g(x)$ for all $x \in \mathbb{R}$.

Exercise 2.4
Determine whether the following functions are continuous on their domains.

(1) $f : x \mapsto \frac{1}{x}$

(2) $f : x \mapsto \frac{1}{x^2 + 1}$

(3) $f : x \mapsto \begin{cases} \frac{1}{x} & \text{if } x > 1 \\ -2x + 1 & \text{if } x \leq 1 \end{cases}$

(4) $f : x \mapsto \begin{cases} \cos(\frac{1}{x}) & \text{if } x \neq 0 \\ 1 & \text{if } x = 0 \end{cases}$

(5) $f : x \mapsto \begin{cases} x^2 + 1 & \text{if } x < 1 \\ 2 & \text{if } x \geq 1 \end{cases}$

Exercise 2.5
Determine the value of c (if one exists) that makes the following functions continuous on their domains.

(1) $f : x \mapsto \begin{cases} \frac{x^2 - 4}{x - 2} & \text{if } x \neq 2 \\ c & \text{if } x = 2 \end{cases}$

(2) $f : x \mapsto \begin{cases} (x+3)^2 & \text{if } x \leq -1 \\ cx + 1 & \text{if } x \geq 1 \end{cases}$

(3) $f : x \mapsto \begin{cases} \frac{x}{x+1} & x \neq 1 \\ c & \text{if } x = 1 \end{cases}$

Exercise 2.6

Assume that $f : [a, b] \to \mathbb{R}$ and $g : [b, c] \to \mathbb{R}$ are continuous and $f(b) = g(b)$. Define $h : [a, c] \to \mathbb{R}$ by $h : x \mapsto \begin{cases} f(x) & \text{if } a \leq x \leq b \\ g(x) & \text{if } b < x \leq c \end{cases}$.

Prove that h is continuous on $[a, c]$.

Exercise 2.7

Let $f, g : I \to \mathbb{R}$ be continuous on I. Define $h : I \to \mathbb{R}$ by $h(x) = \max\{f(x), g(x)\}$ and prove that h is continuous on I.

Exercise 2.8

Prove that if f is continuous at a and $f(a) > 0$, then there is an open interval J containing a such that $f(x) > 0$ for all $x \in J$.

Exercise 2.9

Let f and g be continuous on $[a, b]$, $f(a) \geq g(a)$, $f(b) \leq g(b)$. Prove that there exists $c \in [a, b]$ such that $f(c) = g(c)$.

Exercise 2.10

Prove that every polynomial of odd degree has at least one root.

Exercise 2.11

Let $f : [a, b] \to \mathbb{R}$. Prove that f attains a maximum at $c \in [a, b]$ if and only if $-f$ attains a minimum at c. Prove that f is strictly increasing on $[a, b]$ if and only if $-f$ is strictly decreasing on $[a, b]$.

Exercise 2.12

Prove Theorem 10 for the case $I = (a, b)$.

Exercise 2.13

Show that the function $f(x) = x^n$ ($n > 0$) is strictly decreasing in the interval $(-\infty, 0]$ and strictly increasing in the interval $[0, \infty)$ when n is even. It is strictly increasing in the interval $(-\infty, \infty)$ when n is odd. Repeat the exercise for the function $f(x) = 1/x^n$.

Exercise 2.14

Determine whether the following functions are uniformly continuous.

(1) $f : x \mapsto 3x + 2$ on $\mathbb{R}$.

(2) $f : x \mapsto x^2$ on $\mathbb{R}$.

(3) $f : x \mapsto \dfrac{1}{x^2 + 1}$ on $\mathbb{R}$.

(4) $f : x \mapsto \sin(x)$ on $\mathbb{R}$.

(5) $f : x \mapsto \dfrac{1}{x}$ on $[1, \infty)$.

(6) $f : x \mapsto \dfrac{1}{x}$ on $(0, \infty)$.

(7) $f : x \mapsto \sin\left(\dfrac{1}{x}\right)$ on $(0, 1]$.

(8) $f : x \mapsto x \cdot \sin\left(\dfrac{1}{x}\right)$ on $(0, 1]$.

Exercise 2.15

Let $a < b$; prove that if f is continuous on $[a, b)$ and uniformly continuous on $[b, \infty)$, then f is uniformly continuous on $[a, \infty)$.

Exercise 2.16

Prove Theorem 16 for one-sided limits.

Exercise 2.17

If $\lim\limits_{x \to a} f(x) \in \mathbb{R}$ and $\lim\limits_{x \to a} g(x) = +\infty$, then

(1) $\lim\limits_{x \to a} (f(x) \pm g(x)) = \pm\infty$.

(2) If $\lim\limits_{x \to a} f(x) > 0$, then $\lim\limits_{x \to a} (f(x) \cdot g(x)) = +\infty$.

(3) $\lim\limits_{x \to a} \left(\dfrac{f(x)}{g(x)}\right) = 0$.

(4) If $\lim\limits_{x \to a} f(x) = +\infty$ and $\lim\limits_{x \to a} g(x) = +\infty$, then

$$\lim_{x \to a} (f(x) + g(x)) = +\infty \text{ and } \lim_{x \to a} (f(x) \cdot g(x)) = +\infty.$$

Exercise 2.18

Prove that if $f(x) \le g(x) \le h(x)$ for $x \in [a, b)$ and

$$\lim_{x \to b^-} f(x) = \lim_{x \to b^-} h(x) = L,$$

then

$$\lim_{x \to b^-} g(x) = L.$$

Exercise 2.19
Calculate the following limits. The answer should be a number, $+\infty$, $-\infty$ or "no limit."

(1) $\displaystyle\lim_{x\to\infty} \frac{6x-4}{2x+5}$

(2) $\displaystyle\lim_{x\to\infty} x^3 - 10x^2 - 6x - 2$

(3) $\displaystyle\lim_{x\to\infty} \frac{x^2-x+4}{3x^2+2x-3}$

(4) $\displaystyle\lim_{x\to\infty} \frac{\sqrt{x+2}}{\sqrt{3x+1}}$

(5) $\displaystyle\lim_{x\to\infty} x - \sqrt{x}$

(6) $\displaystyle\lim_{x\to\infty} \sqrt[3]{x+2}$

(7) $\displaystyle\lim_{x\to 0^-} 1 + \frac{1}{x}$

(8) $\displaystyle\lim_{x\to 0} \frac{1}{x^2} - \frac{1}{x}$

(9) $\displaystyle\lim_{x\to 0} \frac{1+2x^{-1}}{7+x^{-1}-5x^{-2}}$

(10) $\displaystyle\lim_{x\to 2} \frac{1-x}{2-x}$

(11) $\displaystyle\lim_{x\to 3^+} \frac{x+1}{(x-2)(x-3)}$

(12) $\displaystyle\lim_{x\to 3} \frac{x+1}{(x-2)(x-3)}$

(13) $\displaystyle\lim_{x\to 1} \frac{3x^2+4}{x^2+x-2}$

(14) $\displaystyle\lim_{x\to 2^+} \frac{x^2+4}{x^2-4}$

(15) $\displaystyle\lim_{x\to\infty} \sqrt{x^2+1} - x$

(16) $\displaystyle\lim_{x\to-\infty} \sqrt{x^2+1} - x$

(17) $\displaystyle\lim_{x\to\infty} \sqrt[3]{x+4} - \sqrt[3]{x}$

(18) $\displaystyle\lim_{x\to 0^+} \frac{1}{\log_a(x)}$ for $a > 1$

(19) $\displaystyle\lim_{x\to\infty} 2^{2x} - 3 \cdot 2^x + 5$

(20) $\displaystyle\lim_{x\to-\infty} 2^{2x} - 3 \cdot 2^x + 5$

(21) $\displaystyle\lim_{x\to\infty} \sqrt{x^2-3x+2} - \sqrt{x^2+1}$

(22) $\displaystyle\lim_{x\to\infty} \log_a(\log_a(x))$ for $a > 1$

Exercise 2.20
Prove: If $\lim_{x\to a} f(x) = L$ and g is continuous at L, then

$$\lim_{x\to a} g(f(x)) = g(L).$$

Exercise 2.21
Let $a > 1$ and consider the function

$$f(x) = \frac{1}{1+a^{\frac{1}{x}}}.$$

Find $\lim_{x\to\infty} f(x)$, $\lim_{x\to-\infty} f(x)$, $\lim_{x\to 0^+} f(x)$ and $\lim_{x\to 0^-} f(x)$.

Exercise 2.22
Prove the remaining properties of Theorem 24.

Exercise 2.23

Let $\varepsilon > 0$ be ultrasmall. Show that

(1) If $n \in \mathbb{N}$ is observable relative to ε, then $\varepsilon^n > \delta$ for every δ ultrasmall relative to ε.

(2) If $n \in \mathbb{N}$ is not observable relative to ε, then ε^n is ultrasmall relative to ε.

Hint: Use the fact that $a^n \simeq 0$ for $0 < a < 1$ and ultralarge n.

Exercise 2.24

Prove that the following statements are equivalent:

(1) L is observable relative to f, and $f(x) \simeq L$ whenever x is ultralarge relative to f.

(2) $f(x) \simeq L$ whenever x is ultralarge relative to f, L.

3

Differentiability

3.1 Derivative

Intuitively, the derivative of a function f at a point a is the instantaneous rate of change of f at a. In general discussions of derivative, we always assume that f is defined in some open interval I containing a; by Closure, we may assume that I is as observable as f and a.

We start with an example. Let $f : x \mapsto x^2$ for all $x \in \mathbb{R}$, and let $a = 1$. Both f and a are standard. If the argument of f changes by an ultrasmall amount h, that is, from 1 to $1 + h$, the value of f changes from $f(1)$ to $f(1 + h)$, that is, by the amount

$$f(1 + h) - f(1) = (1 + h)^2 - 1^2 = 2h + h^2.$$

The average rate of change is

$$\frac{f(1 + h) - f(1)}{h} = 2 + h.$$

The choice of h makes an ultrasmall difference in the computed average rate of change, but the "part" of the rate which is observable, that is, 2, is independent of the choice of h. This is the instantaneous rate of change of f at 1. We note that, in fact,

$$2 = \lim_{h \to 0} \frac{f(1 + h) - f(1)}{h}.$$

This motivates the general definition.

Definition 19. *Let f be a function defined on an open interval containing a. We say that f is **differentiable at** a if there is a real number D such that*

$$\lim_{h \to 0} \frac{f(a + h) - f(a)}{h} = D.$$

*The number D is denoted $f'(a)$ and called the **derivative of f at** a.*

Equivalently, we may define the derivative in the following way.

Definition 20. *Let f be a function defined on an open interval containing a. We say that f is **differentiable at** a if there is an observable number D such that for every ultrasmall h we have*

$$\frac{f(a+h) - f(a)}{h} \simeq D.$$

If such a number exists, we denote it $f'(a)$.

The parameters of this definition are f and a. As for all limits, by Stability, we may also work relative to any context where f and a are observable.

Example.

(1) Consider

$$f : x \mapsto x^2 + 3x \quad \text{at} \quad x = a.$$

The parameters are f and a (since f is standard, the context is in fact a). Let h be ultrasmall; then

$$\frac{f(a+h) - f(a)}{h} = \frac{(a+h)^2 + 3(a+h) - (a^2 + 3a)}{h}$$

$$= \frac{2a \cdot h + 3h + h^2}{h}$$

$$= 2a + 3 + h$$

$$\simeq 2a + 3.$$

Since $2a + 3$ is observable and does not depend on h, it is the limit. This shows that the function is differentiable at a and that its derivative is $f'(a) = 2a + 3$.

(2) Consider the function **absolute value**

$$g : x \mapsto |x|$$

at the point $a = 0$. Let h be ultrasmall. If $h > 0$, we have

$$\frac{g(0+h) - g(0)}{h} = \frac{|h| - |0|}{h} = \frac{h}{h} = 1.$$

and if $h < 0$, we have

$$\frac{g(0+h) - g(0)}{h} = \frac{|h| - |0|}{h} = \frac{-h}{h} = -1.$$

Hence there is no unique real value which satisfies the conditions to be the limit. We conclude that the derivative of g at 0 does not exist.

Letting $x = a+h$, we get an equivalent formulation: f is differentiable at a if and only if

$$\lim_{x \to a} \frac{f(x) - f(a)}{x - a}$$

exists (recall that this means that the limit is a real number); $f'(a)$ is then equal to this limit.

The derivative has a geometric interpretation. For $x \neq a$, the ratio

$$\frac{f(x) - f(a)}{x - a}$$

is the slope of the straight line through the two distinct points $\langle a, f(a) \rangle$ and $\langle x, f(x) \rangle$ on the graph of the function f, that is, the slope of the **secant line**. For $x \simeq a$, all these slopes are ultraclose to $f'(a)$.

Definition 21. *Let f be differentiable at a. The **tangent line to f at** a is the straight line of slope $f'(a)$ going through $\langle a, f(a) \rangle$.*

It is clear that the tangent line to f at a is given by

$$y = f'(a)(x - a) + f(a).$$

Let f be differentiable at a and let $x \simeq a$. Then by definition

$$\frac{f(x) - f(a)}{x - a} \simeq f'(a), \quad \text{so} \quad \frac{f(x) - f(a)}{x - a} - f'(a) = \varepsilon, \quad \text{for some } \varepsilon \simeq 0.$$

We deduce that

$$\frac{f(x) - (f(a) + f'(a)(x - a))}{x - a} = \varepsilon.$$

Thus,

$$f(x) = f(a) + f'(a)(x - a) + \varepsilon \cdot (x - a), \quad \text{for some } \varepsilon \simeq 0.$$

This shows that for $x \simeq a$ the value $f(x)$ is very well approximated by the value of the tangent line. Indeed, the error $\varepsilon \cdot (x - a)$ is a product of ultrasmall numbers. This approximation property of the tangent line singles it out among all straight lines going through $\langle a, f(a) \rangle$.

Exercise 29 (Answer page 248)
Let f be a function defined on an open interval containing a and let m be a real number. Show that the following conditions are equivalent:

(1) f is differentiable at a and $f'(a) = m$.

(2) The line $\ell : x \mapsto m(x - a) + f(a)$ through $\langle a, f(a) \rangle$ satisfies

$$\lim_{x \to a} \frac{f(x) - \ell(x)}{x - a} = 0.$$

Notation: It is customary to indicate an increment of the variable x by dx, assumed to be nonzero.

We rephrase the approximation property of the tangent line in this notation for further reference.

Theorem 25 (Increment Equation). *Let f be a function differentiable at $a \in \mathbb{R}$. If dx is ultrasmall, then*

$$f(a + dx) = f(a) + f'(a) \cdot dx + \varepsilon \cdot dx,$$

where $\varepsilon \simeq 0$.

The converse is also true.

Theorem 26 (Increment Equation: Converse). *Let f be a function and a a real number. Suppose there exists an observable real number L such that, for each ultrasmall dx,*

$$f(a + dx) = f(a) + L \cdot dx + \varepsilon \cdot dx,$$

where $\varepsilon \simeq 0$. Then f is differentiable at a and $f'(a) = L$.

Proof. The hypothesis implies that for any ultrasmall dx we have

$$\frac{f(a + dx) - f(a)}{dx} \simeq L.$$

By definition of the derivative, this means that f is differentiable and that $f'(a) = L$. □

We simply say "by the Increment Equation" when we use one of the previous two theorems.

Exercise 30 (Straddle Version) (Answer page 248)
Let f be differentiable at a. Let $x_1 \leq a \leq x_2$ be such that $x_1 \simeq a$ and $x_2 \simeq a$. Show that

$$f(x_2) - f(x_1) = f'(a)(x_2 - x_1) + \varepsilon \cdot (x_2 - x_1), \quad \text{for some } \varepsilon \simeq 0.$$

Definition 22. *The **increment** of f at a, denoted by $\Delta f(a)$, is*

$$\Delta f(a) = f(a + dx) - f(a).$$

This increment depends on f and a, as indicated by the notation. For fixed f and a, it is a function of dx; a more pedantic (and rarely used) notation would be $\Delta f(a)(dx)$.

If f is differentiable at a and dx is ultrasmall, then

$$\frac{\Delta f(a)}{dx} \simeq f'(a).$$

Notation: The **differential** of f at a, denoted by $df(a)$, is

$$df(a) = f'(a) \cdot dx.$$

It represents the increment along the tangent line to f at a. For fixed f and a, it is a linear function of dx.

The Increment Equation can be rewritten in this notation as follows. *For dx ultrasmall,*

$$\Delta f(x) = df(x) + \varepsilon \cdot dx, \quad \varepsilon \simeq 0.$$

We note that, if f is differentiable at x,

$$f'(x) = \frac{df(x)}{dx}.$$

For this reason, the notation $\frac{df(x)}{dx}$ or, loosely written, $\frac{df}{dx}(x)$, is often used for the derivative of f at x. In the case when $y = f(x)$, one writes $\frac{dy}{dx}$ for y'.

For functions of one variable, derivatives and differentials are just two equivalent ways to describe the same idea. In this book, we generally use the language of derivatives. Differentials become important in the study of functions of two or more variables.

We conclude this section with an important observation.

Theorem 27 (Continuity of a Differentiable Function). *Let f be a function differentiable at a. Then f is continuous at a.*

Proof. The number $f'(a)$ exists by hypothesis and is observable. Let dx be ultrasmall. By definition of $\Delta f(a)$, it is enough to show that $\Delta f \simeq 0$. But, by Rule 5,

$$\Delta f(a) = \frac{\Delta f(a)}{dx} \cdot dx \simeq f'(a) \cdot dx \simeq 0.$$

$\square$

The converse is false. The function $f : x \mapsto |x|$ is not differentiable at $a = 0$ (see Example (2) on page 68). However, f is continuous at $a = 0$, since $h \simeq 0$ implies $f(h) = |h| \simeq 0 = f(0)$.

Exercise 31 (Answer page 249)
Let $f : x \mapsto x^3 - x^2$.
Using the definition, calculate $f'(2)$ and $f'(3 + h)$ for $h \simeq 0$.

Exercise 32 (Answer page 249)
Let $f : x \mapsto \sqrt{x}$. Show that $f'(x) = \frac{1}{2\sqrt{x}}$, for $x > 0$.

3.2   Rules of Differentiation

Recall the notation

$$\Delta f(a) = f(a + dx) - f(a),$$

which implies that $f(a + dx) = f(a) + \Delta f(a)$, and the fact that

$$\frac{\Delta f(a)}{dx} \simeq f'(a),$$

if f is differentiable at a.

We now proceed to give uniform proofs of the usual theorems about derivatives.

Theorem 28 (Derivative of a Constant Function). *Let $\lambda \in \mathbb{R}$ and I be an open interval. Let $f : I \to \mathbb{R}$ be given by $x \mapsto \lambda$. Then*

$$f'(a) = 0, \quad \text{for each } a \in I.$$

Proof. Let dx be ultrasmall.

$$\frac{\Delta f(a)}{dx} = \frac{f(a + dx) - f(a)}{dx} = \frac{\lambda - \lambda}{dx} = 0.$$

But 0 is observable and hence $f'(a) = 0$.   □

Theorem 29 (Derivative of a Product by a Constant). *Let $\lambda \in \mathbb{R}$ and $a \in \mathbb{R}$. Let f be a function differentiable at a. Then $\lambda \cdot f$ is differentiable at a and*

$$(\lambda \cdot f)'(a) = \lambda \cdot f'(a).$$

Proof. Let dx be ultrasmall.

$$
\begin{aligned}
\frac{\Delta(\lambda f)(a)}{dx} &= \frac{(\lambda \cdot f)(a + dx) - (\lambda \cdot f)(a)}{dx} \\
&= \frac{\lambda \cdot f(a + dx) - \lambda \cdot f(a)}{dx} \\
&= \frac{\lambda \cdot (f(a) + \Delta f(a)) - \lambda \cdot f(a)}{dx} \\
&= \lambda \cdot \frac{\Delta f(a)}{dx} \\
&\simeq \lambda \cdot f'(a).
\end{aligned}
$$

But $\lambda \cdot f'(a)$ is observable by Closure, hence $(\lambda \cdot f)'(a) = \lambda \cdot f'(a)$. $\square$

Theorem 30 (Derivative of a Sum). *Let f and g be functions differentiable at a. Then $(f + g)$ is differentiable at a and*

$$
(f + g)'(a) = f'(a) + g'(a).
$$

Proof. Let dx be ultrasmall.

$$
\begin{aligned}
\frac{\Delta(f + g)(a)}{dx} &= \frac{(f + g)(a + dx) - (f + g)(a)}{dx} \\
&= \frac{(f(a + dx) + g(a + dx)) - (f(a) + g(a))}{dx} \\
&= \frac{(f(a) + \Delta f(a) + g(a) + \Delta g(a)) - (f(a) + g(a))}{dx} \\
&= \frac{\Delta f(a)}{dx} + \frac{\Delta g(a)}{dx} \\
&\simeq f'(a) + g'(a).
\end{aligned}
$$

But $f'(a) + g'(a)$ is observable by Closure, hence $(f + g)'(a) = f'(a) + g'(a)$. $\square$

Theorem 31 (Derivative of a Product). *Let f and g be functions differentiable at a. Then $(f \cdot g)$ is differentiable at a and*

$$
(f \cdot g)'(a) = f'(a) \cdot g(a) + f(a) \cdot g'(a).
$$

Proof. Let dx be ultrasmall.

$$
\begin{aligned}
\frac{\Delta(f \cdot g)(x)}{dx} &= \frac{(f \cdot g)(a + dx) - (f \cdot g)(a)}{dx} \\
&= \frac{f(a + dx) \cdot g(a + dx) - f(a) \cdot g(a)}{dx}
\end{aligned}
$$

$$= \frac{(f(a) + \Delta f(a)) \cdot (g(a) + \Delta g(a)) - f(a) \cdot g(a)}{dx}$$

$$= f(a) \cdot \underbrace{\frac{\Delta g(a)}{dx}}_{\simeq g'(a)} + \underbrace{\frac{\Delta f(a)}{dx}}_{\simeq f'(a)} \cdot g(a) + \underbrace{\frac{\Delta f(a)}{dx}}_{\simeq f'(a)} \cdot \underbrace{\Delta g(a)}_{\simeq 0}$$

$$\simeq f(a) \cdot g'(a) + f'(a) \cdot g(a).$$

By Theorem 27 $\Delta g(a) \simeq 0$, and $\frac{\Delta f(a)}{dx}$ is ultraclose to $f'(a)$, which is not ultralarge, so $\frac{\Delta f(a)}{dx} \cdot \Delta g(a) \simeq 0$. But $f'(a) \cdot g(a) + f(a) \cdot g'(a)$ is observable by Closure, hence $(f \cdot g)'(a) = f'(a) \cdot g(a) + f(a) \cdot g'(a)$. $\qquad \square$

Theorem 32 (Derivative of a Quotient of Functions). *Let f and g be functions differentiable at a. Suppose that $g(a) \neq 0$. Then $\frac{f}{g}$ is differentiable at a and*

$$\left(\frac{f}{g}\right)'(a) = \frac{f'(a) \cdot g(a) - f(a) \cdot g'(a)}{g^2(a)}.$$

Proof. Let dx be ultrasmall.

$$\frac{\Delta\left(\frac{f}{g}\right)(a)}{dx} = \frac{\left(\frac{f}{g}\right)(a + dx) - \left(\frac{f}{g}\right)(a)}{dx}$$

$$= \frac{\dfrac{f(a + dx)}{g(a + dx)} - \dfrac{f(a)}{g(a)}}{dx}$$

$$= \frac{\dfrac{f(a) + \Delta f(a)}{g(a) + \Delta g(a)} - \dfrac{f(a)}{g(a)}}{dx}$$

$$= \frac{\dfrac{(f(a) + \Delta f(a)) \cdot g(a) - f(a) \cdot (g(a) + \Delta g(a))}{(g(a) + \Delta g(a)) \cdot g(a)}}{dx}$$

$$= \frac{\dfrac{\Delta f(a)}{dx} \cdot g(a) - f(a) \cdot \dfrac{\Delta g(a)}{dx}}{g^2(a) + \Delta g(a) \cdot g(a)}$$

by Rule 5

$$\simeq \frac{f'(a) \cdot g(a) - f(a) \cdot g'(a)}{g^2(a)}.$$

This last expression is observable by Closure, hence it is the derivative.

$\square$

We next consider the rule for differentiation of the composition of functions.

Theorem 33 (Chain Rule). *Let f and g be functions such that g is differentiable at a and f is differentiable at $g(a)$. Then the composition $f \circ g$ is differentiable at a and*

$$(f \circ g)'(a) = f'(g(a)) \cdot g'(a).$$

Proof. Let dx be ultrasmall.

$$\frac{\Delta(f \circ g)(a)}{dx} = \frac{(f \circ g)(a + dx) - (f \circ g)(a)}{dx}$$

$$= \frac{f(g(a + dx)) - f(g(a))}{dx}$$

$$= \frac{f(g(a) + \Delta g(a)) - f(g(a))}{dx}.$$

We now distinguish two cases:
Case 1: $\Delta g(a) \neq 0$. Then

$$\frac{\Delta(f \circ g)(a)}{dx} = \frac{f(g(a) + \Delta g(a)) - f(g(a))}{dx}$$

$$= \frac{f(g(a) + \Delta g(a)) - f(g(a))}{\Delta g(a)} \cdot \frac{\Delta g(a)}{dx}$$

$$\simeq f'(g(a)) \cdot g'(a),$$

since $\Delta g(a)$ is ultrasmall by Theorem 27, so

$$\frac{f(g(a) + \Delta g(a)) - f(g(a))}{\Delta g(a)} \simeq f'(g(a))$$

by differentiability of f at $g(a)$.
Case 2: $\Delta g(a) = 0$. Then

$$\frac{\Delta(f \circ g)(a)}{dx} = \frac{f(g(a) + \Delta g(a)) - f(g(a))}{dx}$$

$$= \frac{f(g(a)) - f(g(a))}{dx} = 0 = f'(g(a)) \cdot g'(a),$$

since g is differentiable at a, so $g'(a) \simeq \dfrac{\Delta g(a)}{dx} = 0$, which shows that $g'(a) = 0$. Since $f'(g(a)) \cdot g'(a)$ is observable by Closure, we have $(f \circ g)'(a) = f'(g(a)) \cdot g'(a)$. $\square$

We end this section with the rule for the derivative of the inverse function.

Theorem 34 (Derivative of the Inverse). *Let $f : (a, b) \to \mathbb{R}$ be a continuous one-to-one function. Let $y = f(x)$. If f is differentiable at $x \in (a, b)$ with $f'(x) \neq 0$, then f^{-1} is differentiable at y and*

$$(f^{-1})'(y) = \frac{1}{f'(x)} = \frac{1}{f'(f^{-1}(y))}.$$

Proof. Let dy be ultrasmall. Then

$$\Delta f^{-1}(y) = f^{-1}(y + dy) - f^{-1}(y) \text{ is ultrasmall}$$

since $\Delta f^{-1}(y) \simeq 0$ by continuity of f^{-1} (Theorem 12) and $\Delta f^{-1}(y) \neq 0$ since f^{-1} is one-to-one.

Let $dx = \Delta f^{-1}(y)$. Notice that, with this dx, we have

$$\Delta f(x) = f(x + dx) - f(x) = f(f^{-1}(y) + \Delta f^{-1}(y)) - y$$
$$= f(f^{-1}(y + dy)) - y = dy.$$

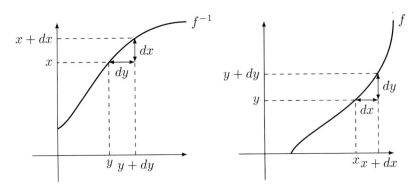

Thus

$$\frac{\Delta f^{-1}(y)}{dy} = \frac{dx}{dy} = \frac{1}{\dfrac{dy}{dx}} = \frac{1}{\dfrac{\Delta f(x)}{dx}} \simeq \frac{1}{f'(x)},$$

by Rule 5, since $\dfrac{\Delta f(x)}{dx} \simeq f'(x) \neq 0$ by assumption. But $\frac{1}{f'(x)}$ is observable by Closure, so $(f^{-1}(y))'$ exists and

$$(f^{-1}(y))' = \frac{1}{f'(x)} = \frac{1}{f'(f^{-1}(y))}.$$

$\square$

Exercise 33 (Answer page 249)

The following example shows a "jump" function with an almost vertical segment.

Let h be ultrasmall relative to 1. Consider the function H defined by

$$H(x) = \begin{cases} 0 & \text{if } x \leq -h; \\ \frac{1}{2h}(x+h) & \text{if } -h < x < h; \\ 1 & \text{if } x \geq h. \end{cases}$$

What are the parameters of H?

Compute $H'(x)$.

Example. We show that if q is rational and $x > 0$, then

$$(x^q)' = q \cdot x^{q-1}.$$

We start with $q = n \in \mathbb{N}$. For $n = 0$, we have $(x^0)' = 0$, so the formula is true. Let $n \geq 1$ and x be arbitrary (and observable). Let h be ultrasmall. To evaluate

$$\frac{(x+h)^n - x^n}{h}$$

we use the binomial formula (see Exercise 26)

$$(x+h)^n = x^n + nx^{n-1}h + \frac{n(n-1)}{2}x^{n-2}h^2 + \cdots + nxh^{n-1} + h^n$$

$$= x^n + nx^{n-1}h + h^2 \cdot \underbrace{\left(\frac{n(n-1)}{2}x^{n-2} + \cdots + nxh^{n-3} + h^{n-2} \right)}_{=p(n,x,h)}.$$

Notice that the number $p(n, x, h)$ is not ultralarge, by Rule 6. Now, using this, we deduce

$$\frac{(x+h)^n - x^n}{h} = \frac{x^n + nx^{n-1}h + h^2 p(n, x, h) - x^n}{h}$$

$$= nx^{n-1} + hp(n, x, h) \simeq nx^{n-1}.$$

(For an alternative proof, see Exercise 53.)

For negative integers we use the rule for the quotient. Let n be a positive integer and $x \neq 0$.

$$(x^{-n})' = \left(\frac{1}{x^n} \right)' = \frac{(1)' \cdot x^n - 1 \cdot (x^n)'}{(x^n)^2} = \frac{0 \cdot x^n - 1 \cdot n \cdot x^{n-1}}{x^{2n}}$$

$$= -nx^{-n-1}.$$

The formula is thus true for all integers.

For $x^{1/n}$ with positive integer n and $x > 0$ we use the fact that $y = x^{1/n}$ is defined as the inverse function to $x = y^n$ on $(0, \infty)$. Hence

$$(x^{1/n})' = \frac{1}{n \cdot y^{n-1}} = \frac{1}{n \cdot (x^{1/n})^{n-1}} = \frac{1}{n} \cdot x^{\frac{1}{n}-1}.$$

Finally, for $q = m/n$, with integers m, n such that $n > 0$, by the Chain Rule:

$$(x^{m/n})' = ((x^m)^{1/n})' = \frac{1}{n} \cdot (x^m)^{\frac{1}{n}-1} \cdot (m \cdot x^{m-1}) = \frac{m}{n} \cdot x^{\frac{m}{n}-1}.$$

3.3 Basic Theorems about Derivatives

Rolle's Theorem, the Mean Value Theorem, Cauchy's Theorem and their consequences constitute the core of differential calculus.

Definition 23.

> *(1) A function f is **increasing** at $a \in \mathbb{R}$ if $f(x) < f(a)$ for all $x \simeq a$, $x < a$, and $f(x) > f(a)$ for all $x \simeq a$, $x > a$.*
>
> *(2) A function f is **decreasing** at $a \in \mathbb{R}$ if $f(x) > f(a)$ for all $x \simeq a$, $x < a$, and $f(x) < f(a)$ for all $x \simeq a$, $x > a$.*

Theorem 35.

> *(1) If $f'(a) > 0$, then f is increasing at a.*
>
> *(2) If $f'(a) < 0$, then f is decreasing at a.*

Proof. Assume $f'(a) > 0$. As $f'(a)$ is observable and $f'(a) \simeq \frac{f(x)-f(a)}{x-a}$ for all $x \simeq a$ ($x \neq a$), we have

$$\frac{f(x) - f(a)}{x - a} > 0, \quad \text{for all } x \simeq a \ (x \neq a).$$

Hence $x - a > 0$ implies $f(x) - f(a) > 0$, and $x - a < 0$ implies $f(x) - f(a) < 0$.

The second claim has a similar proof. $\qquad\square$

Exercise 34 (Answer page 250)
Show directly from the previous theorem and the idea of the proof of the Intermediate Value Theorem that if $f'(x) > 0$ for all x in the interval I then f is strictly increasing on I.

Definition 24.

(1) A function f has **local maximum at** $a \in \mathbb{R}$ if $f(x) \leq f(a)$ for all $x \simeq a$.

(2) A function f has **local minimum at** $a \in \mathbb{R}$ if $f(x) \geq f(a)$ for all $x \simeq a$.

Theorem 36 (Derivative at a maximum or minimum). *If f is differentiable at a and has a local maximum or a local minimum at a, then* $f'(a) = 0$.

Proof. This is an immediate consequence of the preceding theorem. □

A function is **differentiable on an interval** (a, b) if it is differentiable at x for all $x \in (a, b)$.

Theorem 37 (Rolle). *Let f be a function continuous on $[a, b]$ and differentiable on (a, b). Suppose that $f(a) = f(b)$. Then there exists $c \in (a, b)$ such that $f'(c) = 0$.*

Proof. As f is continuous, it reaches its maximum and its minimum (by Theorem 9). If $f(a) = f(b)$ is both maximum and minimum, then the function is constant on $[a, b]$ and all points $c \in (a, b)$ satisfy the conclusion by Theorem 28.

Otherwise, there exists $c \in (a, b)$ such that $f(c)$ is either a maximum or a minimum. By Theorem 36 we have $f'(c) = 0$. □

Theorem 38 (Mean Value). *Let f be a function continuous on $[a, b]$ and differentiable on (a, b). Then there exists $c \in (a, b)$ such that $f(b) - f(a) = f'(c) \cdot (b - a)$.*

Proof. We subtract from f the line connecting $\langle a, f(a) \rangle$ and $\langle b, f(b) \rangle$ to obtain a function satisfying the conditions of Rolle's Theorem.

The connecting line is the graph of the function

$$\ell(x) = f(a) + (x - a) \cdot \frac{f(b) - f(a)}{b - a}.$$

Let

$$h(x) = f(x) - \ell(x) = f(x) - f(a) - (x - a) \cdot \frac{f(b) - f(a)}{b - a};$$

then $\quad h'(x) = f'(x) - \dfrac{f(b) - f(a)}{b - a}.$

The function h is continuous on $[a, b]$ and differentiable on (a, b), and we have $h(a) = h(b) = 0$. By Rolle's Theorem, there is a $c \in (a, b)$ such that $h'(c) = 0$. Hence

$$f'(c) = \frac{f(b) - f(a)}{b - a}.$$

□

Theorem 39 (Cauchy's Mean Value). *Let f and g be continuous on $[a, b]$ and differentiable on (a, b). Then there exists $c \in (a, b)$ such that*

$$\big(f(b) - f(a)\big) \cdot g'(c) = \big(g(b) - g(a)\big) \cdot f'(c).$$

If $g(b) \neq g(a)$ and $g'(c) \neq 0$, then this can be written

$$\frac{f(b) - f(a)}{g(b) - g(a)} = \frac{f'(c)}{g'(c)}.$$

Proof. We define the following auxiliary function

$$h : x \mapsto \big(g(b) - g(a)\big) \cdot f(x) - \big(f(b) - f(a)\big) \cdot g(x).$$

Then h is continuous on $[a, b]$ and differentiable on (a, b) and $h(a) = f(a) \cdot g(b) - g(a) \cdot f(b) = h(b)$.

Hence, by Rolle's Theorem, there is a $c \in (a, b)$ such that $h'(c) = 0$, that is,

$$\big(g(b) - g(a)\big) \cdot f'(c) - \big(f(b) - f(a)\big) \cdot g'(c) = 0,$$

which implies that

$$\big(f(b) - f(a)\big) \cdot g'(c) = \big(g(b) - g(a)\big) \cdot f'(c).$$

□

In the previous theorems the parameters are a, b and f. By Closure, it is always possible to find an observable c satisfying the conditions of the previous three theorems.

We now investigate the link between derivatives and behavior of a function on an interval.

Definition 25.

*(1) A function f is **increasing on an interval** I if $f(x) \leq f(y)$ whenever $x < y$, for $x, y \in I$.*

*(2) A function f is **decreasing on an interval** I if $f(x) \geq f(y)$ whenever $x < y$, for $x, y \in I$.*

*A function is **strictly** increasing/decreasing if (1) or (2) hold for strict inequalities.*

Theorem 40. *Let f be differentiable on $I = (a, b)$.*

    (1) If $f'(x) \geq 0$ for each $x \in I$, then f is increasing on I.

    (2) If $f'(x) \leq 0$ for each $x \in I$, then f is decreasing on I.

    (3) If $f'(x) = 0$ for each $x \in I$, then f is constant on I.

If the inequalities are replaced by strict inequalities, the function is respectively strictly increasing, strictly decreasing.

Proof. (1) Assume that for all $c \in I$ we have $f'(c) \geq 0$. Let $x < y$ be in I. Then by the Mean Value Theorem, there is $c \in (x, y)$ such that

$$f(y) - f(x) = f'(c)(y - x).$$

As $y - x > 0$ and $f'(c) \geq 0$, we have $f(y) \geq f(x)$.

All the other cases are proved similarly. $\qquad\square$

We also prove here a simple version of L'Hôpital's Rule.

Theorem 41 (L'Hôpital's Rule for 0/0 – Simple Form). *Let f and g be differentiable at a, and $f(a) = g(a) = 0$ with $g'(a) \neq 0$. Then*

$$\lim_{x \to a} \left(\frac{f(x)}{g(x)} \right) = \frac{f'(a)}{g'(a)}.$$

Proof. Let dx be ultrasmall and write $x = a + dx$. As $g'(a) \neq 0$, the function g is either increasing or decreasing at a by Theorem 35, and hence $g(x) \neq g(a) = 0$. Then, using the assumption that $f(a) = g(a) = 0$, we have

$$\frac{f(x)}{g(x)} = \frac{f(a + dx)}{g(a + dx)} = \frac{f(a + dx) - f(a)}{dx} \cdot \frac{dx}{g(a + dx) - g(a)} \simeq \frac{f'(a)}{g'(a)},$$

by Rule 5, since $g'(a) \neq 0$ $\qquad\square$

We conclude this section with a definition of one-sided derivatives.

Definition 26. *Let f be a function defined on an interval $[a, b)$. We say that f is **differentiable on the right at** a if*

$$\lim_{h \to 0^+} \frac{f(a + h) - f(a)}{h} \quad \textit{exists.}$$

*If the limit exists, we denote it $f'_+(a)$ and call it the **right derivative of f at** a. Similarly for the left derivative.*

The results of this chapter hold for one-sided derivatives, with obvious modifications. For example, if f is differentiable on the right at a, then f is continuous on the right at a.

Definition 27. *A function f is **differentiable on** I if f is differentiable at each $x \in I$ (one-sided derivatives suffice at endpoints).*

3.4   Smooth Functions

In Definition 19 we define the derivative of f at x as a certain limit. The concept of limit is internal (see Theorem 15) and consequently also the statement $y = f'(x)$ is internal. By the Definition Principle, there is a *function f'*, defined for all x where $f'(x)$ exists, and observable whenever f is observable.

We introduce smooth functions (functions of the class $\mathcal{C}^1$), for which a very useful characterization can be given.

Definition 28. *The function f is **smooth on** I if f is differentiable on I and if f' is continuous on I (the values of f' at endpoints are given by the appropriate one-sided derivatives).*

Functions for which f' is differentiable are smooth, by Theorem 27. Smooth functions are exactly those that satisfy a stronger form of the Increment Equation.

Theorem 42 (Uniform Increment Equation). *Let f be differentiable on $[a, b]$. The following conditions are equivalent.*

> *(1) f is smooth on $[a, b]$.*
>
> *(2) For all $x \in [a, b]$ and all $dx \simeq 0$ such that $x + dx \in [a, b]$,*
>
> $$f(x + dx) - f(x) = f'(x) \cdot dx + \varepsilon \cdot dx,$$
>
> *where $\varepsilon \simeq 0$.*

This version is stronger than the Increment Equation because in the above equation, dx does not have to be ultrasmall relative to x (the parameters of the theorem are f, a, b).

Proof. (1) implies (2): Fix $dx \simeq 0$ relative to f, a, b (but not necessarily relative to x). By the Mean Value Theorem we have

$$f(x + dx) - f(x) = f'(c) \cdot dx,$$

for some c between x and $x + dx$. As f' is continuous on $[a, b]$, it is uniformly continuous, hence (note $c \simeq x$)

$$f'(c) = f'(x) + \varepsilon, \qquad \text{with } \varepsilon \simeq 0.$$

This yields

$$f(x + dx) - f(x) = f'(x) \cdot dx + \varepsilon \cdot dx.$$

(2) implies (1): Let x, $x + dx \in [a, b]$ and dx be ultrasmall (relative to f, a, b). By (2) applied to $x + dx$ in place of x, there is an $\varepsilon \simeq 0$ such that

$$f(x + dx + dx) - f(x + dx) = f'(x + dx) \cdot dx + \varepsilon \cdot dx,$$

so

$$f'(x + dx) \simeq \frac{f(x + 2dx) - f(x + dx)}{dx}.$$

But

$$\frac{f(x + 2dx) - f(x + dx)}{dx} = 2 \cdot \frac{f(x + 2dx) - f(x)}{2dx} - \frac{f(x + dx) - f(x)}{dx}$$

$$\simeq 2f'(x) - f'(x) = f'(x);$$

hence $f'(x + dx) \simeq f'(x)$. The argument proves that f' is uniformly continuous on $[a, b]$, hence f' is continuous on $[a, b]$. $\qquad \square$

3.5 Derivatives of Trigonometric Functions

Here we continue the study of properties of trigonometric functions, based on geometric considerations. Recall that the point on the circle of radius 1 centered at 0, determined by the angle θ, has coordinates $\langle \cos(\theta), \sin(\theta) \rangle$. The point on the tangent line to this circle at $\langle 1, 0 \rangle$, determined by the angle θ, has coordinates $\langle 1, \tan(\theta) \rangle$. This defines the functions

$$\sin : \mathbb{R} \to [-1, 1], \qquad \cos : \mathbb{R} \to [-1, 1] \qquad \text{and}$$

$$\tan : \mathbb{R} \setminus \left\{ \frac{\pi}{2} + k\pi : k \in \mathbb{Z} \right\} \to \mathbb{R}.$$

These functions are standard. The sine and cosine functions are continuous everywhere (see page 38). The continuity of tan follows from the rule for continuity of a quotient, as $\tan(\theta) = \sin(\theta)/\cos(\theta)$, for $\theta \neq \pi/2 + k\pi$, with $k \in \mathbb{Z}$. The value of θ (in radians) is the length of the arc spanning the angle.

We have

Theorem 43.
$$\lim_{\theta \to 0} \frac{\sin(\theta)}{\theta} = 1.$$

Proof. Suppose first that $\theta > 0$ is in the first quadrant.

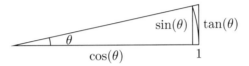

Then by comparing the area of the sector with that of the inside and outside triangles, we obtain

$$\frac{\sin(\theta) \cdot \cos(\theta)}{2} \leq \frac{\theta}{2} \leq \frac{\tan(\theta) \cdot 1}{2} = \frac{\sin(\theta)}{2\cos(\theta)}.$$

We deduce that

$$\cos(\theta) \leq \frac{\sin(\theta)}{\theta} \leq \frac{1}{\cos(\theta)}.$$

By using $-\theta$ if θ is negative, we see that the same inequalities are true for negative θ (in the fourth quadrant).

From continuity of cos at 0 it follows that, for any ultrasmall θ,

$$1 \simeq \cos(\theta) \leq \frac{\sin(\theta)}{\theta} \leq \frac{1}{\cos(\theta)} \simeq 1,$$

which shows that $\lim_{\theta \to 0} \frac{\sin(\theta)}{\theta} = 1$.   □

We also have

Theorem 44.
$$\lim_{\theta \to 0} \frac{1 - \cos(\theta)}{\theta} = 0.$$

Proof. Let θ be ultrasmall. Then

$$\frac{1 - \cos(\theta)}{\theta} = \frac{1 - \cos(\theta)}{\theta} \cdot \frac{1 + \cos(\theta)}{1 + \cos(\theta)}$$

$$= \frac{1 - \cos^2(\theta)}{\theta \cdot (1 + \cos(\theta))} = \frac{\sin^2(\theta)}{\theta \cdot (1 + \cos(\theta))}$$

$$= \underbrace{\frac{\sin(\theta)}{\theta}}_{\simeq 1} \cdot \overbrace{\underbrace{\frac{\sin(\theta)}{1 + \cos(\theta)}}_{\simeq 2}}^{\simeq 0} \simeq 0.$$

We used the Pythagorean Theorem in the form $\sin^2(\theta) + \cos^2(\theta) = 1$, as well as the previous theorem.   □

Exercise 35 (Answer page 251)
Prove that
$$\lim_{\theta \to 0} \frac{\tan(\theta)}{\theta} = 1.$$

Theorem 45.

(1) Let $\theta \in \mathbb{R}$. Then $\sin'(\theta) = \cos(\theta)$.

(2) Let $\theta \in \mathbb{R}$. Then $\cos'(\theta) = -\sin(\theta)$.

(3) Let $\theta \in \mathbb{R}$, $\theta \neq \frac{\pi}{2} + k\pi$, for $k \in \mathbb{Z}$. Then $\tan'(\theta) = \frac{1}{\cos^2(\theta)} = 1 + \tan^2(\theta)$.

Proof.   (1) Let $d\theta$ be ultrasmall (relative to θ). Then using the addition formula for sine, we have

$$
\begin{aligned}
\frac{\Delta \sin(\theta)}{d\theta} &= \frac{\sin(\theta + d\theta) - \sin(\theta)}{d\theta} \\
&= \frac{\sin(\theta)\cos(d\theta) + \cos(\theta)\sin(d\theta) - \sin(\theta)}{d\theta} \\
&= \frac{\sin(\theta) \cdot (\cos(d\theta) - 1) + \cos(\theta) \cdot \sin(d\theta)}{d\theta} \\
&= \sin(\theta) \cdot \underbrace{\frac{\cos(\theta) - 1}{d\theta}}_{\simeq 0} + \cos(\theta) \cdot \underbrace{\frac{\sin(d\theta)}{d\theta}}_{\simeq 1} \simeq \cos(\theta),
\end{aligned}
$$

where we used the last two theorems. But $\cos(\theta)$ is observable, so $\sin'(\theta) = \cos(\theta)$.

(2) Let $d\theta$ be ultrasmall. We give a similar proof using the addition formula for cosine:

$$
\begin{aligned}
\frac{\Delta \cos(\theta)}{d\theta} &= \frac{\cos(\theta + d\theta) - \cos(\theta)}{d\theta} \\
&= \frac{\cos(\theta)\cos(d\theta) - \sin(\theta)\sin(d\theta) - \cos(\theta)}{d\theta} \\
&= \frac{\cos(\theta) \cdot (\cos(d\theta) - 1) - \sin(\theta) \cdot \sin(d\theta)}{d\theta} \\
&= \cos(\theta) \cdot \underbrace{\frac{\cos(\theta) - 1}{d\theta}}_{\simeq 0} - \sin(\theta) \cdot \underbrace{\frac{\sin(d\theta)}{d\theta}}_{\simeq 1} \\
&\simeq -\sin(\theta).
\end{aligned}
$$

But $-\sin(\theta)$ is observable, so $\cos'(\theta) = -\sin(\theta)$.

(3) We use $\tan(\theta) = \dfrac{\sin(\theta)}{\cos(\theta)}$ and the rule for the derivative of a quotient.

$$\tan'(\theta) = \left(\frac{\sin(\theta)}{\cos(\theta)}\right)' = \frac{\cos(\theta) \cdot \cos(\theta) - \sin(\theta) \cdot (-\sin(\theta))}{\cos^2(\theta)}$$

$$= \frac{1}{\cos^2(\theta)} = 1 + \tan^2(\theta).$$

□

Exercise 36 (Answer page 251)
Here is another way to prove that $\sin'(x) = \cos(x)$.

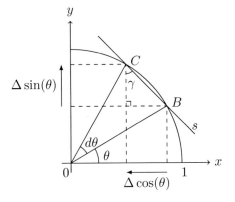

First prove that if $d\theta$ is ultrasmall, there is $\varepsilon \simeq 0$ such that $\overline{BC} = d\theta + \varepsilon \cdot d\theta$. Then deduce that $\sin'(\theta) = \cos(\theta)$ from the equality $\cos(\gamma) = \dfrac{\Delta \sin(\theta)}{\overline{BC}}$.

Since the trigonometric functions are periodic, it is necessary to restrict their domain in order to define their inverse. For sine, we consider

$$\sin : \left[-\frac{\pi}{2}, \frac{\pi}{2}\right] \to [-1, 1],$$

which is a one-to-one correspondence, to define

$$\arcsin : [-1, 1] \to \left[-\frac{\pi}{2}, \frac{\pi}{2}\right]$$

by

$$\arcsin(x) = \theta \quad \text{if} \quad \sin(\theta) = x.$$

We define arccos and arctan similarly, where

$$\arccos : [-1, 1] \to [0, \pi] \quad \text{and} \quad \arctan : \mathbb{R} \to \left(-\frac{\pi}{2}, \frac{\pi}{2}\right).$$

The trigonometric functions are smooth, as are their inverses, except for arcsin and arccos at the endpoints of their domains ($x = \pm 1$), where the tangents to the graphs of these functions are vertical.

Theorem 46.

(1) Let $x \in (-1, 1)$. Then $\arcsin'(x) = \dfrac{1}{\sqrt{1 - x^2}}$.

(2) Let $x \in (-1, 1)$. Then $\arccos'(x) = -\dfrac{1}{\sqrt{1 - x^2}}$.

(3) Let $x \in \mathbb{R}$. Then $\arctan'(x) = \dfrac{1}{1 + x^2}$.

Proof. (1) Assume $\arcsin(x) = y$, that is, $\sin(y) = x$, and recall that $\sin'(y) = \cos(y)$. Then

$$\arcsin'(x) = \frac{1}{\sin'(y)} = \frac{1}{\cos(y)} = \frac{1}{\sqrt{1 - \sin^2(y)}} = \frac{1}{\sqrt{1 - x^2}}.$$

(2) Assume $\arccos(x) = y$, then

$$\arccos'(x) = \frac{1}{\cos'(y)} = -\frac{1}{\sin(y)} = -\frac{1}{\sqrt{1 - \cos^2(y)}} = -\frac{1}{\sqrt{1 - x^2}}.$$

(3) Assume $\arctan(x) = y$, then

$$\arctan'(x) = \frac{1}{\tan'(y)} = \frac{1}{1 + \tan^2(y)} = \frac{1}{1 + x^2}.$$

$\square$

Exercise 37 (Answer page 251)
This exercise is similar to Exercise 33; it is concerned with a smooth function with an almost vertical "jump."
Let ε be ultrasmall relative to 1. Consider $H : x \mapsto \dfrac{1}{2} + \dfrac{1}{\pi} \cdot \arctan\left(\dfrac{x}{\varepsilon}\right)$.
Sketch the graph of H and compute $H'(x)$, for each $x \in \mathbb{R}$.

Exercise 38 (Answer page 252)

(1) Show that f defined by $f(x) = \sin\left(\frac{1}{x}\right)$, for $x \neq 0$, cannot be extended to a function which is continuous at 0.

(2) Let g be defined by

$$g(x) = \begin{cases} x \cdot \sin\left(\frac{1}{x}\right) & \text{if } x \neq 0; \\ 0 & \text{if } x = 0. \end{cases}$$

Show that g is not differentiable at 0.

(3) Show that h defined by

$$h(x) = \begin{cases} x^2 \cdot \sin\left(\frac{1}{x}\right) & \text{if } x \neq 0; \\ 0 & \text{if } x = 0 \end{cases}$$

is continuous at $x = 0$ and differentiable everywhere, but h' is not continuous at $x = 0$. (This provides an example of a differentiable function which is not smooth.)

3.6 Second Order Derivatives

Assume that f is differentiable on I, so that the function f' is defined on I. We can consider the derivative of f' at $a \in I$, that is, $(f')'(a)$. If it exists, we denote it by $f''(a)$, we call $f''(a)$ the **second derivative of** f at a and say that f is **twice differentiable at** a.

To explain the geometric meaning of the second derivative, we consider an approximation of f by a quadratic polynomial. We show in Section 3.1 that the linear polynomial ℓ given by

$$\ell : x \mapsto f(a) + f'(a) \cdot (x - a)$$

is the best approximation of f at a among linear polynomials, in the sense that

$$\lim_{x \to a} \frac{f(x) - \ell(x)}{x - a} = 0.$$

We show now that the quadratic polynomial q given by

$$q : x \mapsto f(a) + f'(a) \cdot (x - a) + \frac{f''(a)}{2} \cdot (x - a)^2$$

is the best approximation of f at a among quadratic polynomials, in the sense that

$$\lim_{x \to a} \frac{f(x) - q(x)}{(x - a)^2} = 0.$$

Theorem 47 (Second Order Increment Equation). *Let f be twice differentiable at a. For every $x \simeq a$ then*

$$f(x) = f(a) + f'(a) \cdot (x - a) + \frac{f''(a)}{2} \cdot (x - a)^2 + \varepsilon \cdot (x - a)^2,$$

where $\varepsilon \simeq 0$.

This equation can be restated in a form similar to the Increment Equation of order one (page 70):

$$f(a + dx) = f(a) + f'(a) \cdot dx + \frac{f''(a)}{2} \cdot (dx)^2 + \varepsilon \cdot (dx)^2, \quad \text{for } dx \simeq 0.$$

Proof. If $f''(a)$ exists, then f' is defined in an interval about a, that is, f is differentiable in an interval about a. We note that $f(a) - q(a) = 0$ and $(a - a)^2 = 0$, therefore we can use L'Hôpital's Rule to compute the limit below; also, $f'(a) = q'(a)$ and $f''(a) = q''(a)$.

$$\lim_{x \to a} \frac{f(x) - q(x)}{(x - a)^2} = \lim_{x \to a} \frac{f'(x) - q'(x)}{2 \cdot (x - a)}$$

$$= \frac{1}{2} \lim_{x \to a} \left(\frac{f'(x) - f'(a)}{x - a} - \frac{q'(x) - q'(a)}{x - a} \right)$$

$$= \frac{1}{2} \left(f''(a) - q''(a) \right) = 0.$$

We conclude that if $x \simeq a$, then $f(x) - q(x) = \varepsilon \cdot (x - a)^2$, for some $\varepsilon \simeq 0$. $\qquad \square$

The following exercise shows that this property characterizes the quadratic polynomial q.

Exercise 39 (Answer page 253)
Let f be twice differentiable at a. Consider the quadratic polynomial q given by

$$q : x \mapsto b_0 + b_1 \cdot (x - a) + b_2 \cdot (x - a)^2,$$

with observable b_0, b_1, b_2. Suppose that for all $x \simeq a$, there is $\varepsilon \simeq 0$ such that

$$f(x) \simeq q(x) + \varepsilon \cdot (x - a)^2.$$

Show that $b_0 = f(a)$, $b_1 = f'(a)$ and $b_2 = \frac{f''(a)}{2}$.

The converse of the Second Order Increment Equation does not hold in general: If b_0, b_1, b_2 are observable and, for all $x \simeq a$, $f(x) = b_0 + b_1 \cdot (x - a) + b_2 \cdot (x - a)^2 + \varepsilon \cdot (x - a)^2$, where $\varepsilon \simeq 0$, then $b_0 = f(a)$ and $b_1 = f'(a)$, but $f''(a)$ need not exist.

Example. For an example, consider f defined by

$$f(x) = \begin{cases} 0 & \text{if } x = 0; \\ x^3 \cdot \sin\left(\frac{1}{x}\right) & \text{otherwise.} \end{cases}$$

The "quadratic" polynomial q which is 0 everywhere, satisfies

$$\frac{f(x) - q(x)}{x^2} = x \cdot \sin\left(\frac{1}{x}\right) \simeq 0 \text{ for } x \simeq 0, x \neq 0,$$

since $\sin\left(\frac{1}{x}\right)$ is bounded (by ± 1). However, $f''(0)$ does not exist. It is easy to check that f' is given by

$$f'(x) = \begin{cases} 0 & \text{if } x = 0; \\ 3x^2 \sin\left(\frac{1}{x}\right) - x\cos\left(\frac{1}{x}\right) & \text{otherwise.} \end{cases}$$

Now let $x \simeq 0$, $x > 0$; we have

$$\frac{f'(x) - f'(0)}{x} = \frac{3x^2 \sin\left(\frac{1}{x}\right) - x\cos\left(\frac{1}{x}\right)}{x} = 3x\sin\left(\frac{1}{x}\right) - \cos\left(\frac{1}{x}\right).$$

Let $n \in \mathbb{N}$ be ultralarge. Then for $x = \frac{1}{n\pi}$ this quotient is $\simeq 1$ if n is odd, and $\simeq -1$ if n is even. So $f''(0)$ does not exist.

Example. Assume that f is twice differentiable at a; then

$$f''(a) = \lim_{h \to 0} \frac{f(a+2h) - 2f(a+h) + f(a)}{h^2}.$$

Proof. The conditions of L'Hôpital's Rule are satisfied. Differentiating numerator and denominator with respect to h yields

$$\frac{f(a+2h) - 2f(a+h) + f(a)}{h^2} = \frac{2f'(a+2h) - 2f'(a+h)}{2h}$$
$$= \frac{f'(a+2h) - f'(a+h)}{h}.$$

For h ultrasmall, this last expression is ultraclose to $f''(a)$:

$$\frac{f'(a+2h) - f'(a+h)}{h} = 2 \cdot \frac{f'(a+2h) - f'(a)}{2h} - \frac{f'(a+h) - f'(a)}{h}$$
$$\simeq 2f''(a) - f''(a)$$
$$= f''(a).$$

$\square$

The converse is false. As a classical example, consider $f : x \mapsto |x|$ and $a = 0$. Then $\lim_{h \to 0} \frac{|2h| - 2|h|}{h^2} = 0$, but $f'(0)$ is not defined, hence $f''(0)$ does not exist.

Definition 29. *Let f be differentiable at a.*

*(1) The function f is **bending upward at** a if $\langle x, f(x) \rangle$ is above the tangent to f at $\langle a, f(a) \rangle$, whenever $x \simeq a$, that is,*

$$f(x) \geq f'(a) \cdot (x - a) + f(a) \quad \text{whenever } x \simeq a.$$

*(2) The function f is **bending downward at** a if $-f$ is bending upward at a.*

Theorem 48. *Let f be twice differentiable at a.*

(1) If $f''(a) > 0$, then f is bending upward at a.

(2) If $f''(a) < 0$, then f is bending downward at a.

(3) If $f'(a) = 0$ and $f''(a) > 0$, then f has local minimum at a.

(4) If $f'(a) = 0$ and $f''(a) < 0$, then f has local maximum at a.

Proof. (1) follows immediately from the Second Order Increment Equation: If $f''(a) > 0$, then

$$\left(\frac{f''(a)}{2} + \varepsilon \right) \cdot (x - a)^2 \geq 0.$$

Again, (2) follows from (1) by considering $-f$. (3) and (4) follow from (1) and (2), respectively. $\qquad\square$

There is also a global version of the preceding theorem.

Definition 30. *Let f be a function defined on an interval I.*

*(1) f is **bending upward on** I if for every $a < b$ in I, and every $x \in (a, b)$,*

$$f(x) \leq \frac{f(b) - f(a)}{b - a} \cdot (x - a) + f(a).$$

This means that the graph of f in the interval (a, b) lies below the secant line connecting the endpoints $\langle a, f(a) \rangle$ and $\langle b, f(b) \rangle$.

*(2) f is **bending downward on** I if $-f$ is bending upward on I.*

Theorem 49. *Let f be twice differentiable on an interval I.*

> *(1) If $f''(x) \geq 0$ for all $x \in I$, then f is bending upward on I.*
>
> *(2) If $f''(x) \leq 0$ for all $x \in I$, then f is bending downward on I.*

Proof. For (1) we have to prove that

$$\Big(f(x) - f(a)\Big) \cdot (b - a) \leq \Big(f(b) - f(a)\Big) \cdot (x - a).$$

We can write $b - a = (b - x) + (x - a)$ and $f(b) - f(a) = (f(b) - f(x)) + (f(x) - f(a))$. This inequality is thus equivalent to the following:

$$\Big(f(x) - f(a)\Big) \cdot (b - x) \leq \Big(f(b) - f(x)\Big) \cdot (x - a),$$

that is,

$$\frac{f(x) - f(a)}{x - a} \leq \frac{f(b) - f(x)}{b - x}.$$

By the Mean Value Theorem, there exist c, d such that $a < c < x < d < b$ and

$$\frac{f(x) - f(a)}{x - a} = f'(c) \quad \text{and} \quad \frac{f(b) - f(x)}{b - x} = f'(d).$$

It follows from $f''(x) \geq 0$ in I, that f' is increasing in I, and in particular, $f'(c) \leq f'(d)$. This proves (1); for the proof of (2) replace f by $-f$. $\quad\square$

  We finish this section with an example of curve sketching. We go through the following steps to systematize the available information about the function.

- Find the domain.

- Find the zeroes and the y-intercept (if any).

- Find the asymptotes (if any).

- Find the derivative (if any).

- Find the zeroes of the derivative (if any).

- Find the second derivative (if any).

- Find the zeroes of the second derivative (if any).

- Put all these values in a table and determine the local maxima and minima and inflexion points.

- Draw arrows which indicate the general direction of the curve.

- Draw "smiles" or "frowns" which indicate the bending of the curve.

- Use this information to choose a convenient scale.

- Sketch the function.

Example. Let

$$f : x \mapsto \frac{6x^2 - 1}{3x^3}.$$

- The function is undefined at $x = 0$, hence the domain is $\mathbb{R} \setminus \{0\}$.

- Because of the domain, there is no y-intercept.
  If $f(x) = 0$, then $x = \pm\sqrt{6}/6 \approx \pm 0.4$.

- Possible vertical asymptote at $x = 0$. For $x \simeq 0$ we have $6x^2 - 1 \simeq -1$ and $3x^3 \simeq 0$, so $f(x)$ is ultralarge. Hence there is a vertical asymptote at 0. Furthermore, $f(x)$ is ultralarge positive if $x < 0$, and $f(x)$ is ultralarge negative if $x > 0$.

  Concerning the horizontal asymptote: If x is ultralarge, then
  $$\frac{6x^2 - 1}{3x^3} = \frac{6 \overbrace{- 1/x^2}^{\simeq 0}}{3x} \simeq \frac{6}{3x} \simeq 0$$ and this is true whether x is positive or negative, hence there is a horizontal asymptote at $y = 0$ on both sides.

- $f'(x) = -\dfrac{2x^2 - 1}{4x^4}.$

- If $f'(x) = 0$, then $x = \pm\sqrt{2}/2 \approx 0.7$.
  $f(\sqrt{2}/2) \approx 1.9.$
  $f(-\sqrt{2}/2) \approx -1.9.$

- $f''(x) = \dfrac{x^2 - 1}{x^5}.$

- If $f''(x) = 0$, then $x = \pm 1$.
  $f(1) = 5/3 \approx 1.7.$
  $f(-1) = -5/3 \approx -1.7.$

-

	$-\infty$	-1	-0.7	-0.4	0		0.4	0.7	1	∞
f	$\simeq 0$	-1.7	-1.9	0	$+\infty$	$\|\| -\infty$	0	1.9	1.7	$\simeq 0$
f'	$-$	$-$	0	$+$		$\|\|$	$+$	0	$-$	$-$
f	$\searrow$	$\searrow$	$\overset{\min}{\longrightarrow}$	$\nearrow$		$\|\|$	$\nearrow$	$\overset{\max}{\longrightarrow}$	$\searrow$	$\searrow$
f''	$-$	0	$+$	$+$		$\|\|$	$-$	$-$	0	$+$
f	$\frown$		$\smile$	$\smile$		$\|\|$	$\frown$	$\frown$		$\smile$

•

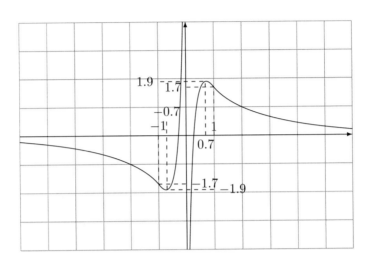

3.7   Additional Exercises

Exercise 3.1
Using Definition 20 calculate the derivatives (if they exist) of the following functions.

(1) $f : x \mapsto 3x^2 + x - 5$ at $x = -2$ and $x = 2$.

(2) $f : x \mapsto 2x^3 - 2$ at $x = 1$ and $x = 0$.

(3) $g : x \mapsto |x|$ at $x = 2$, $x = -2$ and $x = 0$.

(4) $f : x \mapsto \begin{cases} x^2 - 1 & \text{if } x < 0 \\ x - 1 & \text{if } x \geq 0 \end{cases}$ at $x = -3$, $x = 1$ and $x = 0$.

(5) $f : x \mapsto \begin{cases} x^2 - 1 & \text{if } x < 0.5 \\ x - 1.25 & \text{if } x \geq 0.5 \end{cases}$ at $x = 0.5$.

(6) $f : x \mapsto \begin{cases} x^2 & \text{if } x > 0 \\ -x^3 & \text{if } x \leq 0 \end{cases}$ at $x = 0$.

(7) $f : x \mapsto \begin{cases} x^2 - 1 & \text{if } x < 0.5 \\ x + 1 & \text{if } x \geq 0.5 \end{cases}$ at $x = 0.5$.

Exercise 3.2
Calculate the derivatives of the following functions.

(1) $f : x \mapsto 5x^2 - 10x$ at $x = 2$.

(2) $f : x \mapsto 5(x - 10)^2$ at $x = 3$.

(3) $f : x \mapsto x^4 + x^3 + x^2 + x + 1$ at $x = 1$.

(4) $f : x \mapsto 5x^2 + 10$ at $x = 2$.

(5) $f : x \mapsto \frac{1}{x}$ for $x = 1$ and $x = 2$.

(6) $f : x \mapsto \frac{1}{3x+2}$ for $x = 0$ and $x = 1$.

(7) $f : x \mapsto \frac{1}{x^2}$ for $x = 1$ and $x = -1$.

Exercise 3.3
Compute the derivatives of the following functions.

(1) $f : x \mapsto (\sqrt{x} + 1)^4$

(2) $f : x \mapsto \sqrt{5x^3 + 3x^2}$

(3) $f : x \mapsto \sqrt{x^2}$

Exercise 3.4
Use the definition of derivative to show that the function $f(x) = \sqrt[3]{x}$ is not differentiable at 0.

Exercise 3.5
Compute the derivatives of the following functions.

(1) $f : x \mapsto \sqrt{3x^3 + 2x + 1}$

(2) $f : x \mapsto (x^2 + 3)^5$

(3) $f : x \mapsto (ax + b)^n$

(4) $f : x \mapsto \sqrt{x^3 + 1}$

(5) $f : x \mapsto \sin(x^2 + 3x)$

(6) $f : \theta \mapsto \cos^2(3\theta)$

(7) $f : u \mapsto \sin(\sin(u))$

(8) $f : x \mapsto \tan^2(\tan^2(x^2))$

(9) $f : v \mapsto \dfrac{\sin(v)}{\tan(v)}$

(10) $f : x \mapsto \sin^2(x) + \cos^2(x)$

Exercise 3.6
Compute the derivatives of the following functions.

(1) $f : x \mapsto \sin^2(3x + \pi)$

(2) $f : x \mapsto x \cdot \sin(x^2 + 1)$

(3) $f : x \mapsto \sin^2 \left(\dfrac{x}{x^2 + 1} \right) + \cos^2 \left(\dfrac{x}{x^2 + 1} \right)$

(4) $f : x \mapsto 1 + \tan^2(x)$

Exercise 3.7

Use the method of Exercise 36 to show that $\cos'(x) = -\sin(x)$.

Exercise 3.8

(1) Show that $f : x \mapsto \sin^6(x) + \cos^6(x) + 3\sin^2(x) \cdot \cos^2(x)$ is a constant function.

  Hint: Use Theorem 40.

(2) Let $f : x \mapsto \sin(x) + \cos(x)$. Solve $f'(x) = 0$.

(3) What is the equation of the straight line tangent to $y = \sin^2(x)$ at $x = \frac{\pi}{4}$?

Exercise 3.9

Use L'Hôpital's Rule to compute the following limits.

(1) $\displaystyle \lim_{t \to 1_+} \frac{1/t - 1}{t^2 - 2t + 1}$

(2) $\displaystyle \lim_{x \to 1} \frac{\sqrt{x} - 1}{\sqrt[3]{x} - 1}$

(3) $\displaystyle \lim_{x \to 0} \frac{x^2}{\sqrt{2x + 1} - 1}$

(4) $\displaystyle \lim_{x \to 0} \frac{\sqrt{9 + x} - 3}{x}$

(5) $\displaystyle \lim_{x \to 2} \frac{2 - \sqrt{x + 2}}{4 - x^2}$

(6) $\displaystyle \lim_{x \to 0} \frac{(1 - x)^{1/4} - 1}{x}$

(7) $\displaystyle \lim_{u \to 1} \frac{(u - 1)^3}{u^{-1} - u^2 + 3u - 3}$

(8) $\displaystyle \lim_{t \to 0} \left(t + \frac{1}{t} \right) \left((4 - t)^{3/2} - 8 \right)$

(9) $\displaystyle \lim_{t \to 0_+} \left(\frac{1}{t} + \frac{1}{\sqrt{t}} \right) (\sqrt{t + 1} - 1)$

Exercise 3.10

Sketch the following functions.

(1) $f : x \mapsto \dfrac{x^2}{x + 2}$

(2) $f : x \mapsto x - 1 + \dfrac{9}{x + 1}$

(3) $f : x \mapsto \dfrac{-x^2 - 2x - 1}{x + 3}$

(4) $f : x \mapsto x + 3 + \dfrac{1}{2x + 1}$

(5) $f : x \mapsto \dfrac{x^2 - 4x + 6}{(x - 2)^2}$

(6) $f : x \mapsto \dfrac{2x^2 - 3}{x^2 - 1}$

(7) $f : x \mapsto \dfrac{x^2 + 3x - 4}{x^2 - x - 2}$

(8) $f : x \mapsto \dfrac{x^3 + 2}{2x}$

(9) $f : x \mapsto \dfrac{x^3 - 1}{x^2}$

(10) $f : x \mapsto \dfrac{2x - 1}{\sqrt{x^2 + 2}}$

(11) $f : x \mapsto \dfrac{\sqrt{x^2 + 1}}{x + 1}$

(12) $f : x \mapsto \dfrac{\sqrt{x^2 - 4x + 3}}{x + 1}$

(13) $f : x \mapsto \sin(\cos(x))$

(14) $f : x \mapsto \cos(\sin(x))$

Exercises 3.12 through 3.14 are applications where differentiation is used to find an optimal solution to a given problem. The method is to first write down the problem as a function in the appropriate variable, then find its maximum or minimum.

Exercise 3.11
A cylindrical jar has a volume defined by its radius and its height. If it contains one litre (1 dm^3), what are the dimensions that will make it have the least outside area?

Exercise 3.12
Imagine you want to protect a part of a rectangular garden against a wall. You have 100 m of fence. (No fence is needed against the wall.) What is the biggest area that you can protect?

Exercise 3.13
Find the length and width of the rectangle inscribed within the ellipse given by the formula $4x^2 + y^2 = 16$ (sides parallel to the coordinate axes) such that its area is maximal.

Exercise 3.14
Let $\mathcal{P}$ be the parabola given by $x \mapsto x^2$ and A be the point $\langle 0, 5 \rangle$. Find the point(s) on the parabola $\mathcal{P}$ such that its (their) distance to A is minimal.

Exercise 3.15
Let f be differentiable on $(-a, a)$. Show that if f is even [respectively, odd], then f' is odd [respectively, even].

Exercise 3.16
Let f be continuous on $[a, \infty)$, differentiable on (a, ∞), $f(a) = 0$ and such that $\lim_{x \to \infty} f(x) = 0$. Show that there exists $c \in (a, \infty)$ such that $f'(c) = 0$.

Exercise 3.17
Assume that $f'(x) \leq g'(x)$ for all $x \in I$ and $f(a) = g(a)$ for some $a \in I$.
Show that $f(x) \leq g(x)$ for all $x \geq a$, $x \in I$, and $g(x) \leq f(x)$ for all
$x \leq a$, $x \in I$.

Exercise 3.18
Prove that if $|f(x) - f(y)| \leq C \cdot (x - y)^2$ holds for all $x, y \in I$, then f is
a constant function on I.

Exercise 3.19
Let f be continuous on $[a, b)$ and differentiable on (a, b). If $\lim_{x \to a^+} f'(x)$
exists, then $f'_+(a) = \lim_{x \to a^+} f'(x)$.

Exercise 3.20
Show that the function

$$x \mapsto \begin{cases} x + 2x^2 \cdot \sin\left(\frac{1}{x}\right) & \text{if } x \neq 0 \\ 0 & \text{if } x = 0 \end{cases}$$

is increasing at 0 but it is not increasing on any interval $(-a, a)$ about 0.

Exercise 3.21
Let f be differentiable on $[a, b]$ and such that $f'_+(a) < 0 < f'_-(b)$. Then
there is some $c \in (a, b)$ where $f'(c) = 0$.
Hint: Use Theorems 27, 9, 35 and 36.

Exercise 3.22
Prove Darboux's Theorem: If f is differentiable on $[a, b]$ and d is strictly
between $f'_+(a)$ and $f'_-(b)$, then there is some $c \in (a, b)$ where $f'(c) = d$.
Hint: Apply the previous exercise to $h : x \mapsto f(x) - d \cdot x$.

Exercise 3.23
Let f be defined by

$$f(x) = \begin{cases} x^3 \sin\left(\frac{1}{x}\right) & \text{if } x \neq 0 \\ 0 & \text{if } x = 0. \end{cases}$$

Show that f is smooth.

Exercise 3.24
We say that a function $f : I \to \mathbb{R}$ is uniformly differentiable on I if it is
differentiable on I and

$$x \simeq y \quad \text{implies} \quad \frac{f(x) - f(y)}{x - y} \simeq f'(x)$$

for all $x, y \in I$, $x \neq y$. Show that if f is uniformly differentiable on I,
then f' is continuous on I.

4

Integration of Continuous Functions

4.1 Fundamental Theorem of Calculus

The fundamental problem of integration is to find the original function F, given its derivative function $F' = f$. Integration is thus the inverse of differentiation. We study this problem under the assumption that f is continuous, that is, the original function F is smooth. The key to the answer is the uniform version of the Increment Equation: For all dx ultrasmall relative to F, a and b, and for all $x, x + dx \in [a, b]$, we have

$$F(x + dx) = F(x) + f(x) \cdot dx + \varepsilon \cdot dx, \text{ where } \varepsilon \text{ is ultrasmall.}$$

This equation tells us that, if we know $F(x)$ and the derivative $f(x)$ of F at x, we can compute $F(x + dx)$ with an error equal to an ultrasmall number times dx, in other words, very small compared to dx. By starting with a fixed $x = a$ and using the Uniform Increment Equation repeatedly, we can compute $F(x)$ for any x.

We fix $b > a$ and suppose that F is smooth on $[a, b]$ and $F(a)$ is known. Let N be a positive ultralarge integer. We define

$$dx = \frac{b - a}{N} > 0 \quad \text{and} \quad x_i = a + i \cdot dx, \text{ for } i = 0, \dots, N.$$

Notice in particular that $x_0 = a$, $x_N = b$, and dx is ultrasmall.

By the Uniform Increment Equation,

$$F(x_1) - F(x_0) = f(x_0) \cdot dx + \varepsilon_0 \cdot dx$$

$$\cdots$$

$$F(x_{i+1}) - F(x_i) = f(x_i) \cdot dx + \varepsilon_i \cdot dx$$

$$\cdots$$

$$F(x_N) - F(x_{N-1}) = f(x_{N-1}) \cdot dx + \varepsilon_{N-1} \cdot dx,$$

where $\varepsilon_i \simeq 0$, for $i = 0, \dots, N - 1$.

By adding up these equations we obtain

$$F(b) - F(a) = \sum_{i=0}^{N-1} f(x_i) \cdot dx + \sum_{i=0}^{N-1} \varepsilon_i \cdot dx.$$

An important observation is that $\sum_{i=0}^{N-1} \varepsilon_i \cdot dx \simeq 0$.
Indeed, let $\varepsilon = \max\{|\varepsilon_0|, \ldots, |\varepsilon_i|, \ldots, |\varepsilon_{N-1}|\}$; then

$$\left| \sum_{i=0}^{N-1} \varepsilon_i \cdot dx \right| \leq \sum_{i=0}^{N-1} |\varepsilon_i| \cdot dx \leq N \cdot \varepsilon \cdot dx = \varepsilon \cdot (b-a) \simeq 0.$$

This implies that

$$F(b) - F(a) \simeq \sum_{i=0}^{N-1} f(x_i) \cdot dx,$$

from which we conclude that

$$F(b) \text{ is the observable neighbor of } F(a) + \sum_{i=0}^{N-1} f(x_i) \cdot dx.$$

The last formula computes $F(b)$ for any b ($b > a$), given f continuous on $[a, b]$ and the value of F at a single point a. The same formula works for $b < a$, assuming that f is continuous on $[b, a]$ (dx is negative in this case), and for $b = a$ ($dx = 0$ and everything is trivial). We showed that it solves the integration problem for f (the problem of finding F such that f is a derivative of F) *provided* that such F exists. Our ultimate goal in this section is to show that *every* continuous function f is a derivative of some function F.

Theorem 50. *Let a, b be real numbers and let f be a continuous function on the closed interval $[a, b]$. Then there exists an observable real number R such that, for any ultralarge positive integer N,*

$$R \simeq \sum_{i=0}^{N-1} f(x_i) \cdot dx, \quad \text{where } dx = (b-a)/N \text{ and } x_i + a + i \cdot dx.$$

Proof. We have to prove that the value of the sum

$$\sum_{i=0}^{N-1} f(x_i) \cdot dx$$

is not ultralarge and is, up to an ultrasmall amount, independent of the choice of N (provided N is ultralarge).

Let $I = [a, b]$; by Theorem 9, there are observable $c, d \in I$ such that

$$f(c) = \min_{x \in I} f(x) \quad \text{and} \quad f(d) = \max_{x \in I} f(x).$$

It follows that

$$f(c) \cdot (b - a) \leq \sum_{i=0}^{N-1} f(x_i) \cdot dx \leq f(d) \cdot (b - a).$$

Since $f(c) \cdot (b - a)$ and $f(d) \cdot (b - a)$ are observable, the sum is indeed not ultralarge.

We now show that the value of the sum is, up to an ultrasmall amount, independent of N. Let M be another positive ultralarge integer. Let $dy = (b - a)/M$ and $y_j = a + j \cdot dy$, for $j = 0, \dots, M$. We prove that

$$\sum_{j=0}^{M-1} f(y_j) \cdot dy \simeq \sum_{i=0}^{N-1} f(x_i) \cdot dx.$$

It is enough to prove this in the case where N is a multiple of M, that is, in the case when the partition induced by N refines the partition induced by M (to deduce the general case, compare the two sums with that induced by $N \cdot M$). Let us therefore assume that there is an integer k such that $N = k \cdot M$. This implies that $dy = k \cdot dx$ and $y_j = x_{k \cdot j}$, for each $j = 0, \dots, M$.

In the first sum we have

$$f(y_j) \cdot dy = f(x_{k \cdot j}) \cdot k \cdot dx = \underbrace{f(x_{k \cdot j}) \cdot dx + \cdots + f(x_{k \cdot j}) \cdot dx}_{k \text{ times}}$$

and in the second sum

$$\sum_{\ell=0}^{k-1} f(x_{k \cdot j + \ell}) \cdot dx = f(x_{k \cdot j}) \cdot dx + \cdots + f(x_{k \cdot j + k - 1}) \cdot dx.$$

But $y_j = x_{k \cdot j} \leq x_{k \cdot j + \ell} \leq y_{j+1}$ and $y_j \simeq y_{j+1}$, so $x_{k \cdot j} \simeq x_{k \cdot j + \ell}$. Hence, by uniform continuity of f on I, we have

$$f(x_{k \cdot j}) = f(x_{k \cdot j + \ell}) + \varepsilon_{k \cdot j + \ell}, \quad \text{for some } \varepsilon_{k \cdot j + \ell} \simeq 0.$$

This implies that

$$f(y_j) \cdot dy = f(x_{k \cdot j}) \cdot k \cdot dx = \sum_{\ell=0}^{k-1} \left(f(x_{k \cdot j + \ell}) + \varepsilon_{k \cdot j + \ell} \right) \cdot dx$$

$$= \sum_{\ell=0}^{k-1} \left(f(x_{k \cdot j + \ell}) \cdot dx + \varepsilon_{k \cdot j + \ell} \cdot dx \right).$$

Inserting these in the sum induced by M, we obtain

$$\sum_{j=0}^{M-1} f(y_j) \cdot dy = \sum_{j=0}^{M-1}\sum_{\ell=0}^{k-1}\left(f(x_{k\cdot j+\ell}) \cdot dx + \varepsilon_{k\cdot j+\ell} \cdot dx \right)$$

$$= \sum_{i=0}^{N-1}\left(f(x_i) \cdot dx + \varepsilon_i \cdot dx \right)$$

$$\simeq \sum_{i=0}^{N-1} f(x_i) \cdot dx,$$

since $\sum_{i=0}^{N-1} \varepsilon_i \cdot dx \simeq 0$ by a previous observation. $\qquad\qquad \square$

Definition 31. *Let a, b be real numbers. Let f be a function which is continuous on the closed interval whose endpoints are a and b (if $a \neq b$). Let $N \in \mathbb{N}$ be ultralarge, $dx = (b-a)/N$ and $x_i + a + i \cdot dx$, for $i = 0, \dots, N$. The observable number R such that*

$$R \simeq \sum_{i=0}^{N-1} f(x_i) \cdot dx$$

*is called the **integral of f from** a **to** b and is denoted*

$$\int_a^b f(x) \cdot dx.$$

(It does not depend on N.)

The context for this definition is given by a, b and f. Hence,b the integral is observable whenever a, b and f are observable. As in the case of limits, the statement that defines the integral is equivalent to an internal statement; see Exercise 4.8 at the end of this chapter. We can therefore work relative to any extended context.

To summarize: (for any ultralarge $N \in \mathbb{N}$)

$$\int_a^b f(x) \cdot dx \text{ is the observable neighbor of } \sum_{i=0}^{N-1} f(x_i) \cdot dx.$$

Theorem 51. *Let f be continuous on $[a, b]$. Then*

$$\int_a^b f(x) \cdot dx = -\int_b^a f(x) \cdot dx.$$

Proof. Let $N \in \mathbb{N}$ be ultralarge. Let $dx = (b-a)/N$ and $x_i = a + i \cdot dx$. Let $dy = (a-b)/N$ and let $y_j = b + j \cdot dy$. Then $dx = -dy$ and for each i there is a unique j, namely $j = N - i$, such that $x_i = y_j$. Hence

$$\int_a^b f(x) \cdot dx \simeq \sum_{i=0}^{N-1} f(x_i) \cdot dx = -\sum_{j=0}^{N-1} f(y_j) \cdot dy \simeq -\int_b^a f(x) \cdot dx.$$

Since the first and last expressions are observable, they have to be equal. $\qquad \square$

In particular, $\int_a^a f(x) \cdot dx = 0$.

The next theorem is proved in the preamble of this section.

Theorem 52 (Fundamental Theorem of Calculus, First Version). *Let F be a smooth function on I and let f be its derivative. For $a, b \in I$,*

$$F(b) - F(a) = \int_a^b f(x) \cdot dx.$$

Before proceeding further, we prove some important properties of integrals.

Theorem 53 (Linearity of the Integral). *Let f and g be continuous on I, $\lambda \in \mathbb{R}$, $a, b \in I$. Then*

(1) $\displaystyle \int_a^b (f(x) + g(x)) \cdot dx = \int_a^b f(x) \cdot dx + \int_a^b g(x) \cdot dx.$

(2) $\displaystyle \int_a^b \lambda \cdot f(x) \cdot dx = \lambda \cdot \int_a^b f(x) \cdot dx.$

Proof. (1) Let N be a positive ultralarge integer. We have that

$$\underbrace{\sum_{i=0}^{N-1} (f(x_i) + g(x_i)) \cdot dx}_{\simeq \int_a^b (f(x) + g(x)) \cdot dx} = \underbrace{\sum_{i=0}^{N-1} f(x_i) \cdot dx}_{\simeq \int_a^b f(x) \cdot dx} + \underbrace{\sum_{i=0}^{N-1} g(x_i) \cdot dx}_{\simeq \int_a^b g(x) \cdot dx} .$$

By Rule 5 we have

$$\int_a^b (f(x) + g(x)) \cdot dx = \int_a^b f(x) \cdot dx + \int_a^b g(x) \cdot dx.$$

(2) is similar. $\qquad \square$

Exercise 40 (Answer page 253)

Let c be a real number. Show that $\int_a^b c \cdot dx = c \cdot (b - a)$.

Theorem 54 (Monotonicity of the Integral). *Let f be a continuous function on $[a, b]$.*

(1) If $f(x) \geq 0$ (respectively > 0) for all $x \in [a, b]$, then

$$\int_a^b f(x) \cdot dx \geq 0 \text{ (respectively } > 0).$$

(2) If $f(x) = 0$ for all $x \in [a, b]$, then $\int_a^b f(x) \cdot dx = 0$.

(3) If $f(x) \leq 0$ (respectively < 0) for all $x \in [a, b]$, then

$$\int_a^b f(x) \cdot dx \leq 0 \text{ (respectively } < 0).$$

Exercise 41 (Answer page 254)

Prove Theorem 54.

We now prove some more sophisticated properties of the integral. We consider the function

$$F : x \mapsto \int_a^x f(t) \cdot dt,$$

where $a \in I$ and $f : I \to \mathbb{R}$ is a continuous function. For each $x \in I$, the number $F(x)$ is uniquely determined, and it is observable whenever f, a and x are observable. Since the integral has an internal defining statement (see Exercise 4.8), F is a well-defined function, observable whenever f and a are, by the Definition Principle.

Theorem 55 (Continuity). *Let $f : I \to \mathbb{R}$ be continuous. Let $a \in I$. Then*

$$F : x \mapsto \int_a^x f(t) \cdot dt$$

is continuous.

Proof. Fix $x \in I$. The context is given by f, I, a and x. We have to show that if $x \simeq c$, $c \in I$, then

$$\int_a^x f(t) \cdot dt \simeq \int_a^c f(t) \cdot dt.$$

Let B be an observable bound of $|f|$ on the interval with endpoints a and x. We now extend the context to f, I, a, x and also c. Let $N \in \mathbb{N}$ be ultralarge relative to this extended context. Let $dx = \frac{x-a}{N}$ and $x_i = a + i \cdot dx$. Let $dt = \frac{c-a}{N}$ and $t_i = a + i \cdot dt$.

We write $\overset{+}{\simeq}$ to indicate when we work relative to the extended context given by the parameters f, I, a, x and c. By the choice of N we have, in this notation,

$$\int_a^x f(t) \cdot dt \overset{+}{\simeq} \sum_{i=0}^{N-1} f(x_i) \cdot dx \quad \text{and} \quad \int_a^c f(t) \cdot dt \overset{+}{\simeq} \sum_{i=0}^{N-1} f(t_i) \cdot dt.$$

Now, for each $i < N$,

$$x_i - t_i = \left(a + i \cdot \frac{x-a}{N} \right) - \left(a + i \cdot \frac{c-a}{N} \right) = \frac{i}{N} \cdot (x - c) \simeq 0.$$

Hence, by uniform continuity of f (on some observable closed and bounded interval J such that $a, x, c \in J \subseteq I$), we have $f(x_i) = f(t_i) + \varepsilon_i$, where $\varepsilon_i \simeq 0$; in particular $|f(t_i)| = |f(x_i) - \varepsilon_i| < B + 1$. Notice also that $dx = dt + \frac{x-c}{N}$. But then

$$\int_a^x f(t) \cdot dt \overset{+}{\simeq} \sum_{i=0}^{N-1} f(x_i) \cdot dx$$

$$= \sum_{i=0}^{N-1} (f(t_i) + \varepsilon_i) \cdot \left(dt + \frac{x-c}{N} \right)$$

$$= \sum_{i=0}^{N-1} f(t_i) \cdot dt + \sum_{i=0}^{N-1} \varepsilon_i \cdot \left(dt + \frac{x-c}{N} \right)$$

$$+ \sum_{i=0}^{N-1} \frac{f(t_i)}{N} \cdot (x - c)$$

$$\overset{+}{\simeq} \int_a^c f(t) \cdot dt + \underbrace{\sum_{i=0}^{N-1} \varepsilon_i \cdot dx}_{\simeq 0} + \underbrace{\sum_{i=0}^{N-1} \frac{f(t_i)}{N} \cdot (x - c)}_{\simeq 0}$$

$$\simeq \int_a^c f(t) \cdot dt.$$

The first sum is ultraclose to zero because each $\varepsilon_i \simeq 0$ and $\sum_{i=0}^{N-1} dx = x - a$ is not ultralarge. The reason why the second sum is ultraclose to zero is that $x - c \simeq 0$ and $\left| \sum_{i=0}^{N-1} \frac{f(t_i)}{N} \right| < B + 1$ is not ultralarge. $\qquad \square$

We can deduce the additivity of the integral.

Theorem 56 (Additivity). *Let f be continuous on $[a,b]$ and $c \in [a,b]$.*
Then
$$\int_a^b f(x) \cdot dx = \int_a^c f(x) \cdot dx + \int_c^b f(x) \cdot dx.$$

Proof. The context is specified by f, a, b, c. We first show additivity for
the case when $\frac{c-a}{b-a}$ is rational. In this case it is immediate, because there
is $N \in \mathbb{N}$ ultralarge such that $c = a + i \cdot \frac{b-a}{N}$, for some ultralarge integer
i. Then

$$\int_a^b f(x) \cdot dx \simeq \sum_{k=0}^{N-1} f(x_k) \cdot dx$$

$$= \sum_{k=0}^{i-1} f(x_k) \cdot dx + \sum_{k=i}^{N-1} f(x_k) \cdot dx$$

$$\simeq \int_a^c f(x) \cdot dx + \int_c^b f(x) \cdot dx.$$

For the general case, when $c \in [a,b]$ is arbitrary, fix $c' \simeq c$ such that
$\frac{c'-a}{b-a}$ is rational. Then by continuity, we have

$$\int_a^c f(x) \cdot dx \simeq \int_a^{c'} f(x) \cdot dx \quad \text{and} \quad \int_c^b f(x) \cdot dx \simeq \int_{c'}^b f(x) \cdot dx.$$

Together, we have that

$$\int_a^b f(x) \cdot dx = \int_a^{c'} f(x) \cdot dx + \int_{c'}^b f(x) \cdot dx \simeq \int_a^c f(x) \cdot dx + \int_c^b f(x) \cdot dx.$$

The left-hand side and the right-hand side are observable, hence they
have to be equal.   □

A more general version of additivity follows by induction.

Theorem 57. *Let f be continuous on $[a,b]$ and $a = c_0 < c_1 < \ldots <$*
$c_{n-1} < c_n = b$. *Then*

$$\int_a^b f(x) \cdot dx = \sum_{i=0}^{n-1} \int_{c_i}^{c_{i+1}} f(x) \cdot dx.$$

Finally we can prove the ultimate goal of this section.

Theorem 58 (Fundamental Theorem of Calculus, Second Version). *Let* $f : I \to \mathbb{R}$ *be a continuous function and let* $a \in I$. *The function* $F : I \to \mathbb{R}$,

$$F : x \mapsto \int_a^x f(t) \cdot dt,$$

satisfies $F'(x) = f(x)$ *for all* $x \in I$, *and* $F(a) = 0$.
(One-sided derivatives are understood at the endpoints of I.)

Proof. We only need to prove that $F'(x) = f(x)$ for all $x \in I$. Fix $x \in I$. The context is specified by f, I, a and x. Let $h > 0$ be an ultrasmall number (the case $h < 0$ is similar). Using additivity, we have that

$$F(x + h) - F(x) = \int_x^{x+h} f(t) \cdot dt.$$

By the Extreme Value Theorem, there are $c, d \in [x, x + h]$ such that

$$f(c) \le f(t) \le f(d)$$

holds for all $t \in [x, x + h]$. This implies that

$$f(c) \cdot h \le \int_x^{x+h} f(t) \cdot dt \le f(d) \cdot h.$$

By continuity of f at x, we have $f(c) \simeq f(x)$ and $f(d) \simeq f(x)$. Hence

$$\frac{F(x + h) - F(x)}{h} = \frac{\int_x^{x+h} f(t) \cdot dt}{h} \simeq f(x),$$

which is what we had to show. $\qquad \square$

4.2 Antiderivatives

Definition 32. *Let f be defined on an interval I. We say that F is an* **antiderivative of f on** I *if $F'(x) = f(x)$ for all $x \in I$. (Left-hand side and right-hand side derivatives suffice at endpoints.)*

An antiderivative of f is differentiable on I by definition, hence an antiderivative is necessarily a continuous function on I. Note that it is *an* antiderivative and not *the* antiderivative. It is an immediate consequence of the rules of differentiation that if F is an antiderivative of f on I, then for any real number C, the function defined by $x \mapsto F(x) + C$ is also an antiderivative of f on I. In fact, these are the only antiderivatives.

Theorem 59. *Let F and G be antiderivatives of f on the interval I. Then there is a constant $C \in \mathbb{R}$ such that $F(x) = G(x) + C$, for all $x \in I$.*

Proof. As $F'(x) = f(x) = G'(x)$ for all $x \in I$, we have $F'(x) - G'(x) = 0$ for all $x \in I$, hence the derivative of $F - G$ is zero on the interval I. This implies that $F - G = C$, with constant $C \in \mathbb{R}$, by Theorem 40, so $F = G + C$. $\qquad\qquad\qquad\qquad\qquad\qquad\qquad\qquad\qquad\qquad\qquad\square$

The Fundamental Theorem of Calculus can be restated in this terminology.

Theorem 60 (Fundamental Theorem of Calculus, First Version). *Let f be continuous on I and let F be an antiderivative of f on I. Then for all $a, b \in I$,*

$$\int_a^b f(x) \cdot dx = F(b) - F(a).$$

With the notation for increments, we can rewrite the Fundamental Theorem of Calculus as follows:

$$F(b) - F(a) = \sum_{i=0}^{N-1} \Delta F(x_i) \simeq \sum_{i=0}^{N-1} f(x_i) \cdot dx \simeq \int_a^b f(x) \cdot dx.$$

There are two ultraclose approximations on the previous line. As $F(b) - F(a)$ and $\int_a^b f(x) \cdot dx$ both are observable, the approximations cancel each other. The integral is exactly equal to the total variation of the function.
Notation:

(1) We write

$$F(x)\Big|_a^b = F(b) - F(a).$$

Thus

$$\int_a^b f(x) \cdot dx = F(x)\Big|_a^b,$$

where F is an antiderivative of f.

(2) We write

$$\int f(x) \cdot dx$$

for the set of antiderivatives of f. We call this the **indefinite integral**, to distinguish it from $\int_a^b f(x) \cdot dx$, which is usually called the *definite integral*. The former is a set of functions, the latter is a number.

The first version of the Fundamental Theorem tells us how to compute the definite integral of a continuous function f *provided* we know some antiderivative F of f. The second version of the Fundamental Theorem provides for existence of such an antiderivative. We restate it in this language too.

Theorem 61 (Fundamental Theorem of Calculus, Second Version). *Let $f : I \to \mathbb{R}$ be a continuous function, and let $a \in I$. The function*

$$F : x \mapsto \int_a^x f(t) \cdot dt$$

is the only antiderivative of f on I satisfying $F(a) = 0$.

Proof. The uniqueness is a consequence of Theorem 59. $\qquad\square$

4.3 Rules of Integration

The rules of integration

$$\int_a^b (f(x) + g(x)) \cdot dx = \int_a^b f(x) \cdot dx + \int_a^b g(x) \cdot dx$$

and

$$\int_a^b \lambda \cdot f(x) \cdot dx = \lambda \cdot \int_a^b f(x) \cdot dx$$

can be viewed as the integral version of the rules of differentiation:

$$(f + g)' = f' + g' \quad \text{and} \quad (\lambda \cdot f)' = \lambda \cdot f'.$$

Integration by parts is based on the rule for the derivative of the product:

$$(f \cdot g)' = f' \cdot g + f \cdot g'.$$

Theorem 62 (Integration by Parts). *Let f and g be smooth on $[a, b]$. Then*

$$\int_a^b f'(x) \cdot g(x) \cdot dx = f(x) \cdot g(x) \Big|_a^b - \int_a^b f(x) \cdot g'(x) \cdot dx.$$

Proof. Under the assumptions of the theorem, the functions $f' \cdot g$, $f \cdot g'$ and $(f \cdot g)'$ are continuous, so all the integrals that occur in the proof are defined.

$$\int_a^b (f \cdot g)'(x) \cdot dx = \int_a^b (f'(x) \cdot g(x) + f(x) \cdot g'(x)) \cdot dx.$$

The Fundamental Theorem of Calculus can be applied to the left-hand side:

$$\int_a^b (f \cdot g)'(x) \cdot dx = (f \cdot g)(x) \Big|_a^b.$$

Now linearity applied to the right-hand side yields

$$\int_a^b f'(x) \cdot g(x) \cdot dx + \int_a^b f(x) \cdot g'(x) \cdot dx.$$

The formula follows immediately.   □

Example. Consider

$$\int_0^{\pi/2} x \cdot \sin(x) \cdot dx.$$

We use integration by parts to evaluate this integral. We write $f'(x) = \sin(x)$ and $g(x) = x$. Then $f(x) = -\cos(x)$ and $g'(x) = 1$, hence

$$\int_0^{\pi/2} x \cdot \sin(x) \cdot dx = -x \cdot \cos(x) \Big|_0^{\pi/2} + \int_0^{\pi/2} \cos(x) \cdot dx$$

$$= -x \cdot \cos(x) \Big|_0^{\pi/2} + \sin(x) \Big|_0^{\pi/2}$$

$$= 1.$$

We also get

$$\int x \cdot \sin(x) \cdot dx = -x \cdot \cos(x) + \sin(x) + C.$$

Theorem 63 (Integration by Substitution). *If $g : [a, b] \to I$ and $f : I \to \mathbb{R}$ are smooth, then*

$$\int_a^b f'(g(x)) \cdot g'(x) \cdot dx = f(g(x)) \Big|_a^b.$$

Proof. This is the integral version of the Chain Rule.   □

Theorem 64 (Integration by Variable Substitution). *Let f be continuous on $[a, b]$. Let $g : [d, e] \to [a, b]$ be smooth and such that $g(d) = a$ and $g(e) = b$. Then*

$$\int_a^b f(x) \cdot dx = \int_d^e f(g(u)) \cdot g'(u) \cdot du.$$

Proof. Let F be an antiderivative of f. Since g is smooth, g' is continuous, so the function $u \mapsto f(g(u)) \cdot g'(u)$ is continuous, and $F \circ g$ is its antiderivative. By the Fundamental Theorem of Calculus, we have

$$\int_d^e f(g(u)) \cdot g'(u) \cdot du = F(g(u))\Big|_d^e = F(b) - F(a) = \int_a^b f(x) \cdot dx.$$

$\square$

If g is a one-to-one correspondence, then $d = g^{-1}(a)$ and $e = g^{-1}(b)$. Furthermore, if H is an antiderivative of $u \mapsto f(g(u)) \cdot g'(u)$, and g^{-1} is itself differentiable, then $x \mapsto H(g^{-1}(x))$ gives an antiderivative of f.

Exercise 42 (Answer page 254)
Let H be an antiderivative of $u \mapsto f(g(u)) \cdot g'(u)$, where g is a one-to-one correspondence whose inverse is differentiable. Show that

$$(H \circ g^{-1})'(x) = f(x).$$

Example. Consider

$$\int_0^1 \sqrt{1 + \sqrt{x}} \cdot dx.$$

Here $f(x) = \sqrt{1 + \sqrt{x}}$. Let $u = 1 + \sqrt{x}$. If $x = 0$, then $u = 1$ and if $x = 1$, then $u = 2$. Then $x = (u - 1)^2 = g(u)$, and g is indeed a smooth function on $[1, 2]$. Also $f(g(u)) = \sqrt{u}$. Now

$$dx - 2(u - 1) \cdot du.$$

Hence replacing all terms, we get

$$\int_0^1 \sqrt{1 + \sqrt{x}} \cdot dx = 2 \int_1^2 \sqrt{u} \cdot (u - 1) \cdot du = 2 \int_1^2 \left(u^{3/2} - u^{1/2} \right) \cdot du$$

and the integral evaluates to

$$2 \left(\frac{2}{5} u^{5/2} - \frac{2}{3} u^{3/2} \right) \Big|_1^2 = \frac{8 + 8\sqrt{2}}{15}.$$

Since g is, in fact, a one-to-one correspondence between $(1, \infty)$ and $(0, \infty)$, whose inverse $x \mapsto 1 + \sqrt{x}$ is differentiable, we can go back to the variable x to find the antiderivative:

$$\int \sqrt{1 + \sqrt{x}} \cdot dx = \frac{4}{5} \left(\sqrt{1 + \sqrt{x}} \right)^5 - \frac{4}{3} \left(\sqrt{1 + \sqrt{x}} \right)^3 + C.$$

4.4   Geometric Interpretation of Integrals

In this section we give a geometric interpretation of the integral. Let f be a non-negative continuous function on a closed interval I. For $a, b \in I$, $a \leq b$, we let

$$A(a, b)$$

denote the **area** of the region below the graph of f, above the x-axis, and between the straight lines $x = a$ and $x = b$.

It is not our goal to give a rigorous geometric definition of area; instead, we proceed axiomatically. Our intuitive understanding of area suggests that the "area function" A (of two variables a and b) has to have the following two properties:

(1) $A(a, b) = A(a, c) + A(c, b)$, whenever $a \leq c \leq b$;

(2′) $m \cdot (b - a) \leq A(a, b) \leq M \cdot (b - a)$, whenever $m \leq f(x) \leq M$ for all $x \in [a, b]$.

Let $N \in \mathbb{N}$ be ultralarge. We now consider the partition $x_0 < x_1 < \ldots < x_N$ of $[a, b]$, where $x_i = a + i \cdot dx$ and $dx = (b - a)/N$. From (1) it follows by induction that

$$A(a, b) = \sum_{i=0}^{N-1} A(x_i, x_{i+1}).$$

From (2′) we get that

(2) $A(x_i, x_{i+1}) = f(x_i) \cdot dx + \varepsilon_i \cdot dx$ for some $\varepsilon_i \simeq 0$.

Indeed, let $m = f(c)$ and $M = f(d)$ be the minimum and maximum values of f on $[x_i, x_{i+1}]$. By (2′),

$$f(c) \cdot dx \leq A(x_i, x_{i+1}) \leq f(d) \cdot dx.$$

Hence,

$$f(c) \leq \frac{A(x_i, x_{i+1})}{dx} \leq f(d).$$

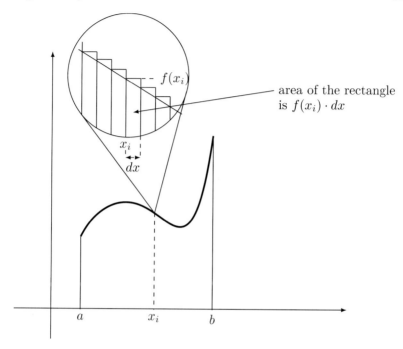

area of the rectangle
is $f(x_i) \cdot dx$

But $c \simeq x_i \simeq d$, so by uniform continuity of f we have

$$f(c) \simeq f(x_i) \simeq f(d).$$

This implies that

$$\frac{A(x_i, x_{i+1})}{dx} \simeq f(x_i),$$

so that $\frac{A(x_i, x_{i+1})}{dx} = f(x_i) + \varepsilon_i$, for some $\varepsilon_i \simeq 0$. This proves claim (2).

Putting these two results together, we get

$$A(a, b) = \sum_{i=0}^{N-1} f(x_i) \cdot dx + \sum_{i=0}^{N-1} \varepsilon_i \cdot dx,$$

The second sum is ultraclose to zero, and we conclude that

$$A(a, b) \text{ is the observable neighbor of } \sum_{i=0}^{N-1} f(x_i) \cdot dx$$

which is

$$\int_a^b f(x) \cdot dx.$$

The above argument shows that, *if* there is a way to assign areas $A(a, b)$ to regions under the graph of f so that properties (1) and (2′) [or (1) and (2)] hold, then $A(a, b)$ is uniquely determined: It has to be equal to the integral of f from a to b. We now reverse the tables and *define* the area by the integral.

Definition 33.

$$A(a, b) = \int_a^b f(x) \cdot dx.$$

The validity of (1) and (2′) follows from the properties of definite integrals established in Section 4.1.

The geometric interpretation of integral as area can be extended to arbitrary continuous functions.

Definition 34. *The **upper part** of f is the function*

$$f^+(x) = \begin{cases} f(x) & \text{if } f(x) > 0; \\ 0 & \text{otherwise.} \end{cases}$$

*The **lower part** of f is the function*

$$f^-(x) = \begin{cases} -f(x) & \text{if } f(x) < 0; \\ 0 & \text{otherwise.} \end{cases}$$

It is immediate to check that

$$f(x) = f^+(x) - f^-(x).$$

Hence by linearity,

$$\int_a^b f(x) \cdot dx = \int_a^b f^+(x) \cdot dx - \int_a^b f^-(x) \cdot dx.$$

So the integral is the difference of the area above the x-axis and the area below the x-axis. It can be positive, negative, or zero, depending on the relative size of the two regions.

To use linearity, we need to know that f^+ and f^- are continuous when f is. The next exercise shows that this is true ($f^+ = \max\{f, 0\}$ and $f^- = \max\{-f, 0\}$).

Exercise 43 (Answer page 254)
Let f, g be continuous functions on I. Define $h : I \to \mathbb{R}$ by

$$h(x) = \max\{f(x), g(x)\}, \quad \text{for each } x \in I.$$

Show that h is continuous on I.

Other applications of integrals can be treated in a similar manner (see the next section).

4.5 Applications of the Integral

Mean Value of a Function

The mean (average) value is unambiguous when we consider n numbers, where n is a positive integer. We now show that the mean value of a continuous function on $[a, b]$ is a natural extension of this concept.

Consider a continuous function f and the interval $[a, b]$. Let N be a positive ultralarge integer. Let $dx = (b - a)/N$ and $x_i = a + i \cdot dx$, for $i = 1, \ldots, N$. Then the mean value of the function can be approximated by the mean value of the N numbers $f(x_i)$, $i = 0, \ldots, N - 1$. But

$$\frac{\sum\limits_{i=0}^{N-1} f(x_i)}{N} = \frac{dx}{b - a} \sum_{i=0}^{N-1} f(x_i) = \frac{1}{b - a} \sum_{i=0}^{N-1} f(x_i) \cdot dx.$$

The mean value of the function should be the observable neighbor of this number, that is, the integral. We therefore define:

Definition 35. *The **mean value** of a function f continuous on $[a, b]$ is*

$$\frac{1}{b - a} \int_a^b f(x) \cdot dx.$$

The mean value is a number μ such that the area under the curve is equal to $\mu \cdot (b - a)$. That is, μ is the height of a rectangle of basis $(b - a)$ whose (oriented) area is equal to the integral.

Theorem 65. *If f is a function continuous on $[a, b]$, then there exists a point $c \in [a, b]$ such that $f(c)$ is the mean value of the function f on $[a, b]$.*

Proof. The mean value cannot be less than the least value nor greater than the greatest value of the function f on $[a, b]$. As f is assumed to be continuous, it reaches all intermediate values, hence also the mean value. □

Note that this theorem is a restatement of the Mean Value Theorem, for the antiderivative of f. When we claim that there is a $c \in [a, b]$ such that

$$f(c) = \frac{1}{b - a} \int_a^b f(x) \cdot dx,$$

we are in fact asserting that there is a $c \in [a, b]$ such that

$$f(c) \cdot (b - a) = \int_a^b f(x) \cdot dx = F(b) - F(a),$$

and as $F'(x) = f(x)$, we conclude that there is a $c \in [a, b]$ such that

$$F'(c) \cdot (b - a) = F(b) - F(a).$$

In fact, the mean value theorem shows that c can be found in (a, b).

Area in Polar Coordinates

We consider the general problem of finding the area $A(\alpha, \beta)$ of a region bounded by the graph of an equation $r = g(\theta)$ in polar coordinates, and the half-rays $\theta = \alpha$ and $\theta = \beta$. We assume that g is nonnegative and continuous on a closed interval $I \subseteq [0, 2\pi]$. Our intuitive understanding of area suggests that

(1) $A(\alpha, \beta) = A(\alpha, \gamma) + A(\gamma, \beta)$, whenever $\alpha \leq \gamma \leq \beta$;

(2) $A(\theta, \theta + d\theta) = \frac{1}{2}[g(\theta)]^2 \cdot d\theta + \varepsilon \cdot d\theta$, with $\varepsilon \simeq 0$, whenever $d\theta \simeq 0$.

To justify (2), let $g(\gamma)$ and $g(\delta)$ be the minimum and maximum values of g on $[\theta, \theta + d\theta]$, with $d\theta > 0$ (the case $d\theta < 0$ is similar).

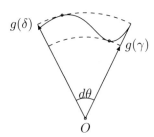

The region under consideration is squeezed between two circular sectors of central angle $d\theta$ and radius $g(\gamma)$ and $g(\delta)$ respectively. Thus

$$\frac{1}{2}[g(\gamma)]^2 \cdot d\theta \leq A(\theta, \theta + d\theta) \leq \frac{1}{2}[g(\delta)]^2 \cdot d\theta$$

and so after dividing by $d\theta$,

$$\frac{1}{2}[g(\gamma)]^2 \leq \frac{A(\theta, \theta + d\theta)}{d\theta} \leq \frac{1}{2}[g(\delta)]^2.$$

The function g is (uniformly) continuous on I so $g(\gamma) \simeq g(\theta) \simeq g(\delta)$ and therefore,

$$A(\theta, \theta + d\theta) = \frac{1}{2}[g(\theta)]^2 \cdot d\theta + \varepsilon \cdot d\theta, \qquad \text{where } \varepsilon \simeq 0.$$

We now follow the same argument as in Section 4.4. Let $N \in \mathbb{N}$ be ultralarge, and let $d\theta = (\beta - \alpha)/N$, and $\theta_i = \alpha + i \cdot d\theta$, for $i = 0, 1, \ldots, N$. From (1) it follows by induction that

$$A(\alpha, \beta) = \sum_{i=0}^{N-1} A(\theta_i, \theta_i + d\theta_i).$$

From (2) we see that

$$A(\theta_i, \theta_i + d\theta_i) = \frac{1}{2}[g(\theta_i)]^2 \cdot d\theta + \varepsilon_i \cdot d\theta, \quad \text{with } \varepsilon_i \simeq 0.$$

Hence

$$A(\alpha, \beta) = \sum_{i=0}^{N-1} \frac{1}{2}[g(\theta_i)]^2 \cdot d\theta + \sum_{i=0}^{N-1} \varepsilon_i \cdot d\theta$$

and, as the second sum is ultraclose to 0,

$$A(\alpha, \beta) \simeq \sum_{i=0}^{N-1} \frac{1}{2}[g(\theta_i)]^2 \cdot d\theta \simeq \int_{\alpha}^{\beta} \frac{1}{2}[g(\theta)]^2 \cdot d\theta.$$

The first and last term are observable, so they are equal.

Formally, we take $A(\alpha, \beta) = \int_{\alpha}^{\beta} \frac{1}{2}[g(\theta)]^2 \cdot d\theta$. as the *definition* of the area of the region under study.

Example. The area of the region bounded by the spiral $r = \theta$ and the half-rays $\theta = 0$ and $\theta = \pi$ is

$$A = \int_0^{\pi} \frac{1}{2}\theta^2 \cdot d\theta = \left[\frac{1}{6}\theta^3\right]_0^{\pi} = \frac{\pi^3}{6}.$$

Our treatment of areas here and in Section 4.4 leaves much to be desired. For example, it is not at all clear that computing the area of a region in rectangular coordinates, as in Section 4.4, will always give the same result as doing so in polar coordinates. The general theory of areas of plane regions requires the introduction of double integrals, and is outside the scope of this book.

Volume, Mass, Force, and other Physical Quantities

When applying mathematics to the real world, we first have to make an idealized mathematical model of the physical reality. In many situations, the model has features similar to those encountered in Section 4.4 and the previous subsections of this section.

The argument already given for the area shows that in general, if a function $F(a, b)$ has the properties

(1) $F(a, b) = F(a, c) + F(c, b)$, whenever $a \leq c \leq b$; and

(2) $F(x, x + dx) = f(x) \cdot dx + \varepsilon \cdot dx$, with $\varepsilon \simeq 0$, whenever $dx \simeq 0$,

for some continuous function f, then

$$F(a, b) \simeq \sum_{i=0}^{N-1} f(x_i) \cdot dx \simeq \int_a^b f(x) \cdot dx,$$

when N is any ultralarge positive integer, leading to $F(a, b) = \int_a^b f(x) \cdot dx$, because the first and last term are both observable.

Volume of a Solid of Revolution

Consider the solid obtained by the revolution around the x-axis of the region under the curve given by $f : x \mapsto f(x)$ between $x = a$ and $x = b$. Assume f is positive and continuous on $[a, b]$.

We let $F(c, d)$ be the volume of the "slice" of the solid between the planes $x = c$ and $x = d$. Let N be an ultralarge positive integer, $dx = (b - a)/N$, $x_i = a + i \cdot dx$, for $i = 0, \ldots, N$. Each slice with $c = x_i$ and $d = x_i + dx$ is between two cylinders with a volume of the form $\pi[f(x)]^2 \cdot dx$, for x in $[x_i, x_i + dx]$ (namely, the one where $f(x)$ is the minimum and the one where it is the maximum, on $[x_i, x_i + dx]$).

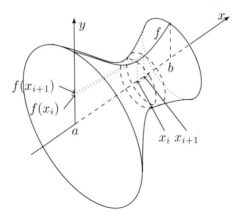

By the Intermediate Value Theorem, there is $c_i \in [x_i, x_i + dx]$ such that the volume of the slice is exactly $\pi[f(c_i)]^2 \cdot dx$. Since $f(c_i) \simeq f(x_i)$ by uniform continuity,

$$F(x_i, x_i + dx) = \pi[f(x_i)]^2 \cdot dx + \varepsilon_i \cdot dx, \qquad \text{with } \varepsilon_i \simeq 0.$$

The total volume is thus

$$V = \sum_{i=0}^{N-1} \left(\pi[f(x_i)]^2 \cdot dx + \varepsilon_i \cdot dx \right) \simeq \sum_{i=0}^{N-1} \pi[f(x_i)]^2 \cdot dx \simeq \int_a^b \pi[f(x)]^2 \cdot dx.$$

As the volume and the integral are both observable,

$$V = \int_a^b \pi[f(x)]^2 \cdot dx.$$

Force

Exercise 44 (Answer on page 254.)
The gravitational force between two masses is given by

$$F = G \cdot \frac{m_1 \cdot m_2}{d^2},$$

where d is the distance between the two masses and G the universal constant of gravitation.
What is the force between objects A and B in the following situation? (For simplicity, the linear mass will be considered to have no width and uniform density, and the other will be considered reduced to a point.)

Length of Curves

Consider a continuous function $f : [a, b] \to \mathbb{R}$. We would like to establish what it means to measure the length of its graph. If the graph of f is a straight line, we can use the Pythagorean Theorem to find the length:

$$\sqrt{(b-a)^2 + (f(b) - f(a))^2}.$$

If the graph of f is a more general curve, then we would like to approximate it with a polygonal line. Let N be an ultralarge positive integer and let $dx = (b - a)/N$ and $x_i = a + i \cdot dx$, for $i = 0, \ldots, N - 1$. We approximate the curve on the interval $[x_i, x_{i+1}]$ by the segment joining the points $\langle x_i, f(x_i) \rangle$ and $\langle x_{i+1}, f(x_{i+1}) \rangle$; its length is

$$\sqrt{(x_{i+1} - x_i)^2 + (f(x_{i+1}) - f(x_i))^2}.$$

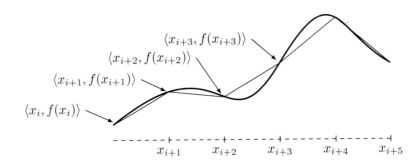

We then sum the lengths of all the segments to obtain an approximation of the length of the curve. Of course, for this approximation to be meaningful, the total sum must not be ultralarge relative to the context f, a, b, and it has to be independent of the choice of the positive integer N, up to an ultrasmall amount.

Definition 36. *Let $f : [a, b] \to \mathbb{R}$ be a continuous function. We say that the graph of f has **length** $L \in \mathbb{R}$ if L is observable and for any positive ultralarge integer N we have*

$$L \simeq \sum_{i=0}^{N-1} \sqrt{(x_{i+1} - x_i)^2 + (f(x_{i+1}) - f(x_i))^2},$$

where $x_i = a + i \cdot \frac{b-a}{N}$, for $i = 0, \ldots, N$.

Theorem 66. *Let $f : [a, b] \to \mathbb{R}$ be smooth. Then the graph of f has length*

$$L = \int_a^b \sqrt{1 + [f'(x)]^2} \cdot dx.$$

Proof. Let $N \in \mathbb{N}$ be ultralarge. Let $dx = (b - a)/N$. By the Uniform Increment Equation (f is smooth) we have

$$f(x_{i+1}) - f(x_i) = f'(x_i) \cdot dx + \varepsilon_i \cdot dx, \quad \text{for some } \varepsilon_i \simeq 0.$$

We deduce that

$$\sqrt{(x_{i+1} - x_i)^2 + (f(x_{i+1}) - f(x_i))^2} = \sqrt{(dx)^2 + (f'(x_i) + \varepsilon_i)^2 \cdot (dx)^2}$$
$$= \sqrt{1 + (f'(x_i) + \varepsilon_i)^2} \cdot dx.$$

Since $f' : [a, b] \to \mathbb{R}$ is continuous, there are observable c, d such that $f' : [a, b] \to [c, d]$. Therefore by uniform continuity of $x \mapsto \sqrt{1 + x^2}$ (on, say, the interval $[c - 1, \ d + 1]$) we have

$$\sqrt{1 + (f'(x_i) + \varepsilon_i)^2} = \sqrt{1 + [f'(x_i)]^2} + \delta_i, \quad \text{for some } \delta_i \simeq 0.$$

Thus

$$\sum_{i=0}^{N-1} \sqrt{(x_{i+1} - x_i)^2 + (f(x_{i+1}) - f(x_i))^2}$$

$$= \sum_{i=0}^{N-1} \sqrt{1 + [f'(x_i)]^2} \cdot dx + \sum_{i=0}^{N-1} \delta_i \cdot dx$$

$$\simeq \sum_{i=0}^{N-1} \sqrt{1 + [f'(x_i)]^2} \cdot dx$$

$$\simeq \int_a^b \sqrt{1 + [f'(x)]^2} \cdot dx,$$

since the function $x \mapsto \sqrt{1 + [f'(x)]^2}$ is continuous on $[a, b]$. As the integral is observable, the graph of f has length equal to the integral. $\square$

4.6 Natural Logarithm and Exponential

Let n be an integer. From $(x^{n+1})' = (n + 1) \cdot x^n$, we deduce

$$\int x^n \cdot dx = \frac{1}{n + 1} \cdot x^{n+1} + C, \quad \text{for } n \neq -1.$$

Hence the antiderivative of $x \mapsto \dfrac{1}{x} = x^{-1}$ cannot be obtained from this formula.

Definition 37. *The **natural logarithm** is the function* $\ln : (0, \infty) \to \mathbb{R}$ *defined by*

$$x \mapsto \int_1^x \frac{1}{t} \cdot dt.$$

The function $t \mapsto \frac{1}{t}$ is continuous on $(0, \infty)$, so, by (the second version of) the Fundamental Theorem of Calculus, the function ln is defined, continuous and differentiable on its domain $(0, \infty)$, and we have

$$\ln(1) = 0 \quad \text{and} \quad (\ln(x))' = \frac{1}{x}.$$

From the definition it is immediate that ln is strictly increasing, hence one-to-one. The function ln is standard.

Our first goal is to show that the natural logarithm is a logarithm in the usual sense.

Theorem 67. *Let $a > 0$ and $b \in \mathbb{Q}$. Then*

$$\ln(a^b) = b \cdot \ln(a).$$

Proof. By definition $\ln(a^b) = \int_1^{a^b} \frac{1}{t} \cdot dt$. If $b = 0$, the equality is clear, so assume that $b \neq 0$. Let $u = t^{1/b}$, so that $t = u^b$. When $t = 1$ we have $u = 1$ and when $t = a^b$ then $u = a$. Since b is rational, we have $dt = b \cdot u^{b-1} \cdot du$ by our rules of differentiation. We thus obtain

$$\ln(a^b) = \int_1^{a^b} \frac{1}{t} \cdot dt = \int_1^a \frac{b \cdot u^{b-1}}{u^b} \cdot du = b \int_1^a \frac{1}{u} \cdot du = b \cdot \ln(a).$$

$\square$

Exercise 45 (Answer page 255)
Let $a, b > 0$. Use $u = \frac{t}{a}$ to show that

$$\int_a^{a \cdot b} \frac{1}{t} \cdot dt = \int_1^a \frac{1}{u} \cdot du.$$

Deduce that $\ln(a \cdot b) = \ln(a) + \ln(b)$.

Theorem 68. *The natural logarithm satisfies*

$$\lim_{x \to 0^+} \ln(x) = -\infty \quad \text{and} \quad \lim_{x \to +\infty} \ln(x) = +\infty.$$

Proof. Let x be positive ultralarge and let N be the largest positive integer such that $2^N \leq x$. Then N is ultralarge and

$$N \cdot \ln(2) = \ln(2^N) \leq \ln(x).$$

As $\ln(2) > 0$ and standard, we have that $N \cdot \ln(2)$ is ultralarge, so $\ln(x)$ is ultralarge, and therefore $\lim_{x \to +\infty} \ln(x) = +\infty$.

To see that $\lim_{x \to 0^+} \ln(x) = -\infty$, let ε be positive and ultrasmall. Then ε^{-1} is positive and ultralarge and $\ln(\varepsilon^{-1}) = -\ln(\varepsilon)$ is a positive ultralarge number, hence $\ln(\varepsilon)$ is ultralarge negative. This implies that $\lim_{x \to 0^+} \ln(x) = -\infty$. $\square$

Theorem 69. *The function* $\ln : (0, \infty) \to \mathbb{R}$ *is a one-to-one correspondence.*

Proof. The function $\ln$ is one-to-one because it is strictly increasing. Its range is an open interval by Theorem 11, because it is continuous. The previous theorem implies that the range is $(-\infty, \infty)$. $\qquad\square$

Definition 38. *We denote by e the unique real number such that*

$$\ln(e) = 1.$$

The number e exists and is unique by the previous theorem, so that, by Closure, e is standard. Approximation of the area under the curve $y = 1/x$ easily shows that $2.5 \le e \le 3$. In fact, e is an irrational number whose first few digits are

$$e = 2.71828\ldots$$

We give two other definitions of e below (Theorems 73 and 108); the second one is very useful for calculating approximations. It turns out that the natural logarithm is the base e logarithm.

Theorem 70. *With e defined as above, we have*

$$\ln(x) = \log_e(x), \quad \text{for all } x > 0.$$

Proof. Let $x > 0$ be given. Let $y = \log_e(x)$, so $x = e^y$. The parameter is x and therefore y is observable. Let $u \simeq y$ be rational. Since the base e exponential function is continuous, we have $e^y \simeq e^u$. As $\ln$ is continuous at e^y, we have

$$\ln(x) = \ln(e^y) \simeq \ln(e^u) = u \cdot \ln(e) = u \simeq y = \log_e(x).$$

But both the left-hand side and the right-hand side are observable, so they have to be equal. $\qquad\square$

It follows that $\ln(a^y) = y \cdot \ln(a)$ for any real y. By letting $x = a^y$ we get $\ln(x) = y \cdot \ln(a) = \log_a(x) \cdot \ln(a)$, so that

$$\log_a(x) = \frac{\ln(x)}{\ln(a)}.$$

Definition 39. *The **exponential function** $\exp : \mathbb{R} \longrightarrow (0, \infty)$ is defined as the inverse function of* $\ln$.

By the previous theorem, we have that

$$\exp(x) = \exp_e(x) = e^x, \quad \text{for all } x \in \mathbb{R}.$$

The following property makes $\exp(x)$ a very special function.

Theorem 71.
$$\exp'(x) = \exp(x).$$

Proof. Let $y = \exp(x)$ and $x = \ln(y)$:

$$\exp'(x) = \frac{1}{\ln'(y)} = \frac{1}{\frac{1}{y}} = y = \exp(x).$$

$\square$

We obtain the following derivatives.

Theorem 72.

(1) Let $b \in \mathbb{R}$. The power function $x \mapsto x^b$ is differentiable and

$$(x^b)' = b \cdot x^{b-1}, \quad \text{for } x > 0.$$

(2) Let $a > 0$. The base a exponential is differentiable and

$$(a^x)' = \ln(a) \cdot a^x, \quad \text{for } x > 0.$$

(3) Let $a > 0$. The base a logarithm is differentiable and

$$\log_a'(x) = \frac{1}{\ln(a) \cdot x}, \quad \text{for } x > 0.$$

Proof. The first two proofs are applications of the Chain Rule after noticing that

$$a^b = \exp(\ln(a^b)) = \exp(b \ln(a)), \quad \text{for all } b \in \mathbb{R}, \ a > 0.$$

For (1) we have

$$\left(x^b\right)' = (\exp(b \cdot \ln(x)))' = \exp(b \cdot \ln(x)) \cdot \frac{b}{x} = x^b \cdot \frac{b}{x} = b \cdot x^{b-1}.$$

For (2) we have

$$(a^x)' = (\exp(x \cdot \ln(a)))' = \exp(x \cdot \ln(a)) \cdot \ln(a) = \ln(a) \cdot a^x.$$

For (3) we have

$$\log_a'(x) = \left(\frac{\ln(x)}{\ln(a)}\right)' = \frac{1}{\ln(a) \cdot x}.$$

$\square$

The following theorem expresses e as a limit.

Theorem 73. *The number e satisfies*

$$e = \lim_{x \to \infty} \left(1 + \frac{1}{x}\right)^x.$$

Proof. Notice that

$$\left(1 + \frac{1}{x}\right)^x = \exp\left(x \cdot \ln\left(1 + \frac{1}{x}\right)\right).$$

Let x be ultralarge and positive. Then $1/x$ is ultrasmall, so by definition of the derivative of ln at 1 we have

$$x \cdot \ln\left(1 + \frac{1}{x}\right) = \frac{\ln\left(1 + \frac{1}{x}\right) - \ln(1)}{\frac{1}{x}} \simeq \ln'(1) = \frac{1}{1} = 1.$$

But exp is continuous at 1, so

$$\exp\left(x \cdot \ln(1 + \frac{1}{x})\right) \simeq \exp(1) = e.$$

□

Exercise 46 (Answer page 255)
Modify the argument in the previous proof to show that for any $z \in \mathbb{R}$ we have

$$e^z = \lim_{x \to \infty} \left(1 + \frac{z}{x}\right)^x.$$

Example. By the Increment Equation applied to ln at 1, we have for dx ultrasmall:

$$\ln(1 + dx) = dx + \varepsilon \cdot dx, \quad \text{for some } \varepsilon \simeq 0.$$

This implies that

$$\frac{\ln(1 + dx)}{dx} \simeq 1,$$

so that

$$\lim_{x \to 0} \frac{\ln(1 + x)}{x} = 1.$$

Similarly, if we apply the Increment Equation to exp at 0, we have for ultrasmall dx:

$$\exp(dx) = 1 + dx + \varepsilon \cdot dx, \quad \text{for some } \varepsilon \simeq 0.$$

We deduce that
$$\frac{\exp(dx) - 1}{dx} \simeq 1,$$

so that
$$\lim_{x \to 0} \frac{\exp(x) - 1}{x} = 1.$$

Example. We showed that
$$e = \lim_{x \to \infty} \left(1 + \frac{1}{x}\right)^x.$$

In Chapter 7 we give another, independent proof that this limit exists (page 172). In this example, we demonstrate directly from this limit, that the function $\exp : x \mapsto e^x$ is its own derivative.

Let x be a real number. We show that $(e^x)' = e^x$. The parameter is x. Let $h > 0$ be ultrasmall (we leave the case when $h < 0$ as an exercise). Then
$$(e^x)' \simeq \frac{e^{x+h} - e^x}{h} = \frac{e^x \cdot e^h - e^x}{h} = e^x \cdot \frac{e^h - 1}{h}.$$

Let $b = \frac{e^h - 1}{h}$. It is enough to show that $b \simeq 1$. But $b \cdot h = e^h - 1$, so $b \cdot h > 0$ is ultrasmall (since $e^h \simeq 1$ and $e^h > 1$). Let $z = 1/bh$. Then z is positive ultralarge, so
$$e \simeq \left(1 + \frac{1}{z}\right)^z = (1 + bh)^{1/bh} = (e^h)^{1/bh} = e^{1/b}.$$

This implies that $1/b \simeq 1$, so $b \simeq 1$.

Exercise 47 (Answer page 255)
Let a be a real number and h be a positive and ultrasmall number relative to a.

(1) Show that
$$\frac{1 - e^{-h}}{h} \simeq 1.$$

Hint: Let $b = \frac{1 - e^{-h}}{h}$, let $x = bh$, and use $e^{-1} \simeq (1 - \frac{1}{x})^x$ (see Exercise 46).

(2) Deduce that $\frac{e^{a+k} - e^a}{k} \simeq e^a$ if k is negative and ultrasmall relative to a.

Theorem 74. *Let f be a smooth function on I, $f(x) \neq 0$ for any $x \in I$. Then*
$$\int \frac{f'(x)}{f(x)} \cdot dx = \ln|f(x)| + C.$$

Proof. The claim is clear if $f(x) > 0$ for all $x \in I$, since $|f(x)| = f(x)$ and, by the Chain Rule,

$$(\ln(f(x)))' = \frac{f'(x)}{f(x)}.$$

If $f(x) < 0$ for all $x \in I$, then $|f(x)| = -f(x)$ and, again by the Chain Rule,

$$(\ln(-f(x)))' = \frac{-f'(x)}{-f(x)} = \frac{f'(x)}{f(x)}.$$

$\square$

Theorem 75. *Let f be a smooth function. Then*

$$\int f'(x) \cdot e^{f(x)} \cdot dx = e^{f(x)} + C.$$

Proof. This is again an immediate consequence of the Chain Rule. $\square$

4.7   Numerical Integration

The Fundamental Theorem of Calculus furnishes an easy way to calculate the definite integral $\int_a^b f(x) \cdot dx$ in those cases when an antiderivative F to f can be found explicitly in terms of familiar functions. It turns out that this is an exception rather than the rule; there are many simple functions whose antiderivatives cannot be expressed in terms of elementary functions such as powers, exponentials, logarithms, trigonometric functions, and their inverses. Functions e^{-x^2} and $\frac{\sin(x)}{x}$ are among the simplest examples. In such cases, one has to resort to calculating an approximation to the exact value of the definite integral. That is, we specify an *error tolerance* $\varepsilon > 0$ and look for a number $\overline{R}$ such that

$$\left| \int_a^b f(x) \cdot dx - \overline{R} \right| < \varepsilon. \tag{4.1}$$

It follows immediately from our definition of the definite integral that

$$\overline{R} = \sum_{i=0}^{n-1} f(x_i) \cdot h = h \cdot (f(x_0) + f(x_1) + \ldots + f(x_{n-1})), \tag{4.2}$$

where $h = \frac{b-a}{n}$ and $x_i = a + i \cdot h$ for $i = 0, 1, \ldots, n$, has the property (4.1) *if $n \in \mathbb{N}$ is ultralarge.* Unfortunately, numerical calculations with ultralarge numbers are beyond the capacity of any physical computer. Fortunately, an ultralarge n is not really necessary; it follows immediately from the Closure Principle that (4.1) holds even for observable n, as long as n is *large enough*.

Theorem 76. *For every observable $\varepsilon > 0$ there exists an observable K such that*

$$n > K \quad implies \quad \left| \int_a^b f(x) \cdot dx - \sum_{i=0}^{n-1} f(x_i) \cdot h \right| < \varepsilon. \qquad (4.3)$$

Proof. The parameters are f, a and b. If K is positive ultralarge, then every $n > K$ is ultralarge and $\int_a^b f(x) \cdot dx \simeq \sum_{i=0}^{n-1} f(x_i) \cdot h$, so K has the property (4.3). By Closure, there is an observable K with the property (4.3). □

For practical purposes we would like to have an idea of the size of K that guarantees (4.3), for a given integral and a given ε. We now derive a theorem that can be used to obtain an explicit value of K.

Theorem 77. *Assume that f' is continuous and $|f'(x)| \leq M$, for all $x \in [a, b]$. Then*

$$\left| \int_a^b f(x) \cdot dx - \sum_{i=0}^{n-1} f(x_i) \cdot h \right| \leq \frac{M \cdot (b-a)^2}{2n}.$$

Proof. We write

$$E = \int_a^b f(x) \cdot dx - \sum_{i=0}^{n-1} f(x_i) \cdot h = \sum_{i=0}^{n-1} \left(\int_{x_i}^{x_{i+1}} f(x) \cdot dx - f(x_i) \cdot (x_{i+1} - x_i) \right),$$

using the additivity of the definite integral (Theorem 57) and the fact that $h = x_{i+1} - x_i$. We wish to estimate the term

$$E_i = \int_{x_i}^{x_{i+1}} f(x) \cdot dx - f(x_i) \cdot (x_{i+1} - x_i).$$

We replace x_{i+1} by a variable z and consider the function

$$F_i(z) = \int_{x_i}^{z} f(x) \cdot dx - f(x_i) \cdot (z - x_i).$$

The function F_i is differentiable in $[x_i, x_{i+1}]$ by the Fundamental Theorem, and $F_i'(z) = f(z) - f(x_i)$. We note that $F_i(x_i) = 0$, $F_i'(x_i) = 0$, and

$F_i(x_{i+1}) = E_i$. By the assumptions of the theorem, $|F_i''(z)| = |f'(z)| \leq M$. We integrate the last inequality over the interval $[x_i, t]$ to obtain

$$|F_i'(t)| = |F_i'(t) - F_i'(x_i)| = \left| \int_{x_i}^{t} F_i''(z) \cdot dz \right| \leq \int_{x_i}^{t} M \cdot dz = M \cdot (t - x_i).$$

Integrating again, over $[x_i, x_{i+1}]$, gives $|E_i| = |F_i(x_{i+1})| = |F_i(x_{i+1}) - F_i(x_i)| =$

$$\left| \int_{x_i}^{x_{i+1}} F_i'(t) \cdot dt \right| \leq \int_{x_i}^{x_{i+1}} M \cdot (t - x_i) \cdot dt = \frac{M}{2} \cdot (x_{i+1} - x_i)^2 = \frac{M}{2} \cdot h^2.$$

Finally,

$$|E| \leq \sum_{i=0}^{n-1} |E_i| \leq \sum_{i=0}^{n-1} \frac{M}{2} \cdot h^2 = n \cdot \frac{M}{2} \cdot \left(\frac{b-a}{n} \right)^2 = \frac{M \cdot (b-a)^2}{2n}.$$

$\square$

Example. Consider computing the integral

$$\int_{1}^{2} \frac{\sin(x)}{x} \cdot dx$$

by this method. From

$$f'(x) = \frac{x \cdot \cos(x) - \sin(x)}{x^2}$$

we get

$$|f'(x)| \leq \frac{|x| \cdot |\cos(x)| + |\sin(x)|}{|x|^2} \leq \frac{1}{|x|} + \frac{1}{|x|^2} \leq 2,$$

for $1 \leq x \leq 2$. The error is guaranteed to be less than ε if

$$\frac{M(b-a)^2}{2n} = \frac{2(2-1)^2}{2n} = \frac{1}{n} < \varepsilon,$$

that is, if $n > \frac{1}{\varepsilon} = K$. For $\varepsilon = 0.0001$ we need to take $n > K = 10000$.

For an increasing function f on $[a, b]$, $f(x_i) \leq f(x) \leq f(x_{i+1})$ holds for all $x \in [x_i, x_{i+1}]$, so it is clear that $f(x_i) \cdot h$ underestimates $\int_{x_i}^{x_{i+1}} f(x) \cdot dx$. On the other hand, $f(x_{i+1}) \cdot h$ would overestimate it. It seems reasonable to expect that the average of the two approximations should work better than either.

This suggests approximating $\int_a^b f(x) \cdot dx$ by

$$T = \sum_{i=0}^{n-1} \frac{f(x_i) + f(x_{i+1})}{2} \cdot h = \frac{h}{2} \cdot (f(x_0) + 2f(x_1) + \ldots + 2f(x_{n-1}) + f(x_n)).$$

As $\frac{f(x_i)+f(x_{i+1})}{2} \cdot h$ represents the area of the trapezoid with the bases $f(x_i)$ and $f(x_{i+1})$ and height h, this technique is called the **trapezoidal method** for numerical integration. We derive a theorem showing that the trapezoidal method in general works better (requires smaller n to achieve the desired accuracy).

Theorem 78. *Assume that f'' is continuous and $|f''(x)| \le M$, for all $x \in [a, b]$. Then*

$$\left| \int_a^b f(x) \cdot dx - T \right| \le \frac{M \cdot (b-a)^3}{12n^2}.$$

Proof. We follow the steps of the proof of Theorem 77. This time we let

$$T_i = \int_{x_i}^{x_{i+1}} f(x) \cdot dx - \frac{f(x_i) + f(x_{i+1})}{2} \cdot (x_{i+1} - x_i),$$

and consider

$$F_i(z) = \int_{x_i}^{z} f(x) \cdot dx - \frac{f(x_i) + f(z)}{2} \cdot (z - x_i).$$

We compute

$$F_i'(z) = f(z) - \frac{f'(z)}{2}(z - x_i) - \frac{f(x_i) + f(z)}{2}$$

$$= \frac{1}{2}(f(z) - f(x_i) - f'(z) \cdot (z - x_i)),$$

$$F_i''(z) = -\frac{1}{2}f''(z) \cdot (z - x_i),$$

and note that $F_i(x_i) = 0$, $F_i'(x_i) = 0$. By the assumptions of the theorem, for $z \in [x_i, x_{i+1}]$, $|F_i''(z)| \le \frac{1}{2}M \cdot (z - x_i)$. We integrate this inequality over $[x_i, t]$ to obtain

$$|F_i'(t)| = \left| \int_{x_i}^{t} F_i''(z) \cdot dz \right| \le \frac{M}{2} \cdot \int_{x_i}^{t} (z - x_i) \cdot dz = \frac{M}{4} \cdot (t - x_i)^2,$$

and integrating again, over $[x_i, x_{i+1}]$, gives $|T_i| = |F_i(x_{i+1})| =$

$$\left| \int_{x_i}^{x_{i+1}} F_i'(t) \cdot dt \right| \le \int_{x_i}^{x_{i+1}} \frac{M}{4} \cdot (t - x_i)^2 \cdot dt = \frac{M}{12} \cdot (x_{i+1} - x_i)^3 = \frac{M}{12} \cdot h^3.$$

Finally,

$$|T| \leq \sum_{i=0}^{n-1} |T_i| \leq \sum_{i=0}^{n-1} \frac{M}{12} \cdot h^3 = n \cdot \frac{M}{12} \cdot \left(\frac{b-a}{n}\right)^3 = \frac{M \cdot (b-a)^3}{12n^2}.$$

□

Example (continued). A computation shows that

$$f''(x) = -\frac{\sin(x)}{x} - \frac{2\cos(x)}{x^2} + \frac{2\sin(x)}{x^3},$$

so $|f''(x)| \leq 5$ for $1 \leq x \leq 2$. To guarantee an error less than ε for the trapezoidal method, we need to make $\frac{M(b-a)^3}{12n^2} = \frac{5}{12n^2} < \varepsilon$, that is, $n > \sqrt{\frac{5}{12\varepsilon}} = K$. For $\varepsilon = 0.0001$ this means $n > \sqrt{\frac{50000}{12}} \approx 64.5 = K$, so $n = 65$ suffices.

Yet better approximation methods can be obtained by similar techniques. We give one well-known formula and its error estimate, without proof.

Theorem 79 (Simpson's Rule). *Assume that $f^{(4)}$ is continuous and $|f^{(4)}(x)| \leq M$, for all $x \in [a, b]$. For even $n = 2k$ and $h = (b-a)/n$, $x_i = a + i \cdot h$, let*

$$S = \frac{1}{3}h \cdot (f(x_0) + 4f(x_1) + 2f(x_2) + 4f(x_3) + 2f(x_4) + \ldots$$

$$+ 2f(x_{n-2}) + 4f(x_{n-1}) + f(x_n))$$

$$= \sum_{i=0}^{k-1} \left(\frac{f(x_{2i}) + 4f(x_{2i+1}) + f(x_{2i+2})}{3} \right) \cdot h.$$

Then

$$\left| \int_a^b f(x) \ dx \quad S \right| \leq \frac{M \cdot (b-a)^5}{180n^4}.$$

Example (continued). For $f(x) = \frac{\sin(x)}{x}$, a further calculation gives $|f^{(4)}(x)| \leq 65$. Using $M = 65$ in the error estimate of Theorem 79 shows that $n = 8$ suffices to guarantee an error less than $\varepsilon = 0.0001$ for the Simpson's method.

4.8 Improper Integrals

In this section, we extend the integration of continuous functions on a closed bounded interval $[a, b]$ to the situations where the interval is not necessarily bounded or not necessarily closed. We call such integrals **improper**.

We first consider the integral of a continuous function over an interval of the form $[a, \infty)$ or $(-\infty, b]$.

Definition 40.

(1) Let $f : [a, \infty) \to \mathbb{R}$ be a continuous function. We say that

$$\int_a^\infty f(x) \cdot dx \text{ \textbf{converges} if}$$

$$\lim_{b \to \infty} \int_a^b f(x) \cdot dx \quad \text{exists.}$$

We define

$$\int_a^\infty f(x) \cdot dx = \lim_{b \to \infty} \int_a^b f(x) \cdot dx.$$

We say that

$$\int_a^\infty f(x) \cdot dx \text{ \textbf{diverges}}$$

if it does not converge.

(2) Let $f : (-\infty, b] \to \mathbb{R}$ be a continuous function. We say that

$$\int_{-\infty}^b f(x) \cdot dx \text{ \textbf{converges} if}$$

$$\lim_{a \to -\infty} \int_a^b f(x) \cdot dx \quad \text{exists.}$$

We define

$$\int_{-\infty}^b f(x) \cdot dx = \lim_{a \to -\infty} \int_a^b f(x) \cdot dx.$$

We say that

$$\int_{-\infty}^b f(x) \cdot dx \text{ \textbf{diverges}}$$

if it does not converge.

Let us unravel these definitions. We consider the case of $\int_a^\infty f(x) \cdot dx$. Let b be positive ultralarge. Then, since f is continuous on $[a, b]$, the number $\int_a^b f(x) \cdot dx$ is defined (by Theorem 50). Saying that

$$\lim_{b \to \infty} \int_a^b f(x) \cdot dx \quad \text{exists}$$

means that $\int_a^b f(x) \cdot dx$ is not ultralarge for any $b \simeq +\infty$ and further, that

$$\int_a^b f(x) \cdot dx \simeq \int_a^c f(x) \cdot dx, \quad \text{whenever } b, c \simeq +\infty. \qquad (4.4)$$

By additivity of the integral (Theorem 56) we have

$$\int_a^c f(x) \cdot dx = \int_a^b f(x) \cdot dx + \int_b^c f(x) \cdot dx,$$

so (4.4) is equivalent to

$$\int_b^c f(x) \cdot dx \simeq 0, \quad \text{whenever } b, c \simeq +\infty.$$

See Exercise 59 for more on this.

Example. We first consider a couple of examples where we can compute the antiderivatives explicitly.

(1) The integral

$$\int_0^\infty \sin(x) \cdot dx \quad \text{diverges, since}$$

$$\int_0^b \sin(x) \cdot dx = -\cos(b) + 1 \text{ and } \lim_{b \to \infty} (1 - \cos(b)) \text{ does not exist.}$$

(2) Let $\alpha \in \mathbb{R}$. Consider $f_\alpha : [1, \infty) \to \mathbb{R}$ given by

$$f_\alpha(x) = \frac{1}{x^\alpha}.$$

If $\alpha = 1$, we have

$$\int_1^b f_1(x) \cdot dx = \ln(x) \Big|_1^b = \ln(b).$$

If $\alpha \neq 1$, then

$$\int_1^b f_\alpha(x) \cdot dx = \frac{x^{-\alpha+1}}{-\alpha+1} \Big|_1^b = \frac{b^{-\alpha+1}}{-\alpha+1} + \frac{1}{\alpha-1}.$$

It follows that

$$\int_1^\infty f_\alpha(x) \cdot dx \begin{cases} \text{diverges} & \text{if } \alpha \le 1; \\ = \dfrac{1}{\alpha - 1} & \text{if } \alpha > 1. \end{cases}$$

Exercise 48 (Answer page 256)
Let $\alpha \in \mathbb{R}$. Let $f_\alpha : [0, \infty) \to \mathbb{R}$ be defined by $f_\alpha(x) = e^{\alpha x}$. For which α does $\int_0^\infty f_\alpha(x) \cdot dx$ converge?

When it is not possible to give an explicit formula for the antiderivative, we can often use the following theorem.

Theorem 80 (Comparison). *Let $f, g : [a, \infty) \to \mathbb{R}$ be continuous functions. Suppose that*

$$|f(x)| \le g(x), \quad \text{for each } x \in [a, \infty).$$

If $\int_a^\infty g(x) \cdot dx$ converges, then $\int_a^\infty f(x) \cdot dx$ converges.

Proof. Suppose that $\int_a^\infty g(x) \cdot dx$ converges. By assumption we have $-g(x) \le f(x) \le g(x)$, for each $x \in [a, \infty)$. Let $b \simeq +\infty$. By monotonicity (Exercise 4.9) of the integral we have

$$-\int_a^b g(x) \cdot dx \le \int_a^b f(x) \cdot dx \le \int_a^b g(x) \cdot dx.$$

Since $\int_a^\infty g(x) \cdot dx$ converges, the numbers $\pm \int_a^b g(x) \cdot dx$ are not ultralarge, and so $\int_a^b f(x) \cdot dx$ is not ultralarge. Now suppose that $b, c \simeq +\infty$, say $b < c$. By monotonicity again, we have

$$-\int_b^c g(x) \cdot dx \le \int_b^c f(x) \cdot dx \le \int_b^c g(x) \cdot dx.$$

But the left-hand side and the right-hand side are ultraclose to 0, because $\int_a^\infty g(x) \cdot dx$ converges, so $\int_b^c f(x) \cdot dx \simeq 0$ and therefore $\int_a^\infty f(x) \cdot dx$ converges. $\qquad\square$

Example. The integral

$$\int_1^\infty \frac{\sin(x)}{x^2} \cdot dx$$

converges. We compare it to the integral

$$\int_1^\infty \frac{1}{x^2} \cdot dx,$$

which converges by the previous example ($|\sin(x)/x^2| \le 1/x^2$, for $x \ge 1$).

It is easy to see that the monotonicity, linearity, continuity and additivity properties are true for improper integrals, provided they converge.

If we want to preserve these properties when integrating over $\mathbb{R}$, it is necessary to reduce the problem to two improper integrals.

Definition 41. *Let $f : \mathbb{R} \to \mathbb{R}$ be a continuous function. We say that $\int_{-\infty}^{\infty} f(x) \cdot dx$ **converges** if both*

$$\int_{-\infty}^{0} f(x) \cdot dx \quad and \quad \int_{0}^{\infty} f(x) \cdot dx \quad converge.$$

We define

$$\int_{-\infty}^{+\infty} f(x) \cdot dx = \int_{-\infty}^{0} f(x) \cdot dx + \int_{0}^{\infty} f(x) \cdot dx.$$

If either integral diverges, we say that $\int_{-\infty}^{+\infty} f(x) \cdot dx$ diverges.

The monotonicity, linearity and additivity properties of the integral extend easily to improper integrals of the previous type, provided they converge.

Finally, we consider the integral of a continuous function over $(a, b]$ or $[a, b)$.

Definition 42.

(1) Let $f : (a, b] \to \mathbb{R}$ be a continuous function. We say that

$$\int_{a}^{b} f(x) \cdot dx \quad \textbf{converges} \quad if$$

$$\lim_{h \to 0^+} \int_{a+h}^{b} f(x) \cdot dx \quad exists.$$

We define

$$\int_{a}^{b} f(x) \cdot dx = \lim_{h \to 0^+} \int_{a+h}^{b} f(x) \cdot dx.$$

We say that

$$\int_{a}^{b} f(x) \cdot dx \quad \textbf{diverges}$$

if it does not converge.

(2) Let $f : [a, b) \to \mathbb{R}$ be a continuous function. We say that

$$\int_a^b f(x) \cdot dx \text{ \textbf{converges} if}$$

$$\lim_{h \to 0^+} \int_a^{b-h} f(x) \cdot dx \quad \text{exists.}$$

We define

$$\int_a^b f(x) \cdot dx = \lim_{h \to 0^+} \int_a^{b-h} f(x) \cdot dx.$$

We say that

$$\int_a^b f(x) \cdot dx \text{ \textbf{diverges}}$$

if it does not converge.

Let us unravel the definitions. We consider the case $\int_a^b f(x) \cdot dx$ for a continuous function $f : (a, b] \to \mathbb{R}$. Suppose that $h > 0$ is ultrasmall. Then $\int_{a+h}^b f(x) \cdot dx$ is defined. Saying that

$$\lim_{h \to 0^+} \int_{a+h}^b f(x) \cdot dx \quad \text{exists}$$

means that for each positive ultrasmall h

$$\int_{a+h}^b f(x) \cdot dx$$

is not ultralarge and, up to a quantity ultraclose to 0, this number is independent of h. The last statement means that for $h, k > 0$ ultrasmall,

$$\int_{a+h}^{a+k} f(x) \cdot dx \simeq 0.$$

It is clear that if $f : [a, b] \to \mathbb{R}$ is continuous on $[a, b]$, then the integral $\int_a^b f(x) \cdot dx$ converges and the value of the limit is the usual integral.

Exercise 49 (Answer page 256)
Let $\alpha \in \mathbb{R}$. Show that

$$\int_0^1 \frac{1}{x^\alpha} \cdot dx \begin{cases} = \frac{1}{1-\alpha} & \text{if } \alpha < 1; \\ \text{diverges} & \text{otherwise.} \end{cases}$$

We leave the proof of the next theorem as an exercise.

Theorem 81 (Comparison). *Let $f, g : (a, b] \to \mathbb{R}$ be continuous functions. Suppose that*

$$|f(x)| \leq g(x), \quad \text{for each } x \in (a, b].$$

If $\int_a^b g(x) \cdot dx$ converges, then $\int_a^b f(x) \cdot dx$ converges.

Exercise 50 (Answer page 256)
Prove the previous theorem.
    The case of continuous functions on $[a, b)$ is similar.

One can finally extend the integral to open unbounded intervals such as (a, ∞). If $f : (a, \infty) \to \mathbb{R}$ is continuous, we say that

$$\int_a^\infty f(x) \cdot dx \quad \text{converges}$$

if, for some $c \in (a, \infty)$, both the integrals $\int_a^c f(x) \cdot dx$ and $\int_c^\infty f(x) \cdot dx$ converge. Notice for example that the integral $\int_0^\infty \frac{1}{x^\alpha} \cdot dx$ diverges for all α. The case of the integral of a continuous function on $(-\infty, b)$ is handled similarly. The appropriate comparison theorems also hold.

4.9 Additional Exercises

Exercise 4.1
Use the definition of the definite integral to compute $\int_0^1 x^2 \cdot dx$.

Exercise 4.2
For each of the following functions, find an antiderivative.

(1) $f : x \mapsto 3x^2 + 1$ (6) $f : x \mapsto |x|$

(2) $f : x \mapsto 4 - 3x^3$ (7) $f : x \mapsto x^2 + x^{-2}$

(3) $f : x \mapsto 7x^{-3}$ (8) $f : x \mapsto 4$

(4) $f : x \mapsto (x - 6)^2$ (9) $f : x \mapsto x$

(5) $f : x \mapsto x^{\frac{3}{2}}$ (10) $f : x \mapsto \frac{2}{x^2}$

Check your results by differentiating them.

Exercise 4.3
Use integration by parts to compute the following integrals.

(1) $\int x \cdot \cos(x) \cdot dx$

(3) $\int x^2 \cdot \sin(x) \cdot dx$

(2) $\int (\cos(x))^2 \cdot dx$

(4) $\int \sin(x) \cdot \cos(x) \cdot dx$

Exercise 4.4
Use variable substitution to evaluate the following integrals.

(1) $\int_0^{10} \frac{1}{(2x+2)^2} \cdot dx$

(8) $\int_0^1 \frac{x}{\sqrt{1-x^2}} \cdot dx$

(2) $\int (3 - 4x)^6 \cdot dx$

(9) $\int_1^2 \frac{x}{\sqrt{1-x^2}} \cdot dx$

(3) $\int_{-1}^1 2x \cdot \sqrt{1-x^2} \cdot dx$

(10) $\int_0^{10} x \cdot (x^2 + 3)^{-2} \cdot dx$

(4) $\int_a^b \sqrt{3x+1} \cdot dx$

(11) $\int_{\sqrt{6}}^5 x \cdot (x^2 + 2)^{\frac{1}{3}} \cdot dx$

(5) $\int \frac{4x}{(2 + 3x^2)^2} \cdot dx$

(12) $\int_{-1}^1 \frac{x^2}{(4 - x^3)^2} \cdot dx$

(6) $\int_{-2}^2 x \cdot (4 - 5x^2)^2 \cdot dx$

(13) $\int_1^2 \frac{1}{x^2 \cdot \sqrt{1 + \frac{1}{x}}} \cdot dx$

(7) $\int (1 - x)^{\frac{3}{2}} \cdot dx$

Exercise 4.5
Starting with $\ln(x) = 1 \cdot \ln(x)$, use integration by parts to compute $\int \ln(x) \cdot dx$.

Exercise 4.6
Find the length of the graph of $f(x) = x^2$ on $[0, 1]$.

Exercise 4.7

(1) Integrate the function $x \mapsto e^x$.
(2) Differentiate the function $x \mapsto \ln(\ln(x))$.
(3) Differentiate the function $x \mapsto \ln(x^a)$.
(4) Differentiate the function $x \mapsto \ln(a^x)$.

(5) Differentiate $x \mapsto e^{x^2}$.

(6) Using the fact that $u = e^{\ln(u)}$ (if $u > 0$) differentiate $x \mapsto a^x$ (for $a > 0$ and $x > 0$).

(7) Same idea: Differentiate the function $x \mapsto x^x$.

Exercise 4.8

Prove that the following statements are equivalent:

(1) R is observable relative to f, a, b, and $R \simeq \sum_{i=0}^{N-1} f(x_i) \cdot dx$ whenever $N \in \mathbb{N}$ is ultralarge, relative to f, a, b.

(2) $R \simeq \sum_{i=0}^{N-1} f(x_i) \cdot dx$ whenever $N \in \mathbb{N}$ is ultralarge, relative to f, a, b, R.

Hint: Imitate the proof of Theorem 15.

Exercise 4.9

Show that if f and g are continuous functions on the interval $[a, b]$ such that $f(x) \leq g(x)$ for all $x \in [a, b]$, then

$$\int_a^b f(x) \cdot dx \leq \int_a^b g(x) \cdot dx.$$

Exercise 4.10

Show that if $f : [a, b] \to \mathbb{R}$ is continuous, then

$$\left| \int_a^b f(x) \cdot dx \right| \leq \int_a^b |f(x)| \cdot dx.$$

Exercise 4.11

Prove: If f is continuous on $[a, b]$, $f(x) \geq 0$ for all $x \in [a, b]$ and $f(x) > 0$ for some $x \in [a, b]$, then $\int_a^b f(x) \cdot dx > 0$.

Exercise 4.12

Assume that $f, g : [a, b] \to \mathbb{R}$ are continuous and $g(x) \geq 0$ for all $x \in [a, b]$. Prove that there is $c \in [a, b]$ such that

$$\int_a^b f(x) \cdot g(x) \cdot dx = f(c) \cdot \int_a^b g(x) \cdot dx.$$

Exercise 4.13
Assuming that f is continuous on $[a, \infty)$ and $\int_a^\infty f(x) \cdot dx$ converges, prove that the function $x \mapsto \int_x^\infty f(x) \cdot dx$ is continuous on $[a, \infty)$.

Exercise 4.14
Compute the following integrals.

(1) $\displaystyle\int_0^\infty e^{-x} \cdot \cos(x) \cdot dx$

(2) $\displaystyle\int_0^\infty x^n \cdot e^{-x} \cdot dx \quad (n \in \mathbb{N})$

Exercise 4.15
Find the values of α for which the integral $\int_0^1 (1 - x)^{-\alpha} \cdot dx$ converges and evaluate it.

Part II

Higher Analysis

5

Basic Concepts Revisited

5.1 Real and Natural Numbers

We assume familiarity with the set of real numbers $\mathbb{R}$ and with the fundamental properties of addition, multiplication, and ordering of real numbers. They are summarized formally below.

(1) $a + (b + c) = (a + b) + c$, for all $a, b, c \in \mathbb{R}$.

(2) $a + b = b + a$, for all $a, b \in \mathbb{R}$.

(3) $a + 0 = a$, for all $a \in \mathbb{R}$.

(4) For every $a \in \mathbb{R}$ there exists an element $-a \in \mathbb{R}$ such that $a + (-a) = 0$.

(5) $a \cdot (b \cdot c) = (a \cdot b) \cdot c$, for all $a, b, c \in \mathbb{R}$.

(6) $a \cdot b = b \cdot a$, for all $a, b \in \mathbb{R}$.

(7) $a \cdot 1 = a$, for all $a \in \mathbb{R}$, and $0 \neq 1$.

(8) For every $a \in \mathbb{R}, a \neq 0$, there exists an element $1/a \in \mathbb{R}$ such that $a \cdot (1/a) = 1$.

(9) $a \cdot (b + c) = a \cdot b + a \cdot c$, for all $a, b, c \in \mathbb{R}$.

(10) If $a \leq b$ and $b \leq c$, then $a \leq c$.

(11) If $a \leq b$ and $b \leq a$, then $a = b$.

(12) For all $a, b \in \mathbb{R}$, either $a \leq b$ or $b \leq a$.

(13) If $a \leq b$, then $a + c \leq b + c$.

(14) If $a \leq b$ and $0 \leq c$, then $a \cdot c \leq b \cdot c$.

Many other familiar facts about real numbers can be deduced from these axioms. However, these axioms are not specific to real numbers. Any set with binary operations $+$ and $\cdot$, a binary relation $\leq$, and two distinguished elements 0 and 1, that satisfies these fourteen axioms, is called an *ordered field*. $\mathbb{R}$ is an ordered field, but so is the set of rational numbers $\mathbb{Q}$ (with $+, \cdot$ and $\leq$ restricted to it), and there are many other examples. The ordered field of real numbers is singled out by an additional property called *completeness*.

Definition 43. *Let A be a subset of* $\mathbb{R}$.

(1) We say that $c \in \mathbb{R}$ *is a* **supremum of** *A if c is observable and*

- *for each* $x \in A$, $c \geq x$;
- *there exists* $x \in A$ *such that* $x \simeq c$.

(2) We say that $c \in \mathbb{R}$ *is an* **infimum of** *A if c is observable and*

- *for each* $x \in A$, $c \leq x$;
- *there exists* $x \in A$ *such that* $x \simeq c$.

Observability and $\simeq$ should be taken relative to A (or, equivalently, relative to any context where A is observable). Notice that the supremum (respectively, infimum) is unique, if it exists. In fact, the supremum is the least upper bound, and the infimum is the greatest lower bound, on the set A. We show that this is the case for the supremum, the case for the infimum is similar.

Let c be the supremum of A and let d be an observable upper bound on A, that is, $x \leq d$ for each $x \in A$. By (2) we can find $x \in A$, $x \simeq c$. Then

$$c \simeq x \leq d, \quad \text{so} \quad c \leq d,$$

since c and d are observable. By Universal Closure, $c \leq d$ holds if d is any upper bound on A.

We write

$$c = \sup A \quad [\text{respectively}, \ c = \inf A]$$

to indicate that the supremum [respectively, infimum] of A is c.

Exercise 51 (Answer page 256)
Find $\sup A$ and $\inf A$ for $A = \left\{ \frac{1}{n} + \frac{1}{m} : n, m \geq 1 \right\}$.

Completeness Axiom

(1) *If* $A \subseteq \mathbb{R}$ *is a nonempty set bounded above, then* $\sup A$ *exists.*

(2) *If* $A \subseteq \mathbb{R}$ *is a nonempty set bounded below, then* $\inf A$ *exists.*

Textbooks of analysis usually postulate the Completeness Axiom, together with the fourteen "algebraic" axioms. We can actually prove it from the Neighbor Principle.

Theorem 82. *The ordered field of real numbers satisfies the Completeness Axiom.*

Proof. We only prove (1), as (2) is similar. In fact, (2) follows from (1) (exercise).

Let A be a nonempty set which is bounded above. Then there exists an observable b such that $x \leq b$, for each $x \in A$. Fix an observable a such that some $x \in A$ satisfies $a < x \leq b$ (this is possible since A is nonempty). Let N be an ultralarge positive number. Let $dx = (b - a)/N$ and $x_i = a + i \cdot dx$, for $i = 0, \dots, N$. There exists a first i such that

$$x_i \geq x \quad \text{for all } x \in A, \tag{5.1}$$

since $x_N = b$ satisfies (5.1), and $i > 0$, since $x_0 = a$ does not satisfy (5.1). As x_i is between a and b, it has an observable neighbor c.

Let $x \in A$ be observable. Then

$$c \simeq x_i \geq x,$$

so $c \geq x$. By the Universal Closure Principle, we have $c \geq x$ for all $x \in A$. Since x_i was the first satisfying (5.1), there exists some $x \in A$ such that

$$x_{i-1} < x \leq x_i,$$

hence $x \simeq c$. This shows that $c = \sup A$. $\qquad\square$

It can be proved that the ordered field satisfying the Completeness Axiom is uniquely determined, up to isomorphism. This is the ordered field of real numbers. We do not give the proof here.

Next, we single out the subset of $\mathbb{R}$ consisting of the natural numbers, and discuss the Principle of Mathematical Induction. The idea is to define $\mathbb{N}$ as the smallest set of real numbers that contains 0, and, with each n, also $n + 1$.

Definition 44. *A set $N \subseteq \mathbb{R}$ is **inductive** if*

(1) $0 \in N$;

(2) if $n \in N$, then $n + 1 \in N$, for every $n \in N$.

It is obvious that, for example, $\mathbb{R}$ and $\{x \in \mathbb{R} : x \geq 0\}$ are inductive sets. We define the *set of natural numbers*:

$$\mathbb{N} = \{x \in \mathbb{R} : x \in N \text{ for every inductive set } N\}.$$

It is easily proved that $\mathbb{N}$ itself is an inductive set (exercise). $\mathbb{N}$ is therefore the smallest inductive set: if N is any set such that $0 \in N$, and

$n + 1 \in N$ whenever $n \in N$, then $\mathbb{N} \subseteq N$; that is, all natural numbers are in N. This is the Principle of Mathematical Induction. We state it in the form in which it is often used when proving theorems.

Principle of Mathematical Induction

Let $\mathcal{P}(n)$ *be an internal statement (it may have additional parameters). If*

(1) $\mathcal{P}(0)$ *is true; and*

(2) $\mathcal{P}(n + 1)$ *is true whenever* $\mathcal{P}(n)$ *is true,*

then $\mathcal{P}(n)$ *is true for all natural numbers* n.

The proof consists in noticing that the internality of $\mathcal{P}$ guarantees the existence of the set $N = \{n \in \mathbb{N} : \mathcal{P}(n) \text{ is true}\}$ (via the Definition Principle), and (1) and (2) guarantee that N is inductive.

We conclude by pointing out that the Principle of Mathematical Induction is not applicable to external statements. For example, let $\mathcal{P}(n)$ be the statement "n is standard." Then (1) and (2) are true (Closure Principle), yet there exist natural numbers n that are not standard (Theorem 2).

We apply mathematical induction to show directly from the definition of continuity that all rational functions are continuous.

Example.   (1) $f_n : x \mapsto x^n$ is a continuous function on $\mathbb{R}$, for all $n \in \mathbb{N}$.

> *Proof.* Let $\mathcal{P}(n)$ be the statement "$f_n(x) = x^n$ is continuous on $\mathbb{R}$." Then $\mathcal{P}(n)$ is an internal statement, and we can proceed by mathematical induction.
>
> It is trivial to verify that $f_0(x) = 1$ and $f_1(x) = x$ are continuous functions. If $\mathcal{P}(n)$ is true, that is, $f_n(x) = x^n$ is continuous, then $f_{n+1}(x) = x^{n+1} = f_1(x) \cdot f_n(x)$ is continuous, so $\mathcal{P}(n+1)$ is true. The Principle of Mathematical Induction tells us that $\mathcal{P}(n)$ is true for all n.
>
> $\square$

(2) Let $p_n(x) = a_0 x^n + a_1 x^{n-1} + \ldots + a_{n-1} x + a_n$ be a polynomial of degree $n \in \mathbb{N}$. Then p_n is continuous on $\mathbb{R}$.

> *Proof.* We again proceed by induction. For $n = 0$, $p_0(x) = a_0$ is continuous. Let $p_{n+1}(x) = a_0 x^{n+1} + a_1 x^n + \ldots + a_n x + a_{n+1}$ be a polynomial of degree $n + 1$. We can write $p_{n+1}(x) =$

$a_0 x^{n+1} + p_n(x)$, where $p_n(x) = a_1 x^n + \ldots + a_{n+1}$. If p_n is continuous, then so is p_{n+1}. The Principle of Mathematical Induction implies that all polynomials are continuous. □

(3) All rational functions $f(x) = \frac{P(x)}{Q(x)}$, where $P(x), Q(x)$ are polynomials, are continuous at every x where $Q(x) \neq 0$.

Proof. Immediate. □

Exercise 52 (Answer page 256)
Show by induction on $n \in \mathbb{N}$ $(n > 0)$ that $\lim_{x \to 0^+} x^n = 0$ and $\lim_{x \to \infty} x^n = \infty$.

Exercise 53 (Answer page 257)
Show by induction on $n \in \mathbb{N}$ that $x \mapsto x^n$ is differentiable everywhere for each $n \in \mathbb{N}$, and $(x^n)' = n \cdot x^{n-1}$.

5.2 Epsilon–Delta Method

In this section we establish the equivalence between our definitions of limit, continuity and integral and those found in traditional textbooks. This material is not used anywhere else in the book.

In order to motivate the traditional approach, we consider the problem of computing values of limits numerically. In Section 4.6 we establish that

$$e = \lim_{x \to \infty} \left(1 + \frac{1}{x} \right)^x .$$

The number e is irrational (see Theorem 101); hence an infinite sequence of digits would have to be exhibited in order to specify it with complete accuracy. In practice we have to make do with *approximations* to the value of e that are *good enough* for the intended purpose. That is, we start with an observable *error tolerance* $\varepsilon > 0$ and we look for a number $\bar{e}$ such that the "error" $|e - \bar{e}|$ is less than this given ε.

The definition of limit suggests a way to obtain such an approximate value: If x is positive ultralarge and $\bar{e}_x = \left(1 + \frac{1}{x} \right)^x$, then $|e - \bar{e}_x|$ is ultrasmall, hence less than ε. This is not very helpful in practice, as we cannot do numerical calculations with ultrasmall numbers, but from the Closure Principle it follows that $|e - \bar{e}_x| < \varepsilon$ holds for all x that are merely *large enough*, that is, larger than some observable K.

We give a general theorem to this effect.

Theorem 83. *Assume that* $\lim_{x \to \infty} f(x) = L$. *Then for every* $\varepsilon > 0$ *there is* K *such that*

$$x > K \quad implies \quad |f(x) - L| < \varepsilon.$$

The number K *can be chosen to be observable relative to* f *and* ε.

Proof. Let ε be given; f and ε specify the context (L is observable relative to f). Let K be any ultralarge positive number. Then $x > K$ implies that x is ultralarge positive, hence $f(x) \simeq L$ by the definition of limit, and $|f(x) - L| \simeq 0 < \varepsilon$.

By Closure, if there is some K such that $x > K$ implies $|f(x) - L| < \varepsilon$ for all x, as we just proved, then there is an observable such K.   □

The converse of Theorem 83 is also true.

Theorem 84. *Assume that for every* $\varepsilon > 0$ *there is* K *such that*

$$x > K \quad implies \quad |f(x) - L| < \varepsilon.$$

Then $\lim_{x \to \infty} f(x) = L$.

Proof. Let z be positive ultralarge relative to f and L. If ε is observable and $\varepsilon > 0$, then, by Closure, there is an observable K such that $x > K$ implies $|f(x) - L| < \varepsilon$. We have $z > K$ because z is ultralarge positive, so $|f(z) - L| < \varepsilon$. As this is true for every observable $\varepsilon > 0$, $|f(z) - L| \simeq 0$, that is, $f(z) \simeq L$. This proves that $\lim_{z \to \infty} f(z) = L$.   □

The number K of course depends on ε; generally, the better accuracy (the smaller ε) one desires, the larger K has to be. Theorem 83 guarantees the *existence* of a suitable K; for practical calculations one would like to be able to determine K as a function of ε. We do not pursue this matter here, but see Section 7.3, and also Section 4.7. Instead, we wish to point out that the two theorems together provide a description of limit at infinity that does not refer to contexts, either explicitly or implicitly:

$$\lim_{x \to \infty} f(x) = L \tag{5.2}$$

holds if and only if

For all $\varepsilon > 0$ there is K such that $x > K$ implies $|f(x) - L| < \varepsilon$. (5.3)

In traditional mathematical textbooks, the statement (5.3) is used as the *definition* of $\lim_{x \to \infty} f(x) = L$, in place of our Definition 14.

Other limits can be handled in a similar way. In the case of $\lim_{x \to a} f(x)$, $|f(x) - L| < \varepsilon$ holds for all x that are *sufficiently close* to a, as measured by the distance $|x - a|$.

Theorem 85. *Assume that* $\lim_{x \to a} f(x) = L$. *Then for every* $\varepsilon > 0$ *there is* $\delta > 0$ *such that*

$$0 < |x - a| < \delta \quad \text{implies} \quad |f(x) - L| < \varepsilon.$$

The number δ *can be chosen to be observable relative to* f, a *and* ε.
Conversely, assume that for every $\varepsilon > 0$ *there is* $\delta > 0$ *such that*

$$0 < |x - a| < \delta \quad \text{implies} \quad |f(x) - L| < \varepsilon.$$

Then $\lim_{x \to a} f(x) = L$.

Proof. Assume that $\lim_{x \to a} f(x) = L$. Let $\delta > 0$ be ultrasmall relative to f, a, ε. Then $0 < |x - a| < \delta$ implies that $x \simeq a$ and $x \neq a$, hence $f(x) \simeq L$ and $|f(x) - L| \simeq 0 < \varepsilon$, so δ has the required properties. By Closure, there is an observable δ with the required properties.

Conversely, let $z \simeq a$, $z \neq a$, relative to f, a, L. For every observable $\varepsilon > 0$ there is $\delta > 0$ such that $0 < |x - a| < \delta$ implies $|f(x) - L| < \varepsilon$; by Closure, we can find an observable such δ. Then $0 < |z - a| < \delta$, so $|f(z) - L| < \varepsilon$. As the last inequality holds for all observable $\varepsilon > 0$, we conclude that $f(z) \simeq L$. This proves that $\lim_{x \to a} f(x) = L$. $\qquad \square$

In summary, the theorem shows that

$$\lim_{x \to a} f(x) = L$$

if and only if

For every $\varepsilon > 0$ *there is* $\delta > 0$ *such that* $0 < |x - a| < \delta$ *implies* $|f(x) - L| < \varepsilon$.

The last statement does not refer to observability; it is a statement of traditional mathematics. This is the notorious epsilon–delta definition of limit, typically used to define limits in contemporary textbooks. We believe that the approach based on ultrasmall numbers is more natural and significantly simpler to work with; this is our main reason for writing this book.

We recall that a function f is continuous at a if and only if $\lim_{x \to a} f(x) = f(a)$. Theorem 85 applied to this limit gives the definition of continuity found in traditional textbooks.

Theorem 86. *A function* f *is continuous at* a *if and only if for every* $\varepsilon > 0$ *there is* $\delta > 0$ *such that* $|x - a| < \delta$ *implies* $|f(x) - f(a)| < \varepsilon$.

It follows that f is continuous on an interval I if and only if for every $a \in I$ and every $\varepsilon > 0$ there is $\delta > 0$ such that, for all $x \in I$, $|x - a| < \delta$ implies $|f(x) - f(a)| < \varepsilon$. We note that the value of δ generally depends on the point $a \in I$, as well as the error tolerance ε. We show next that uniformly continuous functions on I are precisely those functions where δ can be chosen independent of a.

Theorem 87. *A function f is uniformly continuous on I if and only if for every $\varepsilon > 0$ there is $\delta > 0$ such that for all $a \in I$ and all $x \in I$, $|x - a| < \delta$ implies $|f(x) - f(a)| < \varepsilon$.*

Proof. Assume that f is uniformly continuous on I and f, I and ε are observable. Let $\delta > 0$ be ultrasmall. Then for all $a, x \in I$, $|x - a| < \delta$ implies that $x \simeq a$, so $f(x) \simeq f(a)$ by Definition 9, and $|f(x) - f(a)| < \varepsilon$.

For the converse, let f and I be observable, $a, z \in I$, and $z \simeq a$. By Closure, given any observable $\varepsilon > 0$, there is an observable $\delta > 0$ such that $|x - a| < \delta$ implies $|f(x) - f(a)| < \varepsilon$. As $|z - a| < \delta$, we have $|f(z) - f(a)| < \varepsilon$, for every observable $\varepsilon > 0$. It follows that $f(z) \simeq f(a)$. $\square$

We conclude this section with a characterization of the definite integral for continuous functions in a way that does not refer to contexts. The proof is left as an exercise; see Theorem 76 in Chapter 4.

Theorem 88. *Let f be a continuous function on $[a, b]$. The following are equivalent:*

(1) $\displaystyle \int_a^b f(x) \cdot dx = R.$

(2) For every $\varepsilon > 0$ there exists K such that $n > K$ implies

$$\left| \sum_{i=0}^{n-1} f(x_i) \cdot h \; - \; R \right| < \varepsilon,$$

where $h = (b - a)/n$ and $x_i = a + i \cdot h$, for $i = 0, 1, \cdots, n$.

5.3   Alternative Characterization of Limits

It is sometimes helpful to work, in the same proof, relative to two contexts: that given by the parameters of the theorem being proved, and an auxiliary extended context that includes some additional parameters. In this situation, as always, the relative concepts are to be taken relative to the parameters of the theorem; we use $\overset{+}{\simeq}$ to indicate ultracloseness relative to the extended context.

Assume that a function f has a limit L at a. Then L is observable and $f(x) \simeq L$ whenever $x \simeq a$, $x \neq a$ (the context is specified by f and a). It is immediate by Stability that we have $f(x) \overset{+}{\simeq} L$ whenever $x \overset{+}{\simeq} a$,

$x \neq a$. As $f(x) \overset{+}{\simeq} L$ implies $f(x) \simeq L$, we also have

$$f(x) \simeq L \text{ whenever } x \overset{+}{\simeq} a, \; x \neq a.$$

The converse may be less intuitive and is a consequence of Stability due to Péraire [22, Partial Transfer]; it is a unique and very useful feature of the relative framework.

Theorem 89. *A function f has a limit at a if and only if there is an observable L such that*

$$f(x) \simeq L \text{ whenever } x \overset{+}{\simeq} a, \; x \neq a.$$

Proof. It remains to prove that "$f(x) \simeq L$ whenever $x \overset{+}{\simeq} a$, $x \neq a$" implies "$f(x) \simeq L$ whenever $x \simeq a$, $x \neq a$."

The parameters are f, a and L. Assume that $f(x) \simeq L$ whenever $x \overset{+}{\simeq} a$, $x \neq a$, where the symbol $\overset{+}{\simeq}$ refers to some extended context. $f(x) \simeq L$ means that for every observable $d > 0$ we have $|f(x) - L| < d$. Hence the assumption can be restated as

For all observable d and for all $x \neq a$, $x \overset{+}{\simeq} a$ implies $|f(x) - L| < d$.
$$(5.4)$$

For every observable d, the statement

For all $x \neq a$, $x \overset{+}{\simeq} a$ implies $|f(x) - L| < d$

is equivalent to the statement

For all $x \neq a$, $x \simeq a$ implies $|f(x) - L| < d$,

by Stability. Substituting this into (5.4) we obtain

For all observable d and for all $x \neq a$, $x \simeq a$ implies $|f(x) - L| < d$,
$$(5.5)$$

that is,

For all $x \neq a$, $x \simeq a$ implies $f(x) \simeq L$.

$\square$

Notice that Stability is in fact *applied only to a part* of the statement (5.4). Similar arguments work for other types of limits, for example

$$\lim_{x \to +\infty} f(x) = L \text{ if and only if } x \overset{+}{\simeq} +\infty \text{ implies } f(x) \simeq L.$$

5.4   Additional Exercises

Exercise 5.1
Define $|a| = \max\{a, -a\}$ and prove that $|a + b| \leq |a| + |b|$ from the axioms 1–14 in Section 5.1.
Hint: Consider separately the cases $a + b \geq 0$ and $a + b < 0$.

Exercise 5.2
Prove that $|x - a| < \varepsilon$ if and only if $a - \varepsilon < x < a + \varepsilon$.

Exercise 5.3
Prove that the set $\{r + s \cdot \sqrt{2} : r, s \in \mathbb{Q}\}$, with the usual operations and ordering, is an ordered field.
Hint: It suffices to verify that the set is closed under the operations $+$ and $\cdot$ and the axioms (4) and (8) are satisfied.

Exercise 5.4
Find $\sup A$ and $\inf A$ for $A = [a, b]$ and $A = (a, b)$.

Exercise 5.5
Find $\sup A$ and $\inf A$ for $A = \{x : x^2 + 2x < 4\}$.

Exercise 5.6
Prove: If a set A has a least element a, then $\inf A = a$. Conversely, if $\inf A = a$ exists and $a \in A$, then a is the least element of A.

Exercise 5.7
Prove that the following statements are equivalent.

(1) $c \in \mathbb{R}$ is a supremum of A.

(2) For each $x \in A$, $c \geq x$, and for each $\varepsilon > 0$ there exists $x \in A$ such that $c - \varepsilon < x$.

Exercise 5.8
For A nonempty and bounded, prove that $\sup A - \inf A = \sup\{x - y : x, y \in A\}$.

Exercise 5.9
Let A, B be nonempty and bounded. Assume that for every $x \in A$ there is $y \in B$ such that $x < y$. Prove that $\sup A \leq \sup B$. Is it true that $\sup A < \sup B$?

Exercise 5.10
Let A, B be nonempty. Assume that $x < y$ holds for every $x \in A$ and every $y \in B$. Prove that $\sup A \leq \inf B$. Is it true that $\sup A = \inf B$?

Exercise 5.11
Show that $c = \sup A$ is equivalent to the internal statement
"For each $x \in A$, $c \geq x$, and there exists $x \in A$ such that $x \simeq c$, relative to A, c."
Hint: Follow the proof of Theorem 15.

Exercise 5.12
Prove the Archimedean property of $\mathbb{R}$:
Given $\varepsilon, a > 0$, there is $n \in \mathbb{N}$ such that $n \cdot \varepsilon > a$.
Hint: Assume the contrary, consider $\sup\{n \cdot \varepsilon : n \in \mathbb{N}\}$ and deduce a contradiction.

Exercise 5.13
Prove that every nonempty set A of natural numbers has a least element using

(1) The Completeness Axiom.

(2) The Principle of Mathematical Induction.

   Hint: Prove by induction that every nonempty subset of n has a least element, for every $n \in \mathbb{N}$.

Exercise 5.14
Let f be a continuous function on $[a, b]$ and $f(a) < d < f(b)$. Prove that there is a $c \in [a, b]$ such that $f(c) = d$ (Theorem 8) using the notion of supremum.
Hint: Consider $S = \{x \in [a, b] : f(y) \leq d \text{ for all } a \leq y \leq x\}$.
Compare your proof with the one given in Section 2.2.

Exercise 5.15
Prove the Extreme Value Theorem (Theorem 9) using the notion of supremum.

Exercise 5.16
Show that if f is increasing at a for every a in the open interval I, then f is increasing on I.

Exercise 5.17
Prove that $\lim_{x \to \infty} c \cdot f(x) = c \cdot \lim_{x \to \infty} f(x)$ first using Definition 14; then directly from (5.3), page 148.

Exercise 5.18
Repeat Exercises 5.17 and 5.19 for $\lim_{x \to a}$.

Exercise 5.19
Prove that $\lim_{x \to \infty}(f(x) + g(x)) = \lim_{x \to \infty} f(x) + \lim_{x \to \infty} g(x)$

(1) using Definition 14;

(2) directly from (5.3), page 148.

Exercise 5.20
Prove that $\lim_{x \to a} f(x) = \infty$ if and only if $f(x)$ is positive ultralarge whenever $x \overset{+}{\simeq} a$, $x \neq a$.

Exercise 5.21
Prove that the following statements are equivalent.

(1) For every $\varepsilon > 0$ there is $\delta > 0$ such that $|x - a| < \delta$ implies $|f(x) - f(a)| < \varepsilon$.

(2) For every observable $\varepsilon > 0$ there is $\delta > 0$ such that $|x - a| < \delta$ implies $|f(x) - f(a)| < \varepsilon$.

(3) For every observable $\varepsilon > 0$ there is an observable $\delta > 0$ such that $|x - a| < \delta$ implies $|f(x) - f(a)| < \varepsilon$.

6

L'Hôpital's Rule and Higher Order Derivatives

6.1 L'Hôpital's Rule

The following theorems are known collectively as L'Hôpital's Rule. They are extensions of the simple case proved in Section 3.3 (Theorem 41). They are useful for theoretical purposes, and also come in handy to evaluate some limits of indeterminate forms, such as $\frac{0}{0}$ or $\frac{\infty}{\infty}$. After prior preparation, they can also be used for other indeterminate forms, such as $0 \cdot \infty$.

Theorem 90 (L'Hôpital's Rule for 0/0 – General Form). *Let f and g be differentiable in a deleted neighborhood of a. Suppose that $\lim\limits_{x \to a} f(x) = 0$, $\lim\limits_{x \to a} g(x) = 0$, and $\lim\limits_{x \to a} \dfrac{f'(x)}{g'(x)}$ exists. Then*

$$\lim_{x \to a} \frac{f(x)}{g(x)} = \lim_{x \to a} \frac{f'(x)}{g'(x)}.$$

Proof. The parameters are f, g and a. Let L be observable such that

$$\lim_{x \to a} \frac{f'(x)}{g'(x)} = L.$$

We have to show that

$$\frac{f(x)}{g(x)} \simeq L \quad \text{for each } x \simeq a, \ x \neq a.$$

Let $x \simeq a$ and assume that $x > a$ (the case $x < a$ is similar). Consider the extended context of f, g, a and x, and fix $y > a$ such that $y \overset{+}{\simeq} a$ (that is, relative to the extended context). We necessarily have

$$a < y < x.$$

By Cauchy's Theorem (Theorem 39), there is $c \in (y, x)$ such that

$$\Big(f(x) - f(y)\Big) \cdot g'(c) = \Big(g(x) - g(y)\Big) \cdot f'(c).$$

Since $\lim_{x \to a} \frac{f'(x)}{g'(x)} = L$ exists, $g'(z) \neq 0$ for all $z \simeq a$, $z \neq a$. Hence $g'(c) \neq 0$ and $g(x) - g(y) \neq 0$ (otherwise, $g'(z) = 0$ for some $z \in (y, x)$, by Rolle's Theorem). Thus, we have:

$$\frac{f(x) - f(y)}{g(x) - g(y)} = \frac{f'(c)}{g'(c)} \simeq L.$$

But $\lim_{x \to a} f(x) = 0$ and $\lim_{x \to a} g(x) = 0$, so $f(y) \overset{+}{\simeq} 0$ and $g(y) \overset{+}{\simeq} 0$, which implies by Rule 5 that

$$\frac{f(x) - f(y)}{g(x) - g(y)} \overset{+}{\simeq} \frac{f(x)}{g(x)}.$$

Hence

$$\frac{f(x) - f(y)}{g(x) - g(y)} \simeq \frac{f(x)}{g(x)}, \quad \text{so} \quad \frac{f(x)}{g(x)} \simeq L.$$

$\square$

Theorem 90 remains valid if $\lim_{x \to a} \frac{f'(x)}{g'(x)} = \pm\infty$; we leave the proof as an exercise.

In many introductory courses L'Hôpital's Rule in the ∞/∞ case is not proved (or only under restricted conditions). We can in fact prove the more general theorem not assuming that the numerator is ∞.

Theorem 91 (L'Hôpital's Rule for ∞/∞). *Let f and g be differentiable in a deleted neighborhood of a. Suppose that $\lim_{x \to a} |g(x)| = \infty$ and $\lim_{x \to a} \dfrac{f'(x)}{g'(x)}$ exists. Then*

$$\lim_{x \to a} \frac{f(x)}{g(x)} = \lim_{x \to a} \frac{f'(x)}{g'(x)}.$$

Proof. The proof of Theorem 90 refers to two contexts. We start with the one given by f, g, a, then fix x such that $x \simeq a$ and consider the extended context f, g, a, x. The situation here is similar except that we start with the extended context and use Theorem 89 to reach the conclusion.

The parameters are f, g and a. Let L be observable and such that

$$\lim_{x \to a} \frac{f'(x)}{g'(x)} = L.$$

We choose $y > a$, $y \simeq a$, and then extend the context to include y. Let $x \overset{+}{\simeq} a$ ($x \neq a$) (that is, relative to this extended context). Assume that

$x > a$ (the case $x < a$ is similar). By Theorem 89 it is enough to show that

$$\frac{f(x)}{g(x)} \simeq L.$$

Necessarily $a < x < y$. By Cauchy's Theorem, just as in the previous proof, we have

$$\frac{f(x) - f(y)}{g(x) - g(y)} = \frac{f'(c)}{g'(c)}, \quad \text{for some } c \in (x, y).$$

But $c \simeq a$, so

$$\frac{f'(c)}{g'(c)} \simeq L.$$

We obtain

$$\frac{f(x) - f(y)}{g(x) - g(y)} = \frac{f'(c)}{g'(c)} \simeq L.$$

Since $\lim_{x \to a} |g(x)| = +\infty$ and $x \overset{+}{\simeq} a$, we have $g(x) \overset{+}{\simeq} \pm\infty$. Hence we have $f(y)/g(x) \overset{+}{\simeq} 0$ and $g(y)/g(x) \overset{+}{\simeq} 0$. By Rule 5 we deduce that

$$\frac{f(x) - f(y)}{g(x)} = \frac{f(x)}{g(x)} - \frac{f(y)}{g(x)} \overset{+}{\simeq} \frac{f(x)}{g(x)}.$$

For the same reason,

$$\frac{g(x)}{g(x) - g(y)} = \frac{1}{\dfrac{g(x) - g(y)}{g(x)}} = \frac{1}{1 - \dfrac{g(y)}{g(x)}} \overset{+}{\simeq} 1.$$

It follows (see Exercise 6) that

$$L \simeq \frac{f(x) - f(y)}{g(x) - g(y)} = \frac{f(x) - f(y)}{g(x)} \cdot \frac{g(x)}{g(x) - g(y)} \overset{+}{\simeq} \frac{f(x)}{g(x)}.$$

Hence

$$L \simeq \frac{f(x)}{g(x)}.$$

$\square$

This use of Theorem 89 introduces a specific way of working with contexts. Start with an extended context. If the proof refers only to this context, Stability ensures the validity of the result. If an "order of magnitude" is lost during the proof, then it is Theorem 89 which ensures the validity of the result.

Theorem 91 also remains valid if $\lim_{x \to a} \frac{f'(x)}{g'(x)} = \pm\infty$.

Exercise 54 (Answer page 257)
Prove L'Hôpital's Rule for $\lim_{x \to \pm\infty}$.

Exercise 55 (Answer page 257)
Let $z \in \mathbb{R}$. Prove that

$$\lim_{x \to \infty} \left(1 + \frac{z}{x}\right)^x = e^z$$

using L'Hôpital's Rule.

6.2 Higher Order Derivatives

Let f be a function and I an interval; then the collection

$$J = \{x \in I : f \text{ is differentiable at } x\}$$

is a set, and it is observable whenever f and I are observable. This follows from the Definition Principle, since the defining statement "f is differentiable at x" is internal. For the same reason, the rule

$$x \mapsto f'(x)$$

defines a function $f' : J \to \mathbb{R}$, observable in the context of f and I. We can therefore consider its derivative, the derivative of its derivative, and so on. We proceed inductively, which is possible since the definition of differentiability is internal.

Definition 45. *Let f be a continuous function. We write $f^{(0)}$ for f. By induction, we say that f is **differentiable** $n + 1$ **times at** x if the function $f^{(n)}$ is differentiable at x. In particular, the function $f^{(n)}$ has to be defined on some open interval about x. We write $f^{(n+1)}(x) = (f^{(n)})'(x)$. The number $f^{(n)}(x)$ is called the **derivative of order** n at x.*

We found a linear polynomial that best approximates a differentiable function, and a quadratic polynomial that best approximates a twice differentiable function. We now generalize these ideas to polynomials of an arbitrary degree.

Assume that f is n times differentiable at a. We look for a polynomial

$$p_n(x) = b_0 + b_1(x - a) + \ldots + b_n(x - a)^n$$

such that $p_n^{(k)}(a) = f^{(k)}(a)$ for all $k = 0, \ldots, n$. An easy computation gives

$$
\begin{aligned}
p_n^{(k)}(x) = {} & k \cdot (k-1) \cdots 2 \cdot 1 \cdot b_k \\
& + (k+1) \cdot k \cdots 3 \cdot 2 \cdot b_{k+1} \cdot (x-a) + \ldots \\
& + n \cdot (n-1) \cdots (n-k+1) \cdot b_n \cdot (x-a)^{n-k},
\end{aligned}
$$

and thus

$$
p_n^{(k)}(a) = k \cdot (k-1) \cdots 2 \cdot 1 \cdot b_k = k! \cdot b_k.
$$

It follows that necessarily

$$
b_k = \frac{f^{(k)}(a)}{k!}.
$$

Definition 46. *Assume that f is differentiable n times at a. The polynomial*

$$
T_n(x) = \sum_{k=0}^{n} \frac{f^{(k)}(a)}{k!} \cdot (x-a)^k
$$

*is called the **Taylor polynomial (of degree n) for f at a.***

Theorem 92 (Increment Equation of Order n). *Let n be a nonnegative integer and let $a \in \mathbb{R}$. If f is a function differentiable n times at a, then for all $x \simeq a$,*

$$
f(x) = \sum_{k=0}^{n} \frac{f^{(k)}(a)}{k!} \cdot (x-a)^k + \varepsilon \cdot (x-a)^n,
$$

where $\varepsilon \simeq 0$.

Equivalently, we may write

$$
f(a+dx) = \sum_{k=0}^{n} \frac{f^{(k)}(a)}{k!} \cdot dx^k + \varepsilon \cdot dx^n, \quad \text{for } dx \text{ ultrasmall.}
$$

Proof. The theorem asserts that

$$
\lim_{x \to a} \frac{f(x) - T_n(x)}{(x-a)^n} = 0.
$$

We prove this for all functions n-times differentiable at a, by induction on n.

For $n = 0$, $T_0(x) = b_0 = f(a)$, and the theorem reduces to $\lim_{x \to a}(f(x) - f(a)) = 0$; this is equivalent to continuity of f at a. For $n = 1$ the theorem is just the Increment Equation.

Let $n \geq 1$. Assume that the theorem has been proved for n and that f is $(n+1)$-times differentiable at a. Observe that

$$T'_{n+1}(x) = \sum_{k=1}^{n+1} \frac{f^{(k)}(a)}{(k-1)!}(x-a)^{k-1} = \sum_{k=0}^{n} \frac{f^{(k+1)}(a)}{k!}(x-a)^k,$$

so T'_{n+1} is nothing but the Taylor polynomial of order n for the n-times differentiable function f'. By inductive hypothesis,

$$\lim_{x \to a} \frac{f'(x) - T'_{n+1}(x)}{(x-a)^n} = 0.$$

Hence, applying L'Hôpital's Rule, we have

$$\lim_{x \to a} \frac{f(x) - T_{n+1}(x)}{(x-a)^{n+1}} = \lim_{x \to a} \frac{f'(x) - T'_{n+1}(x)}{(n+1)(x-a)^n} = 0.$$

$\square$

The Taylor polynomial is the only polynomial satisfying the Increment Equation of order n.

Exercise 56 (Answer page 258)
Let f be n times differentiable at a. Let $p(x) = \sum_{k=0}^{n} c_k(x-a)^k$ and assume that for all $x \simeq a$ there is $\varepsilon \simeq 0$ such that

$$f(x) = p(x) + \varepsilon \cdot (x-a)^n.$$

Show that $c_k = \frac{f^k(a)}{k!}$, for all $k = 0, \ldots, n$.

Theorem 93. *Let f be differentiable n times at a $(n > 0)$, and*

$$f'(a) = f''(a) = \ldots = f^{(n-1)}(a) = 0.$$

(1) If n is odd and $f^{(n)}(a) > 0$, then f is increasing at a.

(2) If n is odd and $f^{(n)}(a) < 0$, then f is decreasing at a.

(3) If n is even and $f^{(n)}(a) > 0$, then f is bending upward at a.

(4) If n is even and $f^{(n)}(a) < 0$, then f is bending downward at a.

Proof. We prove (1); the other cases are similar.
Let $x \simeq a$. Then by the Increment Equation of order n we have

$$f(x) - f(a) = \left(f^{(n)}(a) + \varepsilon\right) \cdot (x-a)^n,$$

with $\varepsilon \simeq 0$. If $f^{(n)}(a) > 0$, then $f^{(n)}(a) + \varepsilon > 0$. If also n is odd, then $(x-a)^n$, and hence $f(x) - f(a)$, is positive for $x > a$, and negative for $x < a$. $\square$

6.3 Additional Exercises

Exercise 6.1
Calculate the following limits.

(1) $\lim\limits_{x\to 0} \dfrac{x}{e^x - 1}$

(2) $\lim\limits_{x\to 0} \dfrac{\sin(x)}{\arcsin(x)}$

(3) $\lim\limits_{x\to \frac{\pi}{2}+} \dfrac{(\cos(x))^2}{x - \frac{\pi}{2}}$

(4) $\lim\limits_{x\to 0} \dfrac{\sin(x) - x}{\cos(x) - 1}$

(5) $\lim\limits_{x\to \infty} \dfrac{x^2}{e^{2x}}$

(6) $\lim\limits_{x\to 0+} x^3 \cdot \ln(x)$

(7) $\lim\limits_{x\to 0} \left(\dfrac{1}{x} - \dfrac{1}{\ln(x+1)} \right)$

(8) $\lim\limits_{x\to 0+} x^x$

(9) $\lim\limits_{x\to \infty} x^{\sin(\frac{1}{x})}$

(10) $\lim\limits_{x\to 1^-} \left(\cos \left(\dfrac{\pi x}{2} \right) \right)^{\ln(x)}$

(11) $\lim\limits_{x\to \infty} \left(1 + \dfrac{a}{x} \right)^x$

(12) $\lim\limits_{x\to \infty} x \cdot \left(\sqrt{x^2 + 2} - \sqrt{x^2 + 1} \right)$

Exercise 6.2
Prove the original version of L'Hôpital's Rule for ∞/∞, that is, assuming in addition that $\lim_{x\to a} |f(x)| = \infty$.

Exercise 6.3
Prove that both L'Hôpital's Rules remain valid when

$$\lim_{x\to a} \frac{f'(x)}{g'(x)} = \pm\infty.$$

Hint: For the ∞/∞ case use Exercise 5.20.

Exercise 6.4
Let $f(x) = x + x^2 \cdot \sin\left(\frac{1}{x}\right)$ and $g(x) = x + \sin(x)$. Show that $\lim_{x\to 0} \frac{f(x)}{g(x)} = \frac{1}{2}$, but $\lim_{x\to 0} \frac{f'(x)}{g'(x)}$ does not exist.

Exercise 6.5
Assume that f and g are differentiable n times at a, and $f^{(k)}(a) = g^{(k)}(a) = 0$ for all $0 \le k \le n - 1$, while $g^{(n)}(a) \neq 0$. Prove that

$$\lim_{x\to a} \frac{f(x)}{g(x)} = \frac{f^{(n)}(a)}{g^{(n)}(a)}.$$

Exercise 6.6
Show that $(\sin(x))^{(n)} = \sin(x + \frac{n\pi}{2})$, for all $n \in \mathbb{N}$.

Exercise 6.7
Assume that f is differentiable n times. Show that if $g(x) = f(ax + b)$, then $g^{(n)}(x) = a^n \cdot f^{(n)}(ax + b)$.

Exercise 6.8
Prove Leibniz's Rule:

$$(f \cdot g)^{(n)} = \sum_{k=0}^{n} \binom{n}{k} f^{(n-k)} \cdot g^{(k)},$$

provided f and g are differentiable n times.

7

Sequences and Series

7.1 Sequences

Definition 47. *A **sequence** is a function*

$$u : \{k, k+1, \dots\} \subseteq \mathbb{N} \longrightarrow \mathbb{R}.$$

We also use the notation

$$(u_n)_{n \geq k}$$

for the sequence above, with $u_n = u(n)$, for $n \geq k$. We occasionally write (u_n) if the set of indices is obvious or irrelevant. The numbers u_n are called the **terms** of the sequence.

The context of a sequence is the list of parameters used in its definition; in particular it includes the integer k.

Example. Let a and d be two real numbers and let k be a nonnegative integer. We define an **arithmetic progression** (with **common difference** d) as follows:

$$u_k = a \quad \text{and} \quad u_{n+1} = u_n + d \quad \text{for} \quad n \geq k.$$

It is immediate that $u_n = a + (n - k) \cdot d$, for all $n \geq k$. The parameters of this definition are a, d and k.

Example. In a similar way, given $a, r \in \mathbb{R}$ and $k \in \mathbb{N}$, we define a **geometric progression** (with **common ratio** r) by

$$u_k = a \quad \text{and} \quad u_{n+1} = u_n \cdot r \quad \text{for} \quad n \geq k.$$

Then $u_n = a \cdot r^{n-k}$ for all $n \geq k$.

Definition 48. *Let $(u_n)_{n \geq k}$ be a sequence. We say that $(u_n)_{n \geq k}$ **converges** if there is an observable real number L such that, for each ultralarge $N \in \mathbb{N}$, we have $u_N \simeq L$. We then write*

$$\lim_{n \to \infty} u_n = L.$$

*The number L is the **limit** of the sequence.*

*If a sequence does not converge, we say that it **diverges**. In particular, if $u_N \simeq \infty$ for all ultralarge N, we say that (u_n) diverges to ∞ and write $\lim_{n\to\infty} u_n = \infty$. Similarly, if $u_N \simeq -\infty$ for all ultralarge N, we say that it diverges to $-\infty$ and write $\lim_{n\to\infty} u_n = -\infty$.*

The limit L is unique and is observable in the context specified by (u_n). Moreover, by Stability, the sequence $(u_n)_{n \geq k}$ converges to L if and only if relative to any extended context, for each N ultralarge we have $u_N \simeq L$. In particular, the convergence and the value of the limit are independent of the first few terms (see the next exercise). This is the reason why we often omit k.

Exercise 57 (Answer page 258)
Let $(u_n)_{n \geq k}$ and $(u'_n)_{n \geq k'}$ be sequences such that $u_n = u'_n$ for all sufficiently large values of n. Then $(u_n)_{n \geq k}$ converges if and only if $(u'_n)_{n \geq k'}$ converges, and if that is the case, then

$$\lim_{n\to\infty} u_n = \lim_{n\to\infty} u'_n.$$

Here are some results similar to those on limits of functions.

Theorem 94. *Let (u_n) an (v_n) be convergent sequences. Then*

(1) $\lim_{n\to\infty} (u_n + v_n) = \lim_{n\to\infty} u_n + \lim_{n\to\infty} v_n.$

(2) $\lim_{n\to\infty} (u_n - v_n) = \lim_{n\to\infty} u_n - \lim_{n\to\infty} v_n.$

(3) $\lim_{n\to\infty} (u_n \cdot v_n) = (\lim_{n\to\infty} u_n) \cdot (\lim_{n\to\infty} v_n).$

(4) If $\lim_{n\to\infty} v_n \neq 0$*, then* $\lim_{n\to\infty} \dfrac{u_n}{v_n} = \dfrac{\lim_{n\to\infty} u_n}{\lim_{n\to\infty} v_n}.$

Theorem 95. *If $\lim_{n\to\infty} u_n = u$ and f is a function continuous at u, then $\lim_{n\to\infty} f(u_n) = f(u)$.*

Exercise 58 (Answer page 259)
Prove Cèsaro's Theorem:
Let $(u_n)_{n \geq k}$ be a sequence converging to L. Then the sequence $(s_n)_{n \geq k}$, defined by

$$s_n = \frac{u_1 + u_2 + \cdots + u_n}{n}, \quad \text{for } n \geq 1,$$

converges to L.

Example. Let $f : [a, b] \to \mathbb{R}$ be a continuous function. Let n be a positive integer, $dx = (b - a)/n$, and $x_i = a + i \cdot dx$.

Define u_n for $n \geq 1$ by

$$u_n = \sum_{i=0}^{n-1} f(x_i) \cdot dx.$$

Then the sequence (u_n) converges to

$$\int_a^b f(x) \cdot dx.$$

So another way of writing the definition of the integral is

$$\int_a^b f(x) \cdot dx = \lim_{n \to \infty} \sum_{i=0}^{n-1} f(x_i) \cdot dx.$$

Definition 49. *The sequence $(u_n)_{n \geq k}$ is:*

*(1) **increasing** if $u_n \leq u_m$ for all $k \leq n \leq m$;*

*(2) **decreasing** if $u_n \geq u_m$ for all $k \leq n \leq m$;*

*(3) **monotone** if it is either increasing or decreasing;*

*(4) **bounded above** if there is an $M \in \mathbb{R}$ such that $u_n \leq M$ for all $n \geq k$ (the number M is an **upper bound**);*

*(5) **bounded below** if there is an $M \in \mathbb{R}$ such that $u_n \geq M$ for all $n \geq k$ (the number M is a **lower bound**);*

*(6) **bounded** if the sequence is bounded above and also bounded below.*

Let $(u_n)_{n \geq k}$ be a sequence. If it is bounded above, then by Closure there is an observable M which is also an upper bound. Conversely, if there is an observable M such that

$$u_n \leq M \text{ for all observable } n,$$

then by Closure this statement is true for all integers (including ultralarge integers). The same remark holds for lower bounds.

Definition 50. *We say that $(u_n)_{n \geq k}$ is a **Cauchy sequence** if*

$$u_{N'} \simeq u_N \quad \text{for all positive ultralarge integers } N, N'.$$

Theorem 96. *A sequence $(u_n)_{n \geq k}$ converges if and only if it is a Cauchy sequence.*

Proof. Assume first that $\lim_{n \to \infty} u_n = L$. Let N and N' be ultralarge. Then $u_N \simeq L$ and $u'_N \simeq L$, hence $u_N \simeq u_{N'}$. Hence (u_n) is a Cauchy sequence.

For the converse, assume that $(u_n)_{n \geq k}$ is a Cauchy sequence. We first show that this sequence is bounded. Let N be a positive ultralarge integer and let $M = \max\{|u_n| : n = k, \dots, N\}$. For any ultralarge N' we have $u_N \simeq u_{N'}$, hence $|u_{N'}| \leq M + 1$. This implies that the sequence is bounded (by $M + 1$).

By Closure, there is an observable bound, hence the terms of the sequence are not ultralarge. Let L be the observable neighbor of u_N. As (u_n) is a Cauchy sequence, we have $u_{N'} \simeq u_N$, thus $u_{N'} \simeq L$, for all ultralarge N'. This shows that the sequence converges.   □

Exercise 59 (Answer page 259)
Let $f : [a, \infty) \to \mathbb{R}$ be continuous. Show that the improper integral $\int_a^\infty f(x) \cdot dx$ converges if and only if $\int_b^c f(x) \cdot dx \simeq 0$ for all b, c positive ultralarge.
Hint: Use the idea of the previous proof.

Theorem 97 (Monotone Convergence).

(1) All increasing sequences bounded above converge.

(2) All decreasing sequences bounded below converge.

Proof. We prove (1) as (2) is similar. Let $(u_n)_{n \geq k}$ be an increasing sequence which is bounded above. We let $a = u_k$ and $b = M$, where M is some observable upper bound on (u_n). We also fix an ultralarge $N \in \mathbb{N}$ and let $dx = (b - a)/N$ and $x_i = a + i \cdot dx$, as usual. There exists a first i such that x_i is an upper bound on (u_n); we let L be the observable neighbor of x_i and prove that $\lim u_n = L$.

If $n \geq k$ is observable, then u_n is also observable, and from $u_n \leq x_i$ it follows that $u_n \leq L$. Hence $u_n \leq L$ holds for all observable $n \geq k$. By Closure, $u_n \leq L$ holds for all $n \geq k$.

On the other hand, if $L' < L$ is observable, then L' is not an upper bound on (u_n) (otherwise, $x_{i-1} \geq L'$ would be an upper bound, contradicting the choice of i). The least n such that $L' < u_n$ is observable, by Closure. As the sequence is increasing, we have $L' < u_{N'} \leq L$ for every ultralarge N'. This is true for all observable $L' < L$. We conclude that $u_{N'} \simeq L$ for all ultralarge N', and the sequence converges to L.   □

In the terminology of Section 5.1, this argument proves that an increasing bounded sequence converges to the supremum of the bounded set $\{u_n : n \geq k\}$. Theorem 82 (the completeness of the real numbers) can be used to make the proof of the Monotone Convergence Theorem considerably simpler (exercise).

Example. As an illustration, we use this theorem to give an alternative proof that

$$\lim_{n \to \infty} a^n = \infty, \quad \text{if } a > 1,$$

and

$$\lim_{n \to \infty} a^n = 0, \quad \text{if } 0 < a < 1.$$

In both cases we define the sequence (u_n) by $u_n = a^n$, for $n \in \mathbb{N}$. This sequence is easily seen to be monotone, by induction (increasing if $a > 1$ and decreasing if $0 < a < 1$).

Let $a > 1$. Assume that (u_n) is bounded above. By the Monotone Convergence Theorem there is an observable L such that

$$a^N \simeq L \quad \text{for all positive ultralarge integers } N.$$

Let N be a positive ultralarge integer. Then $a^N \simeq L$ and $a^{N+1} \simeq L$. But we also have $a^{N+1} = a \cdot a^N \simeq a \cdot L$. Since $a \cdot L \simeq L$ and both are observable, we have $a \cdot L = L$, so $(a-1) \cdot L = 0$. This shows that $L = 0$, which is a contradiction, since $a^n \geq a > 1$ for all n (another easy induction). So (u_n) is not bounded above and u_N is positive ultralarge for all ultralarge N.

Now if $0 < a < 1$, then (u_n) is decreasing and bounded below by 0. Reasoning similar to the above shows that (u_n) converges to 0.

Example. Consider the sequence (u_n) defined by

$$u_0 = 1 \quad \text{and} \quad u_{n+1} = \sqrt{1 + u_n}, \quad \text{for } n \geq 0.$$

The first few terms of the sequence are

$$u_0 = 1, \quad u_1 = \sqrt{1+1} = \sqrt{2}, \quad u_2 = \sqrt{1 + \sqrt{2}}, \quad \dots$$

This sequence is well defined since all the u_n are positive (easily seen by induction). We show that (u_n) converges by using the Monotone Convergence Theorem.

We first prove that (u_n) is increasing. It suffices to show that $u_n \leq u_{n+1}$, for all $n \geq 0$. For $n = 0$, it is clear. Assume inductively that $u_n \leq u_{n+1}$, for $n \geq 0$. Then $1 + u_n \leq 1 + u_{n+1}$, so

$$u_{n+1} = \sqrt{1 + u_n} \leq \sqrt{1 + u_{n+1}} = u_{n+2},$$

as $x \mapsto \sqrt{x}$ is an increasing function.

We now show that (u_n) is bounded by 2, also by induction. For $n = 0$, it is clear. Assume inductively that $u_n < 2$, for $n \geq 0$. Then $1 + u_n < 3$, so

$$u_{n+1} = \sqrt{1 + u_n} < \sqrt{3} < 2.$$

We conclude that the sequence converges; let $\bar{u} = \lim_{n \to \infty} u_n$ be its limit; notice that necessarily $\bar{u} > 0$. By Exercise 57 also $\lim_{n \to \infty} u_{n+1} = \bar{u}$, and by Theorem 95, $\lim_{n \to \infty} \sqrt{1 + u_n} = \sqrt{1 + \bar{u}}$ (the function $x \mapsto \sqrt{1 + x}$ is continuous on its domain). Applying $\lim_{n \to \infty}$ to both sides of the equation $u_{n+1} = \sqrt{1 + u_n}$ we obtain

$$\bar{u} = \sqrt{1 + \bar{u}} \quad \text{and therefore} \quad \bar{u}^2 = 1 + \bar{u}.$$

Solving this equation, we have that $\bar{u} = \frac{1 \pm \sqrt{5}}{2}$. But $\bar{u} > 0$, so we conclude that

$$\bar{u} = \frac{1 + \sqrt{5}}{2}.$$

7.2   Series

Let $(u_n)_{n \geq k}$ be a sequence. It is possible to define another sequence by considering the **partial sums** $s_k = u_k$ and $s_{n+1} = s_n + u_{n+1}$, for $n \geq k$. In other words, for a positive integer $N \geq k$ we have

$$s_N = u_k + u_{k+1} + \cdots + u_N = \sum_{n=k}^{N} u_n.$$

Definition 51. *Let $(u_n)_{n \geq k}$ be a sequence. A **series** is the sequence*

$$\left(\sum_{n=k}^{N} u_n \right)_{N \geq k}$$

of the partial sums. We denote this series by

$$\sum_{n \geq k} u_n.$$

Definition 52. *Let $\sum_{n \geq k} u_n$ be a series. We say that $\sum_{n \geq k} u_n$ **converges** (to L) if the sequence $\left(\sum_{n=k}^{N} u_n \right)_{N \geq k}$ converges (to L). Otherwise, we say that it **diverges**.*

If the series converges, then the **total sum**

$$u_k + u_{k+1} + u_{k+2} + \ldots$$

is defined to be equal to the limit of the sequence of partial sums. Of course, the total sum is observable (in the context of the series, or sequence).

Example. Consider the **arithmetic series**

$$\sum_{n \geq 1} u_n, \quad \text{with } u_1 = a \text{ and } u_n = a + (n-1) \cdot d,$$

for $a, d \in \mathbb{R}$. To establish the value of $\sum_{n=1}^{N} u_n$, first note that

$$1 + 2 + \cdots + (N-1)$$
$$= \frac{1}{2}([1 + (N-1)] + [2 + (N-2)] + \cdots + [(N-1) + 1])$$
$$= \frac{N(N-1)}{2};$$

thus

$$\sum_{n=1}^{N} a + (n-1) \cdot d = N \cdot a + d \sum_{n=1}^{N} (n-1)$$
$$= N \cdot a + d \cdot \frac{N \cdot (N-1)}{2} = \frac{N}{2}(2a + (N-1)d)$$
$$= \frac{N}{2}(u_1 + u_N).$$

One sees immediately that $\frac{N}{2}(2a + (N-1)d)$ is ultralarge positive if $d > 0$, or $d = 0$ and $a > 0$, and ultralarge negative if $d < 0$, or $d = 0$ and $a < 0$. So the arithmetic series diverges, unless $a = d = 0$.

Example. Consider the **geometric series**

$$\sum_{n \geq 1} u_n, \quad \text{with } u_1 = a \text{ and } u_n = a \cdot r^{n-1},$$

with $a, r \in \mathbb{R}$ ($a \neq 0$). Let

$$s_N = \sum_{n \geq 1}^{N} a \cdot r^{n-1}.$$

Note that

$$s_N + ar^{N+1} = a + ar + \cdots + ar^N + ar^{N+1} = a + r(a + \cdots + ar^N) = a + r \cdot s_N;$$

therefore $s_N \cdot (1 - r) = a \cdot (1 - r^{N+1})$ and

$$s_N = a \cdot \frac{1 - r^{N+1}}{1 - r}, \qquad \text{if } r \neq 1.$$

If $r = 1$, then $s_N = a \cdot N$, and the series diverges.

Using the results on powers with ultralarge exponents (Exercise 26) we obtain

$$\sum_{n \geq 1} a \cdot r^{n-1} \begin{cases} \text{diverges} & \text{if } |r| \geq 1; \\ \text{converges to } \frac{a}{1-r} & \text{if } |r| < 1. \end{cases}$$

The next exercise shows that the initial terms do not influence the convergence of a series (but they do influence the value of the sum).

Exercise 60 (Answer page 260.)
Let $\sum_{n \geq k} u_n$ be a series and m an integer such that $m \geq k$. Show that $\sum_{n \geq k} u_n$ converges if and only if $\sum_{n \geq m} u_n$ converges.

We next formulate a number of criteria for determining convergence or divergence of a series. The first criterion is a reformulation of Theorem 96.

Theorem 98. *Let $\sum_{n \geq k} u_n$ be a series. Then $\sum_{n \geq k} u_n$ converges if and only if*

$$\sum_{n=N}^{N'} u_n \simeq 0 \quad \text{for all ultralarge numbers } N \leq N'.$$

Theorem 99 (Comparison Test). *Let $(u_n)_{n \geq k}$ and $(w_n)_{n \geq k}$ be two sequences with nonnegative terms such that*

$$u_n \geq w_n \quad \text{for each } n \geq k.$$

If the series $\sum_{n \geq k} u_n$ converges, then the series $\sum_{n \geq k} w_n$ converges also.

Proof. If $\sum_{n \geq k} u_n$ converges, then $\sum_{n=N}^{N'} u_n \simeq 0$ for all ultralarge $N \leq N'$. Under the assumptions of the Comparison Test, $0 \leq \sum_{n=N}^{N'} w_n \leq \sum_{n=N}^{N'} u_n$, so also $\sum_{n=N}^{N'} w_n \simeq 0$ for all ultralarge $N \leq N'$, and $\sum_{n=N}^{N'} u_n$ converges. $\qquad\square$

The contrapositive of the previous theorem can be used to prove the divergence of a series.

Theorem 100. *If $\sum_{n \geq k} u_n$ converges, then $\lim_{n \to \infty} u_n = 0$.*

Proof. Let N be ultralarge. By Theorem 98, $u_N = \sum_{n=N}^{N} u_n \simeq 0$. □

Example. The converse of this theorem is false. Consider the **harmonic series**

$$\sum_{n \geq 1} \frac{1}{n}.$$

We have $\lim_{n \to \infty} \frac{1}{n} = 0$. We show now that this series diverges.

We observe that

$$s_2 = 1 + \frac{1}{2}$$

$$s_4 = s_2 + \left(\frac{1}{3} + \frac{1}{4} \right) \geq s_2 + \left(\frac{1}{4} + \frac{1}{4} \right) = s_2 + \frac{1}{2} = 1 + 2 \cdot \frac{1}{2}$$

$$s_8 = s_4 + \left(\frac{1}{5} + \frac{1}{6} + \frac{1}{7} + \frac{1}{8} \right) \geq s_4 + 4 \cdot \frac{1}{8} = s_4 + \frac{1}{2} \geq 1 + 3 \cdot \frac{1}{2}.$$

By induction, we see that

$$s_{2^N} \geq 1 + N \cdot \frac{1}{2}.$$

But this implies that the series diverges, because if N is ultralarge, then 2^N is ultralarge and $s_{2^N} \geq 1 + \frac{N}{2}$ is ultralarge, hence not ultraclose to any observable real number L.

By the same argument as in the proof of Theorem 89, a sequence $(u_n)_{n \geq k}$ converges if there is an observable L such that

$$N \stackrel{+}{\simeq} \infty \quad \text{implies} \quad u_N \simeq L.$$

Similarly, a series converges if there is an observable L such that

$$N \stackrel{+}{\simeq} \infty \quad \text{implies} \quad \sum_{n=k}^{N} u_n \simeq L.$$

We leave the proof as an exercise. This observation is used in the next example.

Example. We prove that

$$\sum_{n \geq 0} \frac{1}{n!} = e.$$

Notice first that the series $\sum_{n \geq 0} \frac{1}{n!}$ converges by the Comparison Test, since for each n we have $0 \leq \frac{1}{n!} \leq \frac{1}{2^{n-1}}$, and the geometric series $\sum_{n \geq 1} \frac{1}{2^{n-1}}$ converges.

We formalize a historic argument due to Euler and show that

$$\lim_{n \to \infty} \left(1 + \frac{1}{n}\right)^n = \sum_{n \geq 0} \frac{1}{n!}.$$

This is enough, by Theorem 73. Note also that this result provides an alternative proof for the existence of the limit, that is, of e. We fix a context and an ultralarge positive integer M. According to the observation above, it suffices to show that

$$\left(1 + \frac{1}{N}\right)^N \simeq \sum_{n=0}^{N} \frac{1}{n!} \tag{7.1}$$

holds for all integers N ultralarge relative to the context extended by M.

By the binomial formula, we have

$$\left(1 + \frac{1}{N}\right)^N = \sum_{n=0}^{N} \frac{N \cdot (N-1) \cdots (N - (n-1))}{n!} \cdot \frac{1}{N^n}$$

$$= \sum_{n=0}^{N} \frac{1 \cdot (1 - 1/N) \cdots (1 - (n-1)/N)}{n!}.$$

Write $\overset{+}{\simeq}$ when working with the context extended by M. We have

$$\left(1 + \frac{1}{N}\right)^N = \sum_{n=0}^{M} \frac{1 \cdot (1 - 1/N) \cdots (1 - (n-1)/N)}{n!}$$

$$+ \sum_{n=M+1}^{N} \frac{1 \cdot (1 - 1/N) \cdots (1 - (n-1)/N)}{n!}.$$

But on the one hand

$$\sum_{n=0}^{M} \frac{1 \cdot (1 - 1/N) \cdots (1 - (n-1)/N)}{n!} \overset{+}{\simeq} \sum_{n=0}^{M} \frac{1}{n!},$$

since for each $n \leq M$ we have $n/N \overset{+}{\simeq} 0$ and since there are only M factors (see Theorem 6(4)). On the other hand

$$0 \leq \sum_{n=M+1}^{N} \frac{1 \cdot (1 - 1/N) \cdot \cdots \cdot (1 - (n-1)/N)}{n!} \leq \sum_{n=M+1}^{N} \frac{1}{n!} \simeq 0,$$

since M is ultralarge and $\sum_{n \geq 0} \frac{1}{n!}$ converges.

We conclude that

$$\left(1 + \frac{1}{N}\right)^N \simeq \sum_{n=0}^{M} \frac{1}{n!} \simeq \sum_{n=0}^{N} \frac{1}{n!}.$$

This establishes (7.1).

Theorem 101. *The number e is irrational.*

Proof. We assume that e is rational, that is, $e = m/n$ for some integers m and $n > 1$, and obtain a contradiction.

We write

$$e = 1 + 1 + \frac{1}{2!} + \frac{1}{3!} + \cdots \frac{1}{n!} + \frac{1}{(n+1)!} + \frac{1}{(n+2)!} + \cdots =$$

$$\frac{k}{n!} + \frac{1}{n!} \left[\frac{1}{n+1} + \frac{1}{(n+1)(n+2)} + \cdots \right] = \frac{k}{n!} + \frac{1}{n!} \cdot r,$$

where k is a positive integer and

$$0 < r = \frac{1}{n+1} + \frac{1}{(n+1)(n+2)} + \cdots < \frac{1}{n+1} + \frac{1}{(n+1)^2} + \cdots = \frac{1}{n} < 1.$$

(The formula for the sum of a geometric series is used in the next-to-last step.) As also

$$e = \frac{m}{n} = \frac{m \cdot (n-1)!}{n!},$$

we have

$$\frac{m \cdot (n-1)!}{n!} = \frac{k}{n!} + \frac{1}{n!} \cdot r.$$

It follows that $r = m \cdot (n-1)! - k$ is an integer, a contradiction. $\qquad \square$

Definition 53. *A series $\sum_{k \geq n} u_n$ (or sequence $(u_n)_{n \geq k}$) is an **alternating series** (or an **alternating sequence**), if $u_n \cdot u_{n+1} < 0$ for each $n \geq k$.*

Theorem 102. *Let $(u_n)_{n \geq k}$ be an alternating sequence decreasing in absolute value. If $\lim_{n \to \infty} u_n = 0$, then $\sum_{n \geq k} u_n$ converges.*

Proof. Without loss of generality we may assume that $u_n = (-1)^n \cdot a_n$, $n \geq k$, with (a_n) a decreasing sequence of positive terms. We may further assume by Exercise 60 that k is even. Then the sequence of partial sums satisfies

$$s_{k+1} \leq s_{k+3} \leq \cdots \leq s_{k+2m+1} \leq \cdots \leq s_{k+2m} \leq \cdots \leq s_{k+2} \leq s_k.$$

In particular, for ultralarge N, N' with $N < N'$ we have

$$|s_{N'} - s_N| \leq |s_{N+1} - s_N| = a_N \simeq 0,$$

because $\lim_{n \to \infty} a_n = 0$. Hence the sequence of partial sums is a Cauchy sequence, and the series converges.   □

Example. This theorem shows that the harmonic alternating series defined by

$$\sum_{n \geq 1} (-1)^{n+1} \frac{1}{n}$$

converges.

Theorem 103 (Integral Test). *Let* $f : [k, \infty) \to \mathbb{R}$ *be a continuous, decreasing and positive function. Then the series* $\sum_{n \geq k} f(n)$ *converges if and only if the improper integral* $\int_k^\infty f(x) \cdot dx$ *converges.*

Proof. The parameters are f and k. As f is decreasing, we have $f(x) \leq f(n)$ for $n \leq x$, and $f(n) \leq f(x)$ for $x \leq n$. Hence

$$\int_n^{n+1} f(x) \cdot dx \leq f(n) \leq \int_{n-1}^n f(x) \cdot dx.$$

Let $N \in \mathbb{N}$ be ultralarge. By additivity of the integral we have

$$\int_k^{N+1} f(x) \cdot dx \leq \sum_{n=k}^N f(n) = f(k) + \sum_{n=k+1}^N f(n) \leq f(k) + \int_k^N f(x) \cdot dx.$$

If $\int_k^\infty f(x) \cdot dx$ converges, then the right-hand side of the inequality shows that the increasing sequence of partial sums is bounded above by $\int_k^\infty f(x) \cdot dx + f(k)$. Hence $\sum_{n \geq k} f(n)$ converges, by the Monotone Convergence Theorem.

On the other hand, if $\int_k^\infty f(x) \cdot dx$ diverges, then

$$\lim_{N \to \infty} \int_k^N f(x) = +\infty$$

(since $f \geq 0$) and the left-hand side inequality shows that the series diverges.   □

Example. Consider the series

$$\sum_{n \geq 1} \frac{1}{n^2}.$$

Let f be defined on $(0, \infty)$ by $x \mapsto \frac{1}{x^2}$. By the Integral Test, the series converges if and only if the improper integral $\int_1^\infty \frac{1}{x^2} \cdot dx$ converges. This integral does converge (see the example on page 133, Section 4.8), and so does the series.

The two following criteria use comparisons with geometric series.

Theorem 104 (Ratio Test). *Let $\sum_{n \geq k} u_n$ be a series with positive terms.*
If

$$\lim_{n \to \infty} \frac{u_{n+1}}{u_n} = L,$$

then

(1) if $L > 1$, the series diverges,

(2) if $L < 1$, the series converges.

Proof. Assume $L < 1$. As L is observable, we can choose an observable r such that $L < r < 1$ (so $r - L \not\simeq 0$). By definition of $\lim_{n \to \infty} \frac{u_{n+1}}{u_n} = L$, we have, for all ultralarge m,

$$\frac{u_{m+1}}{u_m} \simeq L.$$

Hence

$$\frac{u_{m+1}}{u_m} \leq r \quad \text{for all ultralarge } m.$$

Fix an ultralarge m; then the inequality above implies that $u_{m+\ell} \leq u_m \cdot r^\ell$, for each $\ell \geq 0$, that is, $u_n \leq u_m \cdot r^{n-m}$ for each $n \geq m$. Let $N \geq m$. We have

$$\sum_{n=m}^{N} u_n \leq \sum_{n=m}^{N} u_m \cdot r^{n-m} = (r^{-m} \cdot u_m) \cdot \sum_{n=m}^{N} r^n.$$

But the geometric series $\sum_{n \geq m} r^n$ converges because $r < 1$, hence, by Comparison Test, $\sum_{n \geq m} u_n$ converges as well. This implies that $\sum_{n \geq k} u_n$ converges, by Exercise 60.

The proof for divergence is similar (see the proof of the following theorem for more details). $\square$

The ratio test is inconclusive in the case $L = 1$. We have seen that $\sum_{n \geq 1} \frac{1}{n}$ diverges but $\sum_{n \geq 1} \frac{1}{n^2}$ converges, while $L = 1$ in both cases.

Theorem 105 (Root Test). *Let $\sum_{n \geq k} u_n$ be a series with non-negative terms.*

If

$$\lim_{n \to \infty} \sqrt[n]{u_n} = L,$$

then

(1) if $L > 1$, the series diverges,

(2) if $L < 1$, the series converges.

Proof. This time we prove the case $L > 1$. Again we choose an observable r such that $L > r > 1$. Fix an ultralarge positive m; then every $n \geq m$ is ultralarge and

$$\sqrt[n]{u_n} \simeq L, \text{ hence } \sqrt[n]{u_n} \geq r.$$

Therefore

$$u_n \geq r^n \quad \text{for all } n \geq m.$$

Then $\sum_{n=m}^{N} u_n \geq \sum_{n=m}^{N} r^n$ for all $N \geq m$. The geometric series $\sum_{n \geq m} r^n$ diverges because $r > 1$, so the series $\sum_{n \geq m} u_n$ diverges by Comparison Test, hence $\sum_{n \geq k} u_n$ diverges also, by Exercise 60.   □

The root test is inconclusive in the case $L = 1$, as can be seen with the same examples as for the ratio test.

We now examine a stronger form of convergence.

Definition 54. *We say that the series $\sum_{n \geq k} u_n$ is **absolutely convergent** if $\sum_{n \geq k} |u_n|$ converges.*

For example, the alternating harmonic series is convergent but not absolutely convergent.

If a series is absolutely convergent, one may rearrange the terms without changing the sum.

Theorem 106. *Let $\sum_{n \geq k} u_n$ be absolutely convergent. Then $\sum_{n \geq k} u_n$ converges and, moreover,*

$$\sum_{n \geq k} u_n = \sum_{n \geq k} u_{\sigma(n)},$$

for any permutation $\sigma : \{k, k+1, k+2, \dots\} \to \{k, k+1, k+2, \dots\}$.

Proof. Let N and N' be ultralarge. Then

$$\left| \sum_{n=N}^{N'} u_n \right| \leq \sum_{n=N}^{N'} |u_n| \simeq 0$$

by assumption, and therefore $\sum_{n \geq k} u_n$ converges.

To prove the "moreover" part, we use the same technique as page 171. Let M be a positive integer ultralarge relative to the context $\sigma, (u_n), k$. Now let N be a positive integer ultralarge relative to the context extended by M. Then for each $n \leq M$, we have $\sigma(n) < N$ (because $\sigma(n)$ is observable relative to the extended context and N is ultralarge relative to that context). Now let $N' > N$ be such that $\sigma(n) \leq N'$ for all $n \leq N$. Consider

$$\left| \sum_{n=k}^{N} u_n - \sum_{n=k}^{N} u_{\sigma(n)} \right|.$$

Since $\sigma(n) \leq N$ for all $n \leq M$, all the terms up to u_M cancel each other, and as both $n \leq N'$ and $\sigma(n) \leq N'$ if $n \leq N$, we have

$$\left| \sum_{n=k}^{N} u_n - \sum_{n=k}^{N} u_{\sigma(n)} \right| \leq \sum_{n=M+1}^{N'} |u_n| \simeq 0.$$

This implies that both series converge to the same limit. $\qquad\square$

One may determine the absolute convergence of a series $\sum u_n$ by applying the convergence criteria for sequences with positive terms to the series $\sum |u_n|$.

7.3   Taylor Series

The idea of this section is to represent a function by a series $\sum_{n \geq k} a_n \cdot (x-c)^n$ so that the series converges to $f(x)$ for values of x near the point c. This is called the **Taylor series for** f **at** c.

Theorem 107. *Let N be a non-negative integer and let $c \in \mathbb{R}$. Let f be a function that has a continuous derivative of order $N + 1$ on an open interval containing c and let x be in this interval. Then*

$$f(x) = \sum_{n=0}^{N} \frac{(x-c)^n}{n!} \cdot f^{(n)}(c) + \int_{c}^{x} \frac{(x-t)^N}{N!} \cdot f^{(N+1)}(t) \cdot dt.$$

Proof. Let N be a non-negative integer. We use integration by parts N times. By the Fundamental Theorem of Calculus, we have $f(x) - f(c) = \int_c^x f'(t) \cdot dt$, hence

$$f(x) = f(c) + \int_c^x 1 \cdot f'(t) \cdot dt.$$

We integrate by parts, choosing $-(x - t)$ as antiderivative of 1 (x being constant), and obtain:

$$f(x) = f(c) - (x - t)f'(t)\Big|_c^x + \int_c^x (x - t)f''(t) \cdot dt$$

$$= f(c) + (x - c)f'(c) + \int_c^x (x - t)f''(t) \cdot dt.$$

We now integrate $\int_c^x (x - t)f''(t) \cdot dt$ by parts again, choosing $-\frac{(x-t)^2}{2}$ as antiderivative of $(x - t)$, and obtain

$$f(x) = f(c) + (x - c)f'(c) + \frac{(x - c)^2}{2!}f''(c) + \int_c^x \frac{(x - t)^2}{2!}f'''(t) \cdot dt.$$

By repeating this process, eventually choosing $-\frac{(x-t)^N}{N}$ as the antiderivative of $(x - t)^{N-1}$, we get

$$f(x) = \sum_{n=0}^{N} \frac{(x - c)^n}{n!} \cdot f^{(n)}(c) + \int_c^x \frac{(x - t)^N}{N!} \cdot f^{(N+1)}(t) \cdot dt.$$

□

As a first application, we give an alternative argument for the fact proved in the example of page 172.

Theorem 108. *The number e satisfies*

$$e = \sum_{n \geq 0} \frac{1}{n!}.$$

Proof. We use Theorem 107 with $f(x) = \exp(x)$, $c = 0$ and $x = 1$. The crucial points are that $f^{(n)}(x) = \exp(x)$ (for all n) and $\exp(0) = 1$. Let N be an ultralarge positive integer.

$$e = 1 + \frac{1}{1!} + \frac{1}{2!} + \cdots + \frac{1}{N!} + \int_0^1 \frac{(1 - t)^N}{N!} \cdot \exp(t) \cdot dt.$$

It is enough to show that the positive number $\int_0^1 \frac{(1-t)^N}{N!} \exp(t) \cdot dt$ is ultrasmall.

But $0 \leq 1-t \leq 1$ for $t \in [0, 1]$, so $0 \leq (1-t)^N \leq 1^N = 1$. Furthermore, $1 \leq \exp(t) \leq e$, since exp is increasing. Hence

$$0 \leq \int_0^1 \frac{(1-t)^N}{N!} \cdot \exp(t) \cdot dt \leq \int_0^1 \frac{1}{N!} \cdot e \cdot dt = \frac{e}{N!} \cdot t \Big|_0^1 = \frac{e}{N!} \simeq 0,$$

since e is standard and N is ultralarge. $\square$

We already know that the alternating harmonic series converges. Now we can prove more.

Theorem 109. *The alternating harmonic series converges to* $\ln(2)$, *i.e.,*

$$\ln(2) = \sum_{n \geq 1} \frac{(-1)^{n+1}}{n}.$$

Proof. Let $f(x) = \ln(x)$ and $c = 1$. Then

$$f^{(n)}(x) = (-1)^{n+1} \cdot \frac{(n-1)!}{x^n},$$

so $f^{(n)}(1) = (-1)^{n+1}(n-1)!$. Let $x = 2$ and N be an ultralarge positive integer. We have

$$\ln(2) = \sum_{n=1}^{N} (-1)^{n+1} \cdot \frac{1}{n} + (-1)^{N+2} \int_1^2 \frac{(2-t)^N}{t^{N+1}} \cdot dt.$$

Hence, it is enough to show that $\int_1^2 \frac{(2-t)^N}{t^{N+1}} \cdot dt$ is ultraclose to zero. But

$$0 \leq \int_1^2 \frac{(2-t)^N}{t^{N+1}} \cdot dt \leq \int_1^2 \frac{1}{t^{N+1}} \cdot dt = -\underbrace{\frac{1}{N}}_{\simeq 0} \cdot \left(\underbrace{\frac{1}{2^N}}_{\simeq 0} - 1 \right) \simeq 0.$$

$\square$

The integral formula for the remainder is sometimes difficult to use. This is why we prove the next theorem.

Theorem 110. *Let N be a non-negative integer and $c \in \mathbb{R}$. Let f be a function differentiable $N+1$ times on an open interval containing c and let x be in this interval. Then*

$$f(x) = \sum_{n=0}^{N} \frac{(x-c)^n}{n!} \cdot f^{(n)}(c) + \frac{(x-c)^{N+1}}{(N+1)!} \cdot f^{(N+1)}(\xi),$$

with ξ between c and x.

Proof. Fix a non-negative integer N. Define

$$R(x) = f(x) - \sum_{n=0}^{N} \frac{(x-c)^n}{n!} \cdot f^{(n)}(c)$$

and

$$S(x) = \frac{(x-c)^{N+1}}{(N+1)!}.$$

Differentiating these as many times as required, we see that

$$R(c) = R'(c) = R''(c) = \cdots = R^{(N)}(c) = 0$$

and

$$S(c) = S'(c) = S''(c) = \cdots = S^{(N)}(c) = 0.$$

By Cauchy's Theorem (page 80),

$$\frac{R(x)}{S(x)} = \frac{R(x) - R(c)}{S(x) - S(c)} = \frac{R'(\xi_1)}{S'(\xi_1)}, \quad \text{where } \xi_1 \text{ is between } c \text{ and } x.$$

Similarly

$$\frac{R'(\xi_1)}{S'(\xi_1)} = \frac{R'(\xi_1) - R'(c)}{S'(\xi_1) - S'(c)} = \frac{R''(\xi_2)}{S''(\xi_2)}, \quad \text{where } \xi_2 \text{ is between } c \text{ and } \xi_1.$$

Repeating the process till N, we get

$$\frac{R(x)}{S(x)} = \frac{R'(\xi_1)}{S'(\xi_1)} = \frac{R''(\xi_2)}{S''(\xi_2)} = \cdots = \frac{R^{(N+1)}(\xi)}{S^{(N+1)}(\xi)},$$

where ξ is between c and ξ_N, so in particular between c and x. But $S^{(N+1)}(x) = 1$ and $R^{(N+1)}(x) = f^{(N+1)}(x)$, so

$$\frac{R(x)}{S(x)} = f^{(N+1)}(\xi).$$

Substituting their values for R and S, we deduce that

$$f(x) = \sum_{n=0}^{N} \frac{(x-c)^n}{n!} \cdot f^{(n)}(c) + \frac{(x-c)^{N+1}}{(N+1)!} \cdot f^{(N+1)}(\xi).$$

$\square$

Example. Compute the value of e with an error less than $\varepsilon = 0.0001$.
In the previous theorem we let $f(x) = e^x$, $c = 0$, and $x = 1$, to get

$$e = \sum_{n=0}^{N} \frac{1}{n!} + \frac{e^\xi}{(N+1)!}$$

with $0 < \xi < 1$. We approximate e by $\bar{e} = \sum_{n=0}^{N} \frac{1}{n!}$, with N chosen to make the error $|e - \bar{e}| = \frac{e^{\xi}}{(N+1)!} < \varepsilon$. As $e^{\xi} < e < 3$, it suffices to make $\frac{3}{(N+1)!} < \varepsilon$, that is, $(N+1)! > \frac{3}{\varepsilon}$. For $\varepsilon = 0.0001$ we thus need $(N+1)! > 30000$. An easy computation gives $7! = 5040$ and $8! = 40320$, so the best choice is $N + 1 = 8$, hence $N = 7$. The approximate value is $\bar{e} = 1 + 1 + \frac{1}{2!} + \ldots + \frac{1}{7!} = 2.7182534\ldots$, which differs from the exact value $e = 2.7182818\ldots$ by less than 0.00003.

To help us in other applications of this theorem, we show that

$$\frac{(x-c)^{N+1}}{(N+1)!} \simeq 0$$

when $N \in \mathbb{N}$ is ultralarge relative to a context where x and c are observable.

There is an observable integer m such that $|x - c| \leq m$. If n is not ultralarge, then $\frac{|x-c|}{n}$ is observable by Closure, and therefore $\prod_{n=1}^{m} \frac{|x-c|}{n}$ is observable. If $n > m$, then each factor $\frac{|x-c|}{n} \leq 1$ and so $\prod_{n=m+1}^{N} \frac{|x-c|}{n} \leq 1$. Finally, $\frac{|x-c|}{N+1} \simeq 0$, because $|x - c|$ is not ultralarge. Hence

$$\frac{|x-c|^{N+1}}{(N+1)!} = \underbrace{\prod_{n=1}^{m} \frac{|x-c|}{n}}_{\text{observable}} \cdot \underbrace{\prod_{n=m+1}^{N} \frac{|x-c|}{n}}_{\leq 1} \cdot \underbrace{\frac{|x-c|}{N+1}}_{\simeq 0} \simeq 0,$$

since the product of an ultrasmall number with a number that is not ultralarge is ultrasmall.

This observation gives us a convergence criterion for Taylor series.

Theorem 111. *Let f be a function infinitely many times differentiable on an open interval containing c and let x be in that interval. Suppose that there is a number M such that $f^{(n)}$ is bounded by M on $[x, c]$ (or $[c, x]$ if $c < x$), for each integer $n \geq 0$. Then the series*

$$\sum_{n \geq 0} \frac{(x-c)^n}{n!} \cdot f^{(n)}(c)$$

converges to $f(x)$.

Proof. By Closure, we may assume that M is observable. Let $N \in \mathbb{N}$ be ultralarge. The previous theorem gives

$$f(x) = \sum_{n=0}^{N} \frac{(x-c)^n}{n!} \cdot f^{(n)}(c) + \frac{(x-c)^{N+1}}{(N+1)!} \cdot f^{(N+1)}(\xi),$$

where ξ is between c and x. All we need to show is that

$$\frac{(x-c)^{N+1}}{(N+1)!} \cdot f^{(N+1)}(\xi) \simeq 0.$$

But it is proved above that $\frac{(x-c)^{N+1}}{(N+1)!} \simeq 0$, and $|f^{(N+1)}(\xi)| \leq M$, which is not ultralarge, hence the product is ultraclose to 0. $\qquad\square$

Theorem 112. *For each real number x we have*

$$\exp(x) = \sum_{n \geq 0} \frac{x^n}{n!}.$$

Proof. We apply the criterion described above to the function $f(x) = \exp(x)$ at $c = 0$. Calculating the derivatives we obtain the Taylor series for $\exp(x)$:

$$\sum_{n \geq 0} \exp^{(n)}(0) \cdot \frac{x^n}{n!} = \sum_{n \geq 0} \frac{x^n}{n!}.$$

As $f^n(\xi) = \exp(\xi)$ and $\exp(\xi)$ is bounded on the interval with the endpoints c and x (by $\exp(c)$ or $\exp(x)$), this series must converge to $\exp(x)$, for all x. $\qquad\square$

Example. The following series converge for all values of x.

$$(1) \quad \sin(x) = x - \frac{x^3}{3!} + \frac{x^5}{5!} + \cdots + (-1)^N \frac{x^{2N+1}}{(2N+1)!} + \cdots$$

$$(2) \quad \cos(x) = 1 - \frac{x^2}{2!} + \frac{x^4}{4!} + \cdots + (-1)^N \frac{x^{2N}}{(2N)!} + \cdots$$

The derivatives $f^{(n)}$ of the function $f(x) = \sin(x)$ are always bounded by 1 because they are $\pm \sin(x)$ or $\pm \cos(x)$. Using Theorem 111 with $c = 0$, the Taylor series for $f(x)$:

$$\underbrace{\frac{x^0}{0!} \sin(0)}_{=0} + \underbrace{x \sin'(0)}_{=x} + \underbrace{\frac{x^2}{2!} \sin''(0)}_{=0} + \underbrace{\frac{x^3}{3!} \sin'''(0)}_{=-\frac{x^3}{3!}} + \underbrace{\frac{x^4}{4!} \sin^{(4)} 0}_{=0} + \cdots$$

converges to $\sin(x)$ for all $x \in \mathbb{R}$.

For exactly the same reason, the second series converges to $\cos(x)$.

Example. A Taylor series for f may converge everywhere without converging to the function f. Consider f given by

$$f(x) = \begin{cases} e^{-\frac{1}{x^2}} & \text{if } x \neq 0; \\ 0 & \text{otherwise.} \end{cases}$$

One can show that

$$f^{(n)}(0) = 0, \quad \text{for each non-negative integer } n.$$

The power series $\sum_{n \geq 0} 0 \cdot x^n$ converges to the function which is everywhere 0 and not to f. This is not a contradiction with Theorem 111, as for each $x \neq 0$ and each M there exist n and ξ between 0 and x such that $|f^{(n)}(\xi)| > M$, so the assumptions of the theorem are not satisfied.

7.4   Uniform Convergence

Let $(f_n)_{n \geq k}$ be a sequence of functions with a common domain A (that is, each $f_n : A \to \mathbb{R}$).

We say that (f_n) **converges pointwise** to $f : A \to \mathbb{R}$ if for every $x \in A$

$$\lim_{n \to \infty} f_n(x) = f(x).$$

Unraveling the definition of limit, this means the following: For all $x \in A$ and all N ultralarge relative to (f_n) and x we have

$$f_N(x) \simeq f(x).$$

Theorem 15 establishes that the statement $y = \lim_{n \to \infty} f_n(x)$ is internal, and hence the function f is observable whenever the sequence (f_n) is observable.

Pointwise convergence turns out to be too weak for many purposes, so we introduce a stronger form of convergence.

Definition 55. *We say that the sequence (f_n)* **converges uniformly** *to $f : A \to \mathbb{R}$ if for all ultralarge N we have*

$$f_N(x) \simeq f(x), \quad \text{for all } x \subset A.$$

The strength of uniform convergence is due to the fact that it requires $f_N(x) \simeq f(x)$ to hold for all N ultralarge relative to (f_n), independently of x. Of course, uniform convergence implies pointwise convergence, by Stability.

Exercise 61 (Answer page 260)
Let (f_n) be a sequence of functions on A. Assume that $f_N(x) \simeq f_M(x)$ holds for all $x \in A$ and all ultralarge N, M. Prove that the sequence converges uniformly to some function f on A.

Theorem 113. *If $f_n : I \to \mathbb{R}$ are continuous functions converging uniformly to f, then f is continuous on I. Moreover, if $a, b \in I$, then*

$$\int_a^b f(x) \cdot dx = \lim_{n \to \infty} \int_a^b f_n(x) \cdot dx.$$

Proof. Let $x \in I$. Let N be ultralarge relative to (f_n), x and let dx be ultrasmall relative to the context extended by N. As usual, we use $\overset{+}{\simeq}$ when we work relative to the extended context. Then $dx \overset{+}{\simeq} 0$ and we have

$$f(x) \simeq f_N(x) \simeq f_N(x + dx) \simeq f(x + dx).$$

This shows that f is continuous. The first and last $\simeq$ are by uniform convergence. The middle one holds by continuity of f_N: $f_N(x) \overset{+}{\simeq} f_N(x + dx)$, since dx is ultrasmall relative to the extended context where f_N and x are observable.

For the "moreover" part of the theorem, the parameters are (f_n), a and b. Let N be ultralarge. Let $r > 0$ be observable. Then $\frac{r}{b-a}$ is also observable, and since the sequence converges uniformly, we have

$$|f(x) - f_N(x)| \le \frac{r}{b - a}, \quad \text{for all } x \in [a, b].$$

Then

$$\left| \int_a^b f(x) \cdot dx - \int_a^b f_N(x) \cdot dx \right| \le \int_a^b |f(x) - f_N(x)| \cdot dx \le \frac{r}{b - a} \cdot (b - a) \le r.$$

Since r is an arbitrary observable positive number, we conclude that

$$\int_a^b f(x) \cdot dx \simeq \int_a^b f_N(x) \cdot dx.$$

$\square$

Theorem 114. *Let $f_n : (a, b) \to \mathbb{R}$ be a sequence of smooth functions converging (pointwise) to f on (a, b), and such that (f_n') converges uniformly to g on (a, b). Then f is smooth on (a, b) and $f'(x) = g(x)$, for each $x \in (a, b)$.*

Proof. The parameters are (f_n), a and b. Let $x \in (a, b)$ be observable and let N be ultralarge. Since the functions f_n' are continuous, the preceding theorem shows that g is continuous and

$$\int_a^x g(t) \cdot dt \simeq \int_a^x f_N'(t) \cdot dt = f_N(x) - f_N(a) \simeq f(x) - f(a).$$

As both $\int_a^x g(t) \cdot dt$ and $f(x) - f(a)$ are observable, it follows that $f(x) = f(a) + \int_a^x g(t) \cdot dt$ holds for all observable x, hence for all x, by Closure. The Fundamental Theorem of Calculus then implies that f is differentiable and satisfies $f'(x) = g(x)$. Since g is continuous, f is smooth. □

The next theorem allows us to apply this theory to series of functions, in particular to convergent power series.

Theorem 115. *Let $(f_n)_{n \geq k}$ be a sequence of functions $f_n : I \to \mathbb{R}$ and $(M_n)_{n \geq k}$ be such that*

$$|f_n(x)| \leq M_n, \quad \text{for all } x \in I.$$

If $\sum_{n \geq k} M_n$ converges, then $\sum_{n \geq k} f_n$ converges uniformly on I.

Proof. For $x \in I$ the series $\sum_{n \geq k} f_n(x)$ converges, since it converges absolutely (by comparison with $\sum_{n \geq k} M_n$). Let $f(x) = \sum_{n \geq k} f_n(x)$. The parameters are (f_n) and (M_n). Let N be ultralarge. For all $x \in I$,

$$\left| f(x) - \sum_{n \geq k}^{N} f_n(x) \right| = \left| \sum_{n \geq N+1} f_n(x) \right| \leq \sum_{n \geq N+1} |f_n(x)| \leq \sum_{n \geq N+1} M_n \simeq 0,$$

since the series $\sum M_n$ converges. □

7.5 Power Series

A **power series** is a series of the form

$$\sum_{n \geq k} a_n (x - c)^n, \qquad \text{where } a_n \in \mathbb{R} \text{ and } c \in \mathbb{R}.$$

A Taylor series is a power series.

By a change of variable $y = (x - c)$, we can always reduce the study of powers series to those where $c = 0$, and by adding terms with $a_n = 0$, we can further reduce it to the form

$$\sum_{n \geq 0} a_n x^n.$$

Theorem 116. *Suppose that $\sum_{n \geq 0} a_n x^n$ converges for some $x = b$. Then $\sum_{n \geq 0} a_n x^n$ converges absolutely for any x such that $|x| < |b|$.*

Proof. If $\sum_{n \geq 0} a_n b^n$ converges, then in particular, its terms are bounded by some $M \in \mathbb{R}$. Hence

$$|a_n x^n| = |a_n b^n| \cdot \left|\frac{x}{b}\right|^n \leq M \cdot \left|\frac{x}{b}\right|^n.$$

But if $\left|\frac{x}{b}\right| < 1$, the series must converge absolutely, by comparison with the geometric series. $\square$

Theorem 115 implies immediately that, under the assumptions of Theorem 116, the series $\sum_{n \geq 0} a_n x^n$ converges uniformly on any closed interval $I \subseteq (-b, b)$.

Definition 56. *Let*

$$r = \sup\{|x| : \sum_{n \geq 0} a_n x^n \ \text{converges}\}.$$

We call r the **radius of convergence** *of the series.*

In the definition above, we take $r = \infty$ if the set is not bounded. The next theorem is immediate.

Theorem 117. *Let $\sum_{n \geq 0} a_n x^n$ be a power series with radius of convergence r. Then $\sum_{n \geq 0} a_n x^n$ converges on $(-r, r)$ and diverges if $|x| > r$. The convergence is uniform on $[-c, c]$, for any $0 < c < r$.*

The convergence of the series for $x = \pm r$ depends on the particular series.

Theorem 118.

(1) *Suppose that*

$$\lim_{n \to \infty} \left|\frac{a_n}{a_{n+1}}\right| \quad \text{exists (or is } \infty\text{)}.$$

Then the value of the limit is the radius of convergence.

(2) *Suppose that*

$$\lim_{n \to \infty} \frac{1}{\sqrt[n]{|a_n|}} \quad \text{exists (or is } \infty\text{)}.$$

Then the value of the limit is the radius of convergence.

Proof. We apply the ratio test or the root test to the series $\sum_{n \geq 0} |a_n x^n|$. We work out the details of (1). We have

$$\lim_{n \to \infty} \left|\frac{a_{n+1} x^{n+1}}{a_n x^n}\right| = \lim_{n \to \infty} \left|\frac{a_{n+1}}{a_n}\right| \cdot |x|.$$

The series converges absolutely if $|x| < \dfrac{1}{\lim_{n\to\infty}\left|\frac{a_{n+1}}{a_n}\right|} = \lim_{n\to\infty}\left|\dfrac{a_n}{a_{n+1}}\right|$ and diverges if $|x| > \lim_{n\to\infty}\left|\dfrac{a_n}{a_{n+1}}\right|$. This shows that this limit is the radius of convergence. The proof of (2) is similar. $\square$

Theorem 119. *The series $\sum_{n\geq 0} a_n x^n$ and $\sum_{n\geq 1} n a_n x^{n-1}$ have the same radius of convergence.*

Proof. Let r be the radius of convergence of the series $\sum_{n\geq 0} a_n x^n$ and r' that of $\sum_{n\geq 1} n a_n x^{n-1}$. First we note that r' is also the radius of convergence of $\sum_{n\geq 1} n a_n x^n$.

If $0 < x < r'$, then $\sum_{n\geq 1} |n a_n x^n|$ converges, hence $\sum_{n\geq 0} |a_n x^n|$ also converges, by Comparison Test, and $x \leq r$. It follows that $r' \leq r$.

Now assume that $0 < x < r$, pick z such that $x < z < r$, and write $n a_n x^n = n \cdot \left(\frac{x}{z}\right)^n \cdot a_n z^n$. Since $\left|\frac{x}{z}\right| < 1$, the sequence $n \cdot \left(\frac{x}{z}\right)^n$ converges (to 0), and hence it is bounded by some $M \in \mathbb{R}$. Then $|n a_n x^n| \leq M |a_n z^n|$. The series $\sum |a_n z^n|$ converges, hence $\sum |n a_n x^n|$ also converges, by Comparison Test, and $x \leq r'$. It follows that $r \leq r'$. $\square$

A power series defines a function on the interval $(-r, r)$. As a consequence of the fact that each $a_n x^n$ is continuous and differentiable, the preceding theorem and our theorems on uniform convergence, power series can be differentiated and integrated term by term. In particular, the functions defined by power series are differentiable infinitely many times in the interval $(-r, r)$. For completeness, we state these observations as a theorem.

Theorem 120. *Let $\sum_{n\geq 0} a_n x^n$ be a power series with radius of convergence r. Then the function $f : x \mapsto \sum_{n\geq 0} a_n x^n$ is differentiable for $x \in (-r, r)$ and*

$$f'(x) = \sum_{n=1}^{\infty} n a_n x^{n-1}.$$

Moreover, the antiderivative of f which has value 0 at 0 is given by

$$\int_0^x f(t) \cdot dt = \sum_{n=0}^{\infty} \frac{a_n}{n+1} x^{n+1}.$$

7.6   Additional Exercises

Exercise 7.1
Find the limits.

(1) $\displaystyle\lim_{n\to\infty} \frac{n^2 + 3n - 1}{2n^2 + 4}$

(2) $\displaystyle\lim_{n\to\infty} \frac{\sqrt{n^4 + 1}}{n!}$

Exercise 7.2
Prove Theorems 94 and 95.

Exercise 7.3
Let f be uniformly continuous on I and let (x_n) be a Cauchy sequence of elements of I. Show that $(f(x_n))$ is a Cauchy sequence.

Exercise 7.4
Give another proof of the Monotone Convergence Theorem by filling in the details of the following sketch.
Let $(u_n)_{n\geq k}$ be an increasing sequence which is bounded above; it suffices to show that it is Cauchy. If not, then there exist ultralarge $N < N'$ such that $u_N \not\simeq u_{N'}$, that is, $u_{N'} - u_N > \varepsilon$ for some observable $\varepsilon > 0$. Hence, for every observable m there exist n, n' such that $m \leq n < n'$ and $u_{n'} - u_n > \varepsilon$. By Closure, this statement is true for all m. For any $\ell \in \mathbb{N}$, we can now find $n_1 < n_1' < n_2 < n_2' < \ldots < n_\ell < n_\ell'$ such that $u_{n_i'} - u_{n_i} > \varepsilon$, for all $i \leq \ell$. This leads to a contradiction with the assumption that the sequence is bounded.

Exercise 7.5
For the following sequences, find the partial sums, determine whether the series converges and find the sum when it exists.

(1) $1 + \dfrac{1}{3} + \dfrac{1}{9} + \cdots + \left(\dfrac{1}{3}\right)^n + \ldots$

(2) $1 + \dfrac{3}{4} + \dfrac{9}{16} + \cdots + \left(\dfrac{3}{4}\right)^n + \ldots$

(3) $\left(1 - \dfrac{1}{2}\right) + \left(\dfrac{1}{2} - \dfrac{1}{6}\right) + \left(\dfrac{1}{6} - \dfrac{1}{24}\right) + \cdots + \left(\dfrac{1}{n!} - \dfrac{1}{(n+1)!}\right) + \ldots$

(4) $\dfrac{1}{1 \cdot 2} + \dfrac{1}{2 \cdot 3} + \cdots + \dfrac{1}{n(n+1)} + \ldots$

   Hint: $\dfrac{1}{n(n+1)} = \dfrac{1}{n} - \dfrac{1}{n+1}$.

(5) $1 - 2 + 4 - 8 + \cdots + (-2)^n + \ldots$

(6) $\dfrac{3}{1^2 \cdot 2^2} + \dfrac{5}{2^2 \cdot 3^2} + \cdots + \dfrac{2n+1}{n^2(n+1)^2} + \cdots$

Hint: $\dfrac{2n+1}{n^2(n+1)^2} = \dfrac{1}{n^2} - \dfrac{1}{(n+1)^2}.$

(7) $\dfrac{1}{1 \cdot 3} + \dfrac{1}{3 \cdot 5} + \dfrac{1}{5 \cdot 7} + \cdots + \dfrac{1}{(2n-1) \cdot (2n+1)} + \cdots$

(8) $\dfrac{1}{3} - \dfrac{2}{5} + \dfrac{3}{7} - \dfrac{4}{9} + \cdots + \dfrac{(-1)^{n-1} \cdot n}{2n+1}$

(9) $\dfrac{1}{4} + \dfrac{1}{7} + \dfrac{1}{10} + \cdots + \dfrac{1}{3n+1} + \cdots$

(10) $\ln(1) + \ln(2) + \ln(3) + \cdots + \ln(n) + \ldots$

Exercise 7.6

For the following, the general term of the series is given. Test the corresponding series for convergence.

(1) $\dfrac{3n-7}{10n+9}$

(2) $\dfrac{5}{6n^2 + n - 1}$

(3) $\dfrac{\sqrt{n}}{1 + 2\sqrt{n} + n}$

(4) ne^{-n}

(5) $\dfrac{5^n}{3^n + 4^n}$

(6) $\dfrac{n^n}{(n!)^2}$

(7) $\dfrac{2^n \cdot n!}{n^n}$

(8) $\dfrac{1}{\ln(n)}$

(9) $\dfrac{n^2}{2^n}$

(10) $\dfrac{\ln(n)}{n}$

Exercise 7.7

Prove: If $u_n > 0$ for all $n \in \mathbb{N}$ and either $\lim_{n \to \infty} \dfrac{u_{n+1}}{u_n} < 1$ or $\lim_{n \to \infty} \sqrt[n]{u_n} < 1$, then $\lim_{n \to \infty} u_n = 0$.

Exercise 7.8

Show that $\left| \sum_{n \geq k} u_n \right| \leq \sum_{n \geq k} |u_n|$, provided that the sum $\sum_{n \geq k} u_n$ is defined.

Exercise 7.9

Show that

(1) $\lim_{n \to \infty} na^n = 0$ for $|a| < 1$.

(2) $\lim\limits_{n\to\infty} \dfrac{a^n}{n!} = 0$ for all $a \in \mathbb{R}$.

(3) $\lim\limits_{n\to\infty} \dfrac{n!}{n^n} = 0$.

(4) $\lim\limits_{n\to\infty} \sqrt[n]{n!} = \infty$.

Exercise 7.10
The **Riemann series** is

$$\sum_{n\geq 1} \frac{1}{n^p}, \quad \text{with } p \in \mathbb{R}.$$

Show that the Riemann series converges if and only if $p > 1$.

Exercise 7.11
Show that the series

(1) $\sum_{n\geq 0} \frac{x^n}{n!}$ converges for all $x \in \mathbb{R}$;

(2) $\sum_{n\geq 0} n^n x^n$ diverges for all $x \in \mathbb{R}$;

(3) $\sum_{n\geq 0} \frac{n! x^n}{n^n}$ converges for $|x| < e$.

Exercise 7.12
Show that the following sequences converge on $\mathbb{R}$ pointwise, but not uniformly.

(1) $f_n(x) = \dfrac{1}{1 + nx^2}$ (2) $g_n(x) = \dfrac{2}{\pi}\arctan(nx)$

Exercise 7.13
Let $f, f_n, n \geq 1$, be functions from I to $\mathbb{R}$. Let $M_n = \sup\{|f_n(x) - f(x)| : x \in I\}$. Prove that (f_n) converges uniformly to f on I if and only if $\lim_{n\to\infty} M_n = 0$.

Exercise 7.14
Under the assumptions of Theorem 114 deduce that the sequence (f_n) converges to f uniformly.

Exercise 7.15
Prove: If $|f_n(x)| \leq g_n(x)$ for all $x \in I$ and all $n \geq 1$, and the series $\sum_{n\geq 1} g_n(x)$ converges uniformly on I, then the series $\sum_{n\geq 1} f_n(x)$ converges uniformly on I.

Exercise 7.16
Show that the following series converge uniformly on $\mathbb{R}$.

(1) $\displaystyle\sum_{n\geq 0} \frac{\sin(nx)}{n^2}$

(2) $\displaystyle\sum_{n\geq 0} \frac{x}{n(1+nx^2)}$

Exercise 7.17
Give the Taylor series for the following functions. State for which values of x they converge.

(1) $\dfrac{1}{1-x}$

(2) $\dfrac{1}{1+x}$

(3) $\dfrac{1}{1-2x}$

(4) $\ln(1-x)$

(5) $\dfrac{1}{1+x^2}$

(6) e^{-x}

(7) e^{-x^2}

(8) $\displaystyle\int_0^x e^{-t^2}\,dt$

(9) $\ln\left(\dfrac{1+x}{1-x}\right)$

Exercise 7.18
Show that the Taylor series for $(1+x)^p$ is

$$\sum_{n\geq 0} \binom{p}{n} x^n, \qquad \text{with} \quad \binom{p}{n} = \frac{p\cdot(p-1)\cdot\cdots\cdot(p-n+1)}{n!}.$$

Show that it converges for all x if $p \in \mathbb{N}$.
Show that it converges for $|x| < 1$ otherwise.

Exercise 7.19
Prove: If $\lim_{n\to\infty} a_n = a$ and $\lim_{n\to\infty} b_n = b$ $(a, b \in \mathbb{R})$, then

$$\lim_{n\to\infty} \frac{a_1\cdot b_n + a_2\cdot b_{n-1} + \cdots + a_{n-1}\cdot b_2 + a_n\cdot b_1}{n} = a\cdot b.$$

8

First Order Differential Equations

8.1 Solutions of Some Differential Equations

We restrict our study to differential equations of order one.

Definition 57. *Let $\langle t, y \rangle \mapsto f(t, y)$ be a function of two variables, defined for all $\langle t, y \rangle \in I \times J$. A **differential equation** is an equation of the form*

$$y' = f(t, y).$$

The unknown y is a function $t \mapsto y(t)$; it is a solution of the differential equation above if $y'(t) = f(t, y(t))$ holds for all $t \in I$. We restrict our attention to the case when $\langle t, y \rangle \mapsto f(t, y)$ is continuous in each of its variables, that is, for each fixed t in I the function $y \mapsto f(t, y)$ is continuous on J and for each fixed $y \in J$ the function $t \mapsto f(t, y)$ is continuous on I. We consider only solutions whose derivatives are continuous on I (smooth solutions).

Definition 58. *The **general solution** of a differential equation is the set of all functions satisfying this equation. A **particular solution** of a differential equation is simply one function satisfying the equation.*

Definition 59. *A differential equation is said to have **separable variables** if it can be written as*

$$g(y) \cdot y' = f(t),$$

with $t \mapsto f(t)$ and $y \mapsto g(y)$ functions of one variable.

Given our assumption on $\langle y, t \rangle \mapsto f(y, t)$, we always assume that $t \mapsto f(t)$ and $y \mapsto g(y)$ are continuous.

Recall that $\int f(x) \cdot dx$ denotes the set of antiderivatives of f, that is, if F is an antiderivative of f, then

$$\int f(x) \cdot dx = \{F(x) + C \ : \quad C \in \mathbb{R}\}.$$

Theorem 121. *Let f and g be continuous functions on I and J, respectively. Let $g(y) \cdot y' = f(t)$ be an equation with separable variables. Then any solution y (if it exists) satisfies*

$$\int g(y) \cdot dy = \int f(t) \cdot dt.$$

Proof. If the function $y = y(t)$ satisfies $g(y) \cdot y' = f(t)$ for all t in I, then by the integral version of Chain Rule (Theorem 64)

$$\int g(y) \cdot dy = \int g(y(t)) \cdot y'(t) \cdot dt = \int f(t) \cdot dt.$$

$\square$

Solving an equation with separable variables thus amounts to writing y' as dy/dt and to "separating" dy from dt. Integrating both sides of the resulting equation yields an implicit formula for y. If one can solve y in terms of t and if the solution is smooth, then one gets a solution of the differential equation.

Consider the simplest differential equation

$$y' = \frac{dy}{dt} = 1.$$

We get $dy = dt$, which implies that $\int dy = \int dt$, so

$$y = t + C, \quad \text{for any constant } C.$$

Hence there are infinitely many solutions. In order to obtain a particular solution, some additional information is needed, from which the constant C can be uniquely determined. Typically, one specifies the **initial value** of y by the requirement $y(t_0) = y_0$.

Definition 60. *A **first order homogeneous linear differential equation** is an equation of the form*

$$y' + p(t) \cdot y = 0.$$

It is called linear because y' and y appear linearly; it is called homogeneous because the right hand side is 0.

Theorem 122. *Let p be a continuous function and let P be an antiderivative of p. Then the general solution to*

$$y'(t) + p(t) \cdot y(t) = 0$$

is $y : t \mapsto C \cdot e^{-P(t)}$, with $C \in \mathbb{R}$.

Proof. We multiply both sides of the equation by the nonzero function $t \mapsto e^{P(t)}$ (called an *integrating factor*). We obtain

$$y'(t) \cdot e^{P(t)} + p(t) \cdot y(t) \cdot e^{P(t)} = 0.$$

The left-hand side is the derivative of $y \cdot e^P$ by the Product Rule, so we have

$$(y(t) \cdot e^{P(t)})' = 0,$$

which gives $y(t) \cdot e^{P(t)} = C$, for $C \in \mathbb{R}$. This implies

$$y(t) = C \cdot e^{-P(t)}.$$

$\square$

Note that there is a trivial solution: $y(t) = 0$ for all t (corresponding to $C = 0$).

Definition 61. *A **first order linear differential equation** is an equation of the form*

$$y' + p(t) \cdot y = f(t).$$

Theorem 123. *Let $y(t)$ be a particular solution of*

$$y' + p(t) \cdot y = f(t)$$

and let $u(t)$ be a nonzero solution of the corresponding homogeneous equation

$$u' + p(t) \cdot u = 0.$$

Then the general solution of $y' + p(t) \cdot y = f(t)$ is

$$y(t) + C \cdot u(t),$$

with C in $\mathbb{R}$.

Proof. First we check that $z(t) = y(t) + C \cdot u(t)$ is a solution of the equation. We have the following chain of equalities:

$$\begin{aligned} z' + p(t) \cdot z &= (y' + C \cdot u') + p(t) \cdot (y + C \cdot u) \\ &= (y' + p(t) \cdot y) + C \cdot (u' + p(t) \cdot u) \\ &= f(t) + 0 = f(t). \end{aligned}$$

We now check that all solutions are of this form. Assume z satisfies $z' + p \cdot z = f$. Then $z - y$ satisfies $(z' - y') + p \cdot (z - y) = f - f = 0$, hence $z - y$ is a solution of the corresponding homogeneous linear differential equation, hence necessarily $z - y$ is of the form $C \cdot u$ for some constant C. Thus we have $z(t) = y(t) + C \cdot u(t)$. $\square$

Theorem 124. *The general solution of a differential equation of the form*

$$y' + p(t) \cdot y = f(t)$$

is

$$y(t) = v(t) \cdot u(t) + C \cdot u(t),$$

where u is a nontrivial solution of the homogeneous equation and $v(t)$ is an antiderivative of $\frac{f(t)}{u(t)}$.

Proof. We use a method called *variation of parameters.*

We replace C by $v(t)$ in the general solution $y = C \cdot u(t)$ of the corresponding homogeneous equation $u' + p \cdot u = 0$ ($u(t) = e^{-P(t)}$; see Theorem 122), and look for a particular solution of the given equation in the form $y(t) = v(t) \cdot u(t)$. In simplified notation we have $y = v \cdot u$. After substituting we obtain

$$
\begin{aligned}
y' + p \cdot y &= (v \cdot u)' + p \cdot v \cdot u \\
&= v' \cdot u + v \cdot \underbrace{u'}_{=-p \cdot u} + p \cdot v \cdot u \\
&= v' \cdot u.
\end{aligned}
$$

The equation $v' \cdot u = f$, that is, $v' = \frac{f}{u}$, is satisfied if $v(t)$ is taken to be an antiderivative of $\frac{f(t)}{u(t)}$. By Theorem 123, the general solution is then

$$y(t) = v(t) \cdot u(t) + C \cdot u(t).$$

$\square$

8.2   Existence and Uniqueness of a Solution

In this section we prove two classical results: the global existence of a solution under the assumption of continuity and boundedness (from which we deduce the local existence under continuity) and the uniqueness of the solution under the assumption of the Lipschitz condition. The first proof uses the Standardization Principle, which is stated in the Appendix.

Definition 62.

(1) *Let I, J be intervals and $F : I \times J \to \mathbb{R}$ a function. Let $t \in I$ and $y \in J$. We say that F is **continuous at** $\langle t, y \rangle$ if*

$$F(t, y) \simeq F(s, z) \quad \text{whenever } s \simeq t, z \simeq y, s \in I, z \in J.$$

*(2) We say that $F : I \times J \to \mathbb{R}$ is **continuous** if F is continuous at each $\langle t, y \rangle \in I \times J$.*

The parameters of the first definition are F, I, J, t and y, and the parameters of the second definition are F, I, J.

Exercise 62 (Answer page 260)
Show that if F is continuous on $I \times J$ and $y : I \to J$ is continuous on I, then $f : I \to \mathbb{R}$ defined by $f(t) = F(t, y(t))$ is continuous on I.

Let y be a solution of the differential equation $y' = F(t, y)$ satisfying the initial condition $y(a) = y_0$. By integrating both sides of $y'(t) = F(t, y(t))$ from a to t we get

$$y(t) - y(a) = \int_a^t y'(s) \cdot ds = \int_a^t F(s, y(s)) \cdot ds \simeq \sum_{i=0}^{N} F(t_i, y(t_i)) \cdot dt,$$

where $t_i = a + i \cdot dt$ and dt is ultrasmall. Hence $y(t)$ is the observable neighbor of the sum $y(a) + \sum_{i=0}^{N} F(t_i, y(t_i)) \cdot dt$. In the proof of the next theorem we "reverse the tables": We *define* the function y as the "observable part" of the sum(s), and then *prove* that it is a solution of $y' = F(t, y)$.

Theorem 125. *Assume that $F : [a, b] \times \mathbb{R} \to \mathbb{R}$ is continuous and bounded. For every $y_0 \in \mathbb{R}$ there is a function $y : [a, b] \to \mathbb{R}$ such that*

$$y(a) = y_0 \quad and \quad y'(t) = F(t, y(t)) \quad for\ all\ t \in [a, b].$$

Proof. The context is given by F, a, b and y_0. By Closure, we can find an observable M such that $|F(t, y)| \le M$ for all $t \in [a, b]$ and all $y \in \mathbb{R}$.

Let N be a positive ultralarge integer. Let $dt = (b - a)/N$ and $t_k = a + k \cdot dt$ for $k = 0, \ldots, N$. We define y_k (for $k \ge 1$) by induction as follows:

$$y_{k+1} = y_k + F(t_k, y_k) \cdot dt.$$

Observe that

$$y_{\ell+1} = y_k + \sum_{i=k}^{\ell} F(t_i, y_i) \cdot dt, \quad \text{for any } k \le \ell \le N,$$

and hence

$$|y_{\ell+1} - y_k| = \left| \sum_{i=k}^{\ell} F(t_i, y_i) \cdot dt \right| \le M \cdot (t_{\ell+1} - t_k) \le M \cdot (b - a).$$

Let $\widetilde{y} : [a, b] \to \mathbb{R}$ be the function that linearly interpolates between the points $\langle t_0, y_0 \rangle, \langle t_1, y_1 \rangle, \ldots, \langle t_N, y_N \rangle$; that is,

$$\widetilde{y}(t) = y_k + F(t_k, y_k)(t - t_k) \quad \text{for } t_k \le t \le t_{k+1}.$$

We note that for any $t \le t'$ in $[a, b]$ there are $k \le \ell$ such that $t_k \le t \le t_{k+1}$, $t_\ell \le t' \le t_{\ell+1}$. Then $\widetilde{y}(t) \simeq y_k$, $\widetilde{y}(t') \simeq y_{\ell+1}$, and $t \simeq t'$ implies $\widetilde{y}(t) \simeq \widetilde{y}(t')$. Note that this is not just a consequence of the continuity of $\widetilde{y}$, because $\widetilde{y}$ is not observable. Also, $\widetilde{y}$ is bounded, by the observable bound $|y_0| + M \cdot (b - a)$.

By the Standardization Principle (see Theorem 161 in the Appendix), there is an observable function $y : [a, b] \to \mathbb{R}$ such that

$$y(t) \simeq \widetilde{y}(t) \quad \text{for all observable } t.$$

We complete the proof by showing that y has the required properties.

As $y(a) \simeq \widetilde{y}(a) = y_0$ and both $y(a)$ and y_0 are observable, we have $y(a) = y_0$.

Let $t \le t'$ be observable; then $t_k \le t \le t_{k+1}$, $t_\ell \le t' \le t_{\ell+1}$, for some $k \le \ell$. We have $y(t) \simeq \widetilde{y}(t) \simeq y_k$, $y(t') \simeq \widetilde{y}(t') \simeq y_{\ell+1}$, and $|y_{\ell+1} - y_k| \le M \cdot (t_{\ell+1} - t_k)$, where $t_{\ell+1} - t_k \simeq t' - t$. It follows that $|y(t') - y(t)| \le M \cdot |t' - t|$ holds for all observable t, t'. By Universal Closure the inequality $|y(t') - y(t)| \le M \cdot |t' - t|$ holds for all $t, t' \in [a, b]$. In particular, the function y is continuous on $[a, b]$.

Let us now fix an observable $t \in [a, b]$; say $t_k \le t \le t_{k+1}$. We have

$$y_{k+1} = y_0 + \sum_{i=0}^{k} F(t_i, y_i) \cdot dt. \tag{8.1}$$

We need one more observation. Fix i and let $\bar{t}$ be the observable neighbor of t_i. Then

$$y(t_i) \simeq y(\bar{t}) \simeq \widetilde{y}(\bar{t}) \simeq \widetilde{y}(t_i) = y_i$$

and, by continuity of F, $F(t_i, y_i) \simeq F(t_i, y(t_i))$. Thus (8.1) gives

$$y_{k+1} \simeq y_0 + \sum_{i=0}^{k} F(t_i, y(t_i)) \cdot dt. \tag{8.2}$$

(Recall that $\sum_{i=0}^{k} \varepsilon_i \cdot dt \simeq 0$, if $\varepsilon_i \simeq 0$ for each $i = 0, \ldots, k$.)

The sum in (8.2) resembles the sum in the definition of the integral of the continuous function $s \mapsto F(s, y(s))$ from a to t (it is exactly right when $t = t_{k+1}$). Exercise 63 (or Theorem 130 in Chapter 9) shows that indeed $\sum_{i=0}^{k} F(t_i, y(t_i)) \cdot dt \simeq \int_a^t F(s, y(s)) \cdot ds$. As also $y(t) \simeq y_{k+1}$, we conclude that

$$y(t) = y_0 + \int_a^t F(s, y(s)) \cdot ds$$

for all observable t, and hence, by Closure, for all t. The Fundamental Theorem of Calculus then gives $y'(t) = F(t, y(t))$ for all $t \in [a, b]$. □

Exercise 63 (Answer page 260)
Let f be continuous on $[a, b]$. Given an ultrasmall $dz > 0$, define $z_i = a + i \cdot dz$, for $i \in \mathbb{N}$. Let N be such that $z_{N-1} \le b \le z_N$. Prove that

$$\sum_{i=0}^{N-1} f(z_i) \cdot dz \simeq \sum_{i=0}^{N-1} f(x_i) \cdot dx \simeq \int_a^b f(x) \cdot dx,$$

where $dx = (b - a)/N$ and $x_i = a + i \cdot dx$, as usual.

The above proof of this theorem goes through under the more technical assumption that $F : [a, a+c] \times [y_0 - Mc, \ y_0 + Mc] \to \mathbb{R}$ is continuous and bounded by M (because the values of y_k, $k = 0, \ldots, N$, are bounded by Mc). From this version one can deduce immediately the usual Peano Existence Theorem.

Theorem 126 (Peano Existence Theorem). *Let I, J be open intervals and $F : I \times J \to \mathbb{R}$ a continuous function. Let $t_0 \in I$ and $y_0 \in J$. There is an open interval $I' \subseteq I$ with $t_0 \in I'$ and a function $y : I' \to J$ such that*

$$y(t_0) = y_0 \quad and \quad y'(t) = F(t, y(t)) \quad for \ all \ t \in I'.$$

In general, the solution satisfying a given initial condition is not uniquely determined; different choices of the ultralarge N in the proof of Theorem 125 may yield different solutions. An additional condition on F guarantees uniqueness.

Definition 63. *A function $F : I \times J \to \mathbb{R}$ satisfies the **Lipschitz condition** if there is a constant $K > 0$ such that, for all $t \in I$ and all $y_1, y_2 \in J$, we have*

$$|F(t, y_1) - F(t, y_2)| \le K \cdot |y_1 - y_2|.$$

Readers familiar with partial derivatives will recognize that a function F that is continuous and has partial derivative with respect to y bounded by K, satisfies the Lipschitz condition with the constant K (use the Mean Value Theorem).

The uniqueness of the solution satisfying a given initial condition $y(a) = y_0$ is a corollary of the next theorem.

Theorem 127. *Let $F : [a, a+c] \times [y_0 - Mc, \ y_0 + Mc] \to \mathbb{R}$ be continuous, bounded by M, and satisfy the Lipschitz condition. Let $\delta < \min\{1/K, c\}$, $\delta > 0$. If $y_1(t)$, $y_2(t)$ are solutions of $y' = F(t, y)$ on $[a, a+c]$ and satisfy $y_1(0) = y_2(0) = y_0$, then $y_1(t) = y_2(t)$ for all $t \in [a, a + \delta]$.*

Proof. Let y_1, y_2 be solutions of $y' = F(t, y)$ on $[a, a + c]$ with $y_1(0) = y_2(0) = y_0$. By integration,

$$y_1(t) = \int_a^t F(s, y_1(s)) \cdot ds \quad \text{and} \quad y_2(t) = \int_a^t F(s, y_2(s)) \cdot ds.$$

We have

$$|y_1(t) - y_2(t)| \leq \int_a^t |F(s, y_1(s)) - F(s, y_2(s))| \cdot ds$$

$$\leq \int_a^t K \cdot |y_1(s) - y_2(s)| \cdot ds.$$

The functions y_1 and y_2 are continuous on $[a, a + \delta]$, so there is some $\bar{t} \in [a, a + \delta]$ where $|y_1(\bar{t}) - y_2(\bar{t})| = \max\{|y_1(s) - y_2(s)| : s \in [a, a + \delta]\} = B$. Applying the last inequality to $\bar{t}$ we obtain

$$B = |y_1(\bar{t}) - y_2(\bar{t})| \leq \int_a^{\bar{t}} K \cdot |y_1(s) - y_2(s)| \cdot ds \leq K \cdot B \cdot \delta.$$

If $B > 0$, we get $1 \leq K \cdot \delta$, a contradiction with the choice of δ. Hence $B = 0$ and $y_1(s) = y_2(s)$ for all $s \in [a, a + \delta]$.   □

Corollary. *Let I, J be open intervals and $F : I \times J \to \mathbb{R}$ a continuous function satisfying the Lipschitz condition. Let $t_0 \in I$ and $y_0 \in J$. If y_1 and y_2 are two solutions of $y' = F(t, y)$ and satisfy $y_1(t_0) = y_2(t_0) = y_0$, then $y_1(t) = y_2(t)$ holds for all $t \in I$ where both sides are defined.*

Proof. We assume that the domains of y_1 and y_2 are intervals. Let $a = \sup\{t \in I : t \geq t_0 \text{ and } y_1(t) = y_2(t)\}$. If both $y_1(a)$ and $y_2(a)$ are defined, then also $y_1(a) = y_2(a)$, by continuity. Suppose that both y_1 and y_2 are defined on some interval $[a, a + c] \subseteq I$. Then $y_1(t) = y_2(t)$ for some $t > a$, by Theorem 127, contradicting the definition of a. For $t \leq t_0$ use the obvious modification of Theorem 127, with $[a, a + c]$ replaced by $[a - c, a]$.   □

8.3   Additional Exercises

Given a differential equation $y' = f(t, y)$, it is convenient to visualize the function f as a **slope field**, that is, by representing the value $f(t, y)$ at each point $\langle t, y \rangle$ of a grid by the slope of a small vector. The graph of a solution has to be tangent to this vector at every point through which the solution passes.

Exercise 8.1
The following slope field is for $y' = y^2 \cdot t$

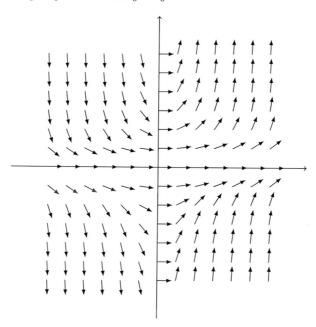

(1) Draw the solutions for different initial values.

(2) Solve the equation algebraically.

Exercise 8.2
Draw the slope field and some of the solutions for

$$y' - 2t\sqrt{1 - y^2}, \quad \text{for } t \in [-1, 1], y \in [-1, 1] \quad \text{(grid every 0.5)}.$$

Exercise 8.3
Draw the slope field and some of the solutions for

$$y' = (1 + y^2)te^t \quad \text{for } t \in [-1, 1], y \in [-1, 1] \quad \text{(grid every 0.5)}.$$

Exercise 8.4
Find the general solutions of the following differential equations.

(1) $y' = y(y + 1)$

(2) $y' = t \sin(t^2)$

(3) $y' = e^{-y}$

(4) $y' = yt \tan(t)$

(5) $(t - 1)y' - 2y = 0$

(6) $y' = ty \ln(t)$

Exercise 8.5
Solve the initial value problems.
Hint: These differential equations have separable variables.

(1) $y' = y^2 t^2$,   initial value $y(1) = 2$.

(2) $y' = t\sqrt{y}$,   initial value $y(0) = 3$.

(3) $y' = \dfrac{\ln(t)}{y}$,   initial value $y(1) = -2$.

(4) $y' = (y^2 - 3y + 2)\sqrt{t}$,   initial value $y(1) = 2$.

Exercise 8.6
Solve the initial value problems.
Hint: These differential equations are linear.

(1) $y' + y = 0$,   initial value $y(0) = 4$.

(2) $y' + y \sin(t) = 0$,   initial value $y(\pi) = 1$.

(3) $y' + y\sqrt{1 + t^2} = 0$,   initial value $y(0) = 0$.

(4) $y' + y \cos(e^t) = 0$,   initial value $y(0) = 0$.

(5) $ty' - 2y = 0$,   initial value $y(1) = 4$, $t > 0$.

(6) $t^2 y' + y = 0$,   initial value $y(1) = -2$, $t > 0$.

(7) $t^3 y' = 2y$,   initial value $y(1) = 1$, $t > 0$.

(8) $t^3 y' = 2y$,   initial value $y(1) = 0$, $t > 0$.

(9) $y' - 3y = 0$,   initial value $y(1) = -2$.

(10) $y' + ye^t = 0$,   initial value $y(0) = e$.

Exercise 8.7
Find the general solution of the differential equations.

(1) $y' + 4y = 8$

(2) $y' - 2y = 6$

(3) $y' + ty = 5t$

(4) $y' + e^t y = -2e^t$

(5) $y' - y = t^2$

(6) $2y' + y = t$

(7) $ty' + 2y = 1/t,$ for $t > 0$.

(8) $ty' + y = \sqrt{t},$ for $t > 0$.

Exercise 8.8
Compute the equation of the curve y_k whose tangent line at t intersects the y-axis at $k \cdot y_k(t)$.

Exercise 8.9
Show that the equation $y' = \sqrt{y}$ has two solutions satisfying the initial condition $y(0) = 0$.

Hint: Consider the functions $y_1(t) = 0$ $(t \in \mathbb{R})$ and

$$y_2(t) = \begin{cases} 0 & \text{if } t \le 0 \\ t^2/4 & \text{otherwise.} \end{cases}$$

9

Integration

9.1 Riemann Integral

In order to extend the integral to functions which are not necessarily continuous, it is convenient to consider "approximations" of intervals by finite sets of points where the distance between two consecutive points is not fixed.

Definition 64. *A **partition** of $[a, b]$ is a finite set $\mathcal{P} = \{x_0, x_1, \ldots, x_n\}$ where*

$$a = x_0 < x_1 < \ldots < x_{i-1} < x_i < \ldots < x_n = b.$$

The partition $\mathcal{P}$ thus divides the interval $[a, b]$ into n non-overlapping closed subintervals

$$[x_0, x_1], [x_1, x_2], \ldots, [x_{n-1}, x_n].$$

It is often convenient to refer to this set of intervals as the partition $\mathcal{P}$. We let

$$dx_i = x_{i+1} - x_i$$

be the length of the subinterval $[x_i, x_{i+1}]$, for $i = 0, \ldots, n - 1$. Clearly

$$\sum_{i=0}^{n-1} dx_i = b - a.$$

Definition 65. *A **tagged partition** consists of a partition $\mathcal{P}$ and a set $\mathcal{T} = \{t_0, \ldots, t_{n-1}\}$ where*

$$x_i \leq t_i \leq x_{i+1}, \quad for \ i = 0, \ldots, n - 1.$$

*The elements of $\mathcal{T}$ are called **tags**. The number t_i is the tag attached to the subinterval $[x_i, x_{i+1}]$.*

Definition 66. *Let f be a function defined on $[a, b]$. The **Riemann sum** $\sum(f; \mathcal{P}, \mathcal{T})$ associated with the function f and the tagged partition $(\mathcal{P}, \mathcal{T})$ is defined as*

$$\sum(f; \mathcal{P}, \mathcal{T}) = \sum_{i=0}^{n-1} f(t_i) \cdot dx_i.$$

Example. Let n be a positive integer and $dx = (b-a)/n$. Let $\mathcal{P}$ be given by

$$a = x_0 < x_1 = a + dx < \cdots < x_i = a + i \cdot dx < \cdots < x_n = b.$$

We refer to this as an **even** partition.

(1) The choice of $t_i = x_i$ gives the left endpoint sums

$$\sum(f; \mathcal{P}, \mathcal{T}) = \sum_{i=0}^{n-1} f(x_i) \cdot dx$$

used in the definition of integral given in Chapter 4. We call this tagging the *left tagging* of the partition $\mathcal{P}$.

(2) The choice of $t_i = x_{i+1}$ gives the right endpoint sums

$$\sum(f; \mathcal{P}, \mathcal{T}) = \sum_{i=0}^{n-1} f(x_{i+1}) \cdot dx.$$

We call this the *right tagging* of the partition $\mathcal{P}$.

(3) The choice of $t_i = \frac{x_i + x_{i+1}}{2}$ gives the *midpoint sums*

$$\sum(f; \mathcal{P}, \mathcal{T}) = \sum_{i=0}^{n-1} f\left(\frac{x_i + x_{i+1}}{2}\right) \cdot dx.$$

(4) Let f be continuous on $[a, b]$. The choice of $t_i \in [x_i, x_{i+1}]$ such that $f(t_i) = \frac{f(x_i) + f(x_{i+1})}{2}$ gives an approximation of the integral using *trapezoids*:

$$\sum(f; \mathcal{P}, \mathcal{T}) = \sum_{i=0}^{n-1} f(t_i) \cdot dx$$

$$= \frac{1}{2} \cdot \left(\sum_{i=0}^{n-1} f(x_i) \cdot dx + \sum_{i=0}^{n-1} f(x_{i+1}) \cdot dx \right).$$

Definition 67. *A partition is **fine** if, relative to the context, all dx_i are ultrasmall.*

We first show that, for continuous functions, all fine tagged partitions give Riemann sums ultraclose to the integral as defined in Chapter 4.

Definition 68. *A partition $\mathcal{P}_2$ **refines** $\mathcal{P}_1$ if $\mathcal{P}_1 \subseteq \mathcal{P}_2$.*

Theorem 128. *Let f be continuous on $[a,b]$. Let $(\mathcal{P}_1, \mathcal{T}_1)$ and $(\mathcal{P}_2, \mathcal{T}_2)$ be fine tagged partitions of $[a,b]$. Then*

$$\sum(f; \mathcal{P}_1, \mathcal{T}_1) \simeq \sum(f; \mathcal{P}_2, \mathcal{T}_2).$$

Proof. With a partition $\mathcal{P} = \{x_0, \ldots, x_n\}$ we associate a minimal and a maximal tagging as follows. By the Extreme Value Theorem, we choose $c_i, d_i \in [x_i, x_{i+1}]$ such that

$$f(c_i) \leq f(x) \leq f(d_i), \quad \text{for all } x \in [x_i, x_{i+1}].$$

We then define $s(\mathcal{P})$ and $S(\mathcal{P})$ as

$$s(\mathcal{P}) = \sum_{i=0}^{n-1} f(c_i) \cdot dx_i \quad \text{and} \quad S(\mathcal{P}) = \sum_{i=0}^{n-1} f(d_i) \cdot dx_i.$$

Then, regardless of $\mathcal{T}$, we have

$$s(\mathcal{P}) \leq \sum(f; \mathcal{P}, \mathcal{T}) \leq S(\mathcal{P}).$$

Now, if $\mathcal{P}$ is fine, then $dx_i \simeq 0$, $c_i \simeq d_i$, and since f is uniformly continuous, there is $\varepsilon_i \simeq 0$ such that $f(d_i) = f(c_i) + \varepsilon_i$. This implies that

$$S(\mathcal{P}) = \sum_{i=0}^{n-1} f(d_i) \cdot dx_i = \sum_{i=0}^{n-1} (f(c_i) + \varepsilon_i) \cdot dx_i = s(\mathcal{P}) + \sum_{i=0}^{n-1} \varepsilon_i \cdot dx_i \simeq s(\mathcal{P}),$$

since $\sum_{i=0}^{n-1} \varepsilon_i \cdot dx_i \simeq 0$. Thus we have, for all $\mathcal{T}$,

$$s(\mathcal{P}) \simeq \sum(f; \mathcal{P}, \mathcal{T}) \simeq S(\mathcal{P}).$$

Now let $(\mathcal{P}_1, \mathcal{T}_1)$ and $(\mathcal{P}_2, \mathcal{T}_2)$ be fine tagged partitions of $[a,b]$. Let $\mathcal{P}^*$ be the union of $\mathcal{P}_1$ and $\mathcal{P}_2$. Since the maximum value of f taken over a smaller interval can only decrease, and the minimum increase, we have

$$s(\mathcal{P}_1) \leq s(\mathcal{P}^*) \leq S(\mathcal{P}^*) \leq S(\mathcal{P}_1)$$

and

$$s(\mathcal{P}_2) \leq s(\mathcal{P}^*) \leq S(\mathcal{P}^*) \leq S(\mathcal{P}_2).$$

But then
$$s(\mathcal{P}_1) \simeq s(\mathcal{P}^*) \quad \text{and} \quad s(\mathcal{P}_2) \simeq s(\mathcal{P}^*).$$

We conclude that
$$\sum(f; \mathcal{P}_1, \mathcal{T}_1) \simeq s(\mathcal{P}_1) \simeq s(\mathcal{P}^*) \simeq s(\mathcal{P}_2) \simeq \sum(f; \mathcal{P}_2, \mathcal{T}_2).$$

$\square$

We want to consider functions which are not necessarily continuous. We always assume that f is bounded on $[a, b]$.

Definition 69. *A bounded function f is **Riemann integrable on** $[a, b]$ if there is an observable real number R such that*

$$\sum(f; \mathcal{P}, \mathcal{T}) \simeq R,$$

for all fine tagged partitions $\mathcal{P}, \mathcal{T}$ of $[a, b]$.
  If this is the case, we let

$$\int_a^b f(x) \cdot dx = R.$$

The parameters are f, a and b, so of course any context where f, a, b are observable can be used in this definition. By the usual argument (see Exercise 4.8), the defining statement is equivalent to an internal one. If f is Riemann integrable, then $\int_a^b f(x) \cdot dx$ is the observable neighbor of $\sum_{i=0}^{n-1} f(t_i) \cdot dx_i$, for any fine tagged partition of $[a, b]$.
  If f is bounded on $[a, b]$, then by Closure there is an observable B such that $|f(x)| \leq B$, for all $x \in [a, b]$. Let B be such a number. Then

$$\left| \sum(f; \mathcal{P}, \mathcal{T}) \right| = \left| \sum_{i=0}^{n-1} f(t_i) \cdot dx_i \right| \leq \sum_{i=0}^{n-1} |f(t_i)| \cdot dx_i \leq B \cdot (b - a),$$

where $B \cdot (b - a)$ is observable. Thus $\sum(f; \mathcal{P}, \mathcal{T})$ is never ultralarge, and the observable neighbor of $\sum_{i=0}^{n-1} f(t_i) \cdot dx_i$ always exists. The issue is therefore only whether different fine partitions might yield sums which are not ultraclose to each other. These considerations prove the following theorem.

Theorem 129. *A bounded function f is Riemann integrable on $[a, b]$ if and only if*
$$\sum(f; \mathcal{P}_1, \mathcal{T}_1) \simeq \sum(f; \mathcal{P}_2, \mathcal{T}_2),$$

for all fine tagged partitions $(\mathcal{P}_1, \mathcal{T}_1)$ and $(\mathcal{P}_2, \mathcal{T}_2)$ of $[a, b]$.

Theorem 128 can be restated as follows.

Theorem 130. *Every f continuous on $[a, b]$ is Riemann integrable on $[a, b]$ and the value of the integral is the same as the one obtained with left-tagged even partitions, that is, partitions where the subintervals are of equal length and the tags are chosen to be the left endpoints.*

This justifies our use of the same notation for the integral here as in Chapter 4.

Theorem 131. *Every f monotonic on $[a, b]$ is Riemann integrable on $[a, b]$.*

Proof. We prove it in the case when f is increasing; the other case is similar. Clearly f is bounded on $[a, b]$ by $f(a)$ below and by $f(b)$ above. Let $\mathcal{P}$ be a fine partition of $[a, b]$:

$$a = x_0 < x_1 < \cdots < x_n = b.$$

Since f is increasing, we have

$$f(x_i) \leq f(t) \leq f(x_{i+1}), \quad \text{for all } t \in [x_i, x_{i+1}].$$

We consider an arbitrary tagging

$$\mathcal{T} = \{t_0, \ldots, t_{n-1}\}, \quad \text{with } t_i \in [x_i, x_{i+1}].$$

With our previous definitions of $s(\mathcal{P})$ and $S(\mathcal{P})$ (see the proof of Theorem 128), we have

$$s(\mathcal{P}) = \sum_{i=0}^{n-1} f(x_i) \cdot dx_i \leq \sum (f; \mathcal{P}, \mathcal{T}) \leq \sum_{i=0}^{n-1} f(x_{i+1}) \cdot dx_i = S(\mathcal{P}).$$

It therefore suffices to show that $s(\mathcal{P}) \simeq S(\mathcal{P})$.

Let $\delta = \max\{dx_0, \ldots, dx_{n-1}\}$, where $dx_i = x_{i+1} - x_i$. Then δ is ultrasmall, since $\mathcal{P}$ is fine. Hence,

$$S(\mathcal{P}) - s(\mathcal{P}) - \sum_{i=0}^{n-1} f(x_{i+1}) \cdot dx_i - \sum_{i=0}^{n-1} f(x_i) \cdot dx_i$$

$$= \sum_{i=0}^{n-1} \Big(f(x_{i+1}) - f(x_i) \Big) \cdot dx_i$$

$$\leq \delta \cdot \sum_{i=0}^{n-1} \Big(f(x_{i+1}) - f(x_i) \Big)$$

$$= \delta \cdot \Big(f(b) - f(a) \Big) \simeq 0,$$

because $f(b) - f(a)$ is observable and δ is ultrasmall. $\qquad \square$

Example. The Dirichlet function

$$f(x) = \begin{cases} 1 & \text{if } x \in \mathbb{Q}; \\ 0 & \text{otherwise} \end{cases}$$

is not Riemann integrable over any $[a, b]$, $a < b$.

Proof. Let $\mathcal{P}$ be any fine partition of $[a, b]$. Let $\mathcal{T}_1$ be a set of tags where each t_i is rational, and let $\mathcal{T}_2$ be a set of tags where each t_i is irrational. Then $\sum(f; \mathcal{P}, \mathcal{T}_1) = b - a$ and $\sum(f; \mathcal{P}, \mathcal{T}_2) = 0$. $\qquad\square$

In particular, consider the interval $[0, 1]$. With equal partitions, the sum $\sum_{i=0}^{n-1} f(x_i) \cdot dx = 1$ as all the x_i are rational. But for an irrational $r \simeq 1$ ($dy = r/N$, $y_i = i \cdot dy$, $N \in \mathbb{N}$ ultralarge relative to r) the sum $\sum_{i=0}^{N-1} f(y_i) \cdot dy = 0$. If we had defined Riemann integral using even partitions and left taggings only, as in Chapter 4, then $\int_0^1 f(x) \cdot dx = 1$ and $\int_0^r f(y) \cdot dy = 0$, so the analog of Theorem 55 would fail.

We now extend our theorems about the integral of continuous functions to the case of Riemann integrable functions.

Theorem 132 (Linearity). *Let f and g be Riemann integrable on $[a, b]$. Let λ, μ be real numbers. Then $\lambda \cdot f + \mu \cdot g$ is Riemann integrable and*

$$\int_a^b (\lambda \cdot f(x) + \mu \cdot g(x)) \cdot dx = \lambda \cdot \int_a^b f(x) \cdot dx + \mu \cdot \int_a^b g(x) \cdot dx.$$

Proof. Let $(\mathcal{P}, \mathcal{T})$ be a fine tagged partition of $[a, b]$. The theorem follows from the fact that

$$\sum(\lambda \cdot f + \mu \cdot g; \mathcal{P}, \mathcal{T}) = \lambda \cdot \sum(f; \mathcal{P}, \mathcal{T}) + \mu \cdot \sum(g; \mathcal{P}, \mathcal{T}).$$

$\qquad\square$

Theorem 133 (Monotonicity). *Let f and g be Riemann integrable on $[a, b]$. Assume $f(x) \le g(x)$ for all $x \in [a, b]$. Then*

$$\int_a^b f(x) \cdot dx \le \int_a^b g(x) \cdot dx.$$

Proof. Let $(\mathcal{P}, \mathcal{T})$ be a fine tagged partition of $[a, b]$. The result follows immediately from the fact that

$$\sum(f; \mathcal{P}, \mathcal{T}) \le \sum(g; \mathcal{P}, \mathcal{T}).$$

$\qquad\square$

Theorem 134 (Additivity). *Assume that $a \leq b \leq c$. Then f is Riemann integrable on $[a, c]$ if and only if f is Riemann integrable on $[a, b]$ and on $[b, c]$. Moreover,*

$$\int_a^b f(x) \cdot dx + \int_b^c f(x) \cdot dx = \int_a^c f(x) \cdot dx.$$

Proof. Assume that f is Riemann integrable on $[a, c]$. Let $(\mathcal{P}_1, \mathcal{T}_1)$ and $(\mathcal{P}_2, \mathcal{T}_2)$ be fine tagged partitions of $[a, b]$. We extend them to fine tagged partitions $(\mathcal{P}_1', \mathcal{T}_1')$ and $(\mathcal{P}_2', \mathcal{T}_2')$ of $[a, c]$ in such a way that the extensions coincide on $[b, c]$. In more detail: Fix a fine tagged partition $(\mathcal{P}_3, \mathcal{T}_3)$ of $[b, c]$ and let $\mathcal{P}_1' = \mathcal{P}_1 \cup \mathcal{P}_3$, $\mathcal{T}_1' = \mathcal{T}_1 \cup \mathcal{T}_3$, and similarly for $(\mathcal{P}_2', \mathcal{T}_2')$.

As f is Riemann integrable on $[a, c]$, we have

$$\sum(f; \mathcal{P}_1', \mathcal{T}_1') \simeq \sum(f; \mathcal{P}_2', \mathcal{T}_2').$$

But

$$\sum(f; \mathcal{P}_1, \mathcal{T}_1) - \sum(f; \mathcal{P}_2, \mathcal{T}_2) = \sum(f; \mathcal{P}_1', \mathcal{T}_1') - \sum(f; \mathcal{P}_2', \mathcal{T}_2'),$$

hence

$$\sum(f; \mathcal{P}_1, \mathcal{T}_1) \simeq \sum(f; \mathcal{P}_2, \mathcal{T}_2).$$

Therefore f is Riemann integrable on $[a, b]$. In a similar way, one shows that f is Riemann integrable on $[b, c]$.

For the converse, assume that f is Riemann integrable on $[a, b]$ and on $[b, c]$. Let $(\mathcal{P}, \mathcal{T})$ be a fine tagged partition of $[a, c]$, with $x_i \leq b < x_{i+1}$. Let $(\mathcal{P}_1, \mathcal{T}_1)$ be the restriction of $(\mathcal{P}, \mathcal{T})$ to $[a, b]$, obtained by adding b as the final element of $\mathcal{P}_1$ and $\mathcal{T}_1$, if necessary. In more detail: If $x_i = b$, let $\mathcal{P}_1 = \mathcal{P} \cap [a, b]$, $\mathcal{T}_1 = \mathcal{T} \cap [a, b]$, and similarly for $\mathcal{P}_2$, $\mathcal{T}_2$. If $x_i < b < x_{i+1}$, let $\mathcal{P}_1 = (\mathcal{P} \cap [a, b]) \cup \{b\}$ and $\mathcal{T}_1 = \mathcal{T} \cap [a, b]$ or $\mathcal{T}_1 = (\mathcal{T} \cap [a, b]) \cup \{b\}$, depending on whether or not $t_i \in [x_i, b]$. Similarly, let $(\mathcal{P}_2, \mathcal{T}_2)$ be the restriction of $(\mathcal{P}, \mathcal{T})$ to $[b, c]$, obtained by adding b as the initial element of $\mathcal{P}_2$ and $\mathcal{T}_2$, if necessary. The partition $(\mathcal{P}_1 \cup \mathcal{P}_2, \mathcal{T}_1 \cup \mathcal{T}_2)$, obtained by concatenating $(\mathcal{P}_1, \mathcal{T}_1)$ and $(\mathcal{P}_2, \mathcal{T}_2)$, refines $(\mathcal{P}, \mathcal{T})$, and as $f(b)$ is not ultralarge, the contributions $f(t_i) \cdot (x_{i+1} - x_i)$, $f(b) \cdot (b - x_i)$ and $f(b) \cdot (x_{i+1} - b)$ are ultrasmall, and we have

$$\sum(f; \mathcal{P}, \mathcal{T}) \simeq \sum(f; \mathcal{P}_1 \cup \mathcal{P}_2, \mathcal{T}_1 \cup \mathcal{T}_2) = \sum(f; \mathcal{P}_1, \mathcal{T}_1) + \sum(f; \mathcal{P}_2, \mathcal{T}_2).$$

Therefore

$$\sum(f; \mathcal{P}, \mathcal{T}) \simeq \int_a^b f(x) \cdot dx + \int_b^c f(x) \cdot dx,$$

which shows that f is integrable on $[a, c]$ and

$$\int_a^c f(x) \cdot dx = \int_a^b f(x) \cdot dx + \int_b^c f(x) \cdot dx.$$

□

Theorem 135 (Continuity). *Let f be Riemann integrable on $[a, b]$. Let $F : [a, b] \to \mathbb{R}$ be defined by*

$$F : x \mapsto \int_a^x f(t) \cdot dt.$$

Then F is a continuous function on $[a, b]$.

Proof. Let $x \in I$; we show the continuity of F at x. The parameters are f, a, b and x. Let B be an observable bound on f. Let h be ultrasmall. Then by additivity, we have

$$F(x + h) = F(x) + \int_x^{x+h} f(t) \cdot dt.$$

But

$$\left| \int_x^{x+h} f(t) \cdot dt \right| \leq B \cdot |h| \simeq 0,$$

which shows that $F(x + h) \simeq F(x)$.   □

Theorem 136 (Fundamental Theorem of Calculus). *Let f be Riemann integrable on $[a, b]$. Let $F : [a, b] \to \mathbb{R}$ be defined by*

$$x \mapsto \int_a^x f(t) \cdot dt.$$

If f is continuous at $c \in [a, b]$, then $F'(c) = f(c)$.

Proof. The parameters are f, a, b and c. Let $h > 0$ be ultrasmall. We no longer have continuity of f on $[c, c + h]$ to conclude that f attains a minimum value and a maximum value on $[c, c + h]$, but f is bounded on $[c, c + h]$ and so it has an infimum and a supremum. Let

$$m = \inf\{f(x) : x \in [c, c + h]\} \quad \text{and} \quad M = \sup\{f(x) : x \in [c, c + h]\}.$$

Then

$$m \cdot h \leq \int_c^{c+h} f(t) \cdot dt \leq M \cdot h,$$

and

$$m \le \frac{F(c+h) - F(c)}{h} = \frac{1}{h} \int_c^{c+h} f(t) \cdot dt \le M.$$

By definition of the infimum and supremum, there are $d, e \in [c, c+h]$ such that

$$m \simeq f(d) \quad \text{and} \quad f(e) \simeq M.$$

But f is continuous at c and $c \simeq d$, $c \simeq e$, thus $m \simeq f(d) \simeq f(c) \simeq f(e) \simeq M$, and therefore

$$\frac{F(c+h) - F(c)}{h} \simeq f(c).$$

The argument for $h < 0$ is similar. $\qquad\square$

Theorem 137. *Let $g : [a, b] \to [g(a), g(b)]$ be smooth and strictly increasing and let f be Riemann integrable on $[g(a), g(b)]$. Then $x \mapsto f(g(x)) \cdot g'(x)$ is Riemann integrable on $[a, b]$ and*

$$\int_a^b f(g(x)) \cdot g'(x) \cdot dx = \int_{g(a)}^{g(b)} f(y) \cdot dy.$$

Proof. The parameters are a, b, g, and f. Let $(\mathcal{P}, \mathcal{T})$ be a fine tagged partition of $[a, b]$, where $\mathcal{P} = \{x_0, \ldots, x_n\}$ and $\mathcal{T} = \{t_0, \ldots, t_{n-1}\}$. Let $y_i = g(x_i)$, $s_i = g(t_i)$, and $dy_i = y_{i+1} - y_i$, for $i = 0, \ldots, n - 1$. As g is strictly increasing, $(\{y_0, \ldots, y_n\}, \{s_0, \ldots, s_{n-1}\})$ is a tagged partition of $[g(a), g(b)]$. It is also fine, because g is uniformly continuous, so all dy_i are ultrasmall. Since g is smooth, we have

$$dy_i = g(x_{i+1}) - g(x_i) = g'(t_i) \cdot dx_i + \varepsilon_i \cdot dx_i,$$

with $\varepsilon_i \simeq 0$. Hence

$$\sum_{i=0}^{n-1} f(g(t_i)) \cdot g'(t_i) \cdot dx_i = \sum_{i=0}^{n-1} f(s_i) \cdot (dy_i - \varepsilon_i \cdot dx_i)$$

$$= \sum_{i=0}^{n-1} f(s_i) \cdot dy_i - \sum_{i=0}^{n-1} f(s_i) \cdot \varepsilon_i \cdot dx_i.$$

But since f is bounded, there is an observable M such that $|f(s_i)| \le M$, for all $i = 0, \ldots, n - 1$. So

$$\left| \sum_{i=0}^{n-1} f(s_i) \cdot \varepsilon_i \cdot dx_i \right| \le \sum_{i=0}^{n-1} |f(s_i) \cdot \varepsilon_i \cdot dx_i| \le \underbrace{M}_{\text{not ultralarge}} \cdot \underbrace{\sum_{i=0}^{n-1} |\varepsilon_i| \cdot dx_i}_{\simeq 0} \simeq 0.$$

As f is integrable on $[g(a), g(b)]$, we have

$$\sum_{i=0}^{n-1} f(g(t_i)) \cdot g'(t_i) \cdot dx_i \simeq \sum_{i=0}^{n-1} f(s_i) \cdot dy_i \simeq \int_{g(a)}^{g(b)} f(y) \cdot dy.$$

Hence $x \mapsto f(g(x)) \cdot g'(x)$ is integrable on $[a, b]$ and

$$\int_a^b f(g(x)) \cdot g'(x) \cdot dx = \int_{g(a)}^{g(b)} f(y) \cdot dy.$$

$\square$

9.2   Darboux Integral

We now consider a second definition of Riemann integral, due to Darboux, and prove that it is equivalent to the one we adopted. Let $\mathcal{P}$ be a partition of $[a, b]$. Let f_i be the infimum of $\{f(x) : x_i \leq x \leq x_{i+1}\}$ and let F_i be the supremum of $\{f(x) : x_i \leq x \leq x_{i+1}\}$. The infimum and supremum exist because we assume that f is bounded. We define the following sums:

- The **lower Darboux sum of** $\mathcal{P}$ is

$$s(\mathcal{P}) = \sum_{i=0}^{n-1} f_i \cdot dx_i.$$

- The **upper Darboux sum of** $\mathcal{P}$ is

$$S(\mathcal{P}) = \sum_{i=0}^{n-1} F_i \cdot dx_i.$$

It is easy to see that if $\mathcal{P}'$ is a refinement of $\mathcal{P}$, then

$$s(\mathcal{P}) \leq s(\mathcal{P}') \leq S(\mathcal{P}') \leq S(\mathcal{P}).$$

We consider the supremum $\sup_{\mathcal{P}} s(\mathcal{P})$ over all partitions of $[a, b]$. This supremum exists because

$$\{s(\mathcal{P}) \ : \ \mathcal{P} \text{ is a partition of } [a, b]\} \subseteq \mathbb{R}$$

is a nonempty set bounded above. Similarly, we consider the infimum $\inf_{\mathcal{P}} S(\mathcal{P})$ over all partitions of $[a, b]$. It follows from the remarks above that

$$\sup_{\mathcal{P}} s(\mathcal{P}) \leq \inf_{\mathcal{P}} S(\mathcal{P}).$$

Definition 70. *Let f be bounded on $[a, b]$. We say that f is **Darboux integrable on** $[a, b]$ if*

$$\sup_{\mathcal{P}} s(\mathcal{P}) = \inf_{\mathcal{P}} S(\mathcal{P}).$$

*The **Darboux integral of** f is $D_f = \sup_{\mathcal{P}} s(\mathcal{P})$.*

Theorem 138. *Let f be a bounded function on $[a, b]$. The following conditions are equivalent:*

(1) There exists a fine partition $\mathcal{P}$ such that

$$s(\mathcal{P}) \simeq S(\mathcal{P}).$$

(2) For all fine partitions $\mathcal{P}$ we have

$$s(\mathcal{P}) \simeq S(\mathcal{P}).$$

Proof. (2) implies (1) is clear.

We prove (1) implies (2). The parameters are f, a and b. By assumption, there is an observable B such that $|f(x)| \leq B$ for all $x \in [a, b]$. By (1) there exists a fine partition $\mathcal{P}^*$ such that

$$s(\mathcal{P}^*) \simeq S(\mathcal{P}^*).$$

We first prove that

$$s(\mathcal{P}) \simeq S(\mathcal{P})$$

for all partitions that are fine relative to the extended context specified by f, a, b and $\mathcal{P}^*$. We deduce the general case in the end.

Consider a partition $\mathcal{P} = \{x_0, \ldots, x_n\}$ of $[a, b]$, which is fine relative to the extended context. We write $\overset{+}{\simeq}$ when working relative to the extended context.

Let $\delta = \max\{dx_0, \ldots, dx_{n-1}\}$; $\delta \overset{+}{\simeq} 0$. Since $\mathcal{P} \cup \mathcal{P}^*$ refines $\mathcal{P}^*$, we must have

$$s(\mathcal{P} \cup \mathcal{P}^*) \simeq S(\mathcal{P} \cup \mathcal{P}^*).$$

Let m be the number of points by which $\mathcal{P}$ differs from $\mathcal{P} \cup \mathcal{P}^*$. Since m is at most the number of points in the partition $\mathcal{P}^*$, this m is not ultralarge (relative to extended context). Then

$$\left| s(\mathcal{P} \cup \mathcal{P}^*) - s(\mathcal{P}) \right| \leq m \cdot 2B \cdot \delta \overset{+}{\simeq} 0.$$

This shows that

$$s(\mathcal{P}) \simeq s(\mathcal{P} \cup \mathcal{P}^*).$$

The same argument shows that

$$S(\mathcal{P}) \simeq S(\mathcal{P} \cup \mathcal{P}^*),$$

and hence

$$s(\mathcal{P}) \simeq S(\mathcal{P}).$$

We now use Stability to show that the same is true for all fine partitions. Note that, given any observable $\varepsilon > 0$, there is $\delta > 0$ such that for any partition $\mathcal{P}$ with $\max\{dx_0, \ldots, dx_{n-1}\} \leq \delta$ we have

$$0 \leq S(\mathcal{P}) - s(\mathcal{P}) < \varepsilon.$$

(Take any $\delta \overset{+}{\simeq} 0$; any such partition is fine relative to the extended context.) By Closure, there exists an observable δ with the same property. This implies that any fine partition $\mathcal{P}$ satisfies $0 \leq S(\mathcal{P}) - s(\mathcal{P}) < \varepsilon$. As this is true for every observable $\varepsilon > 0$, the conclusion $s(\mathcal{P}) \simeq S(\mathcal{P})$ follows. $\qquad\square$

Theorem 139. *A bounded function f on $[a, b]$ is Riemann integrable if and only if it is Darboux integrable. In this case, the values of the Riemann and Darboux integrals are equal.*

Proof. The parameters are a, b and f.

Assume f is Riemann integrable. Let R be the value of the Riemann integral; then R is observable. Let $\mathcal{P}^* = \{x_0, \ldots, x_n\}$ be a fine partition of $[a, b]$. Fix an ultrasmall $\varepsilon > 0$. By definition of f_i and F_i, there are numbers $t_i, s_i \in [x_i, x_{i+1}]$ such that

$$f_i \leq f(t_i) < f(t_i) + \varepsilon \quad \text{and} \quad F_i \geq f(s_i) > F_i - \varepsilon.$$

It follows that

$$f_i = f(t_i) - \varepsilon_i \quad \text{and} \quad f(s_i) + \delta_i = F_i,$$

where ε_i, δ_i are positive and ultrasmall (or 0). But $\sum_{i=0}^{n-1} \varepsilon_i \cdot dx_i \simeq 0$ and $\sum_{i=0}^{n-1} \delta_i \cdot dx_i \simeq 0$. Hence,

$$s(\mathcal{P}^*) = \sum_{i=0}^{n-1} f_i \cdot dx_i = \sum_{i=0}^{n-1} (f(t_i) - \varepsilon_i) \cdot dx_i \simeq \sum_{i=0}^{n-1} f(t_i) \cdot dx_i \simeq R$$

and similarly

$$R \simeq \sum_{i=0}^{n-1} f(s_i) \cdot dx_i \simeq \sum_{i=0}^{n-1} (f(s_i) + \delta_i) \cdot dx_i = \sum_{i=0}^{n-1} F_i \cdot dx_i = S(\mathcal{P}^*).$$

Hence,

$$\sup_{\mathcal{P}} \ s(\mathcal{P}) \geq s(\mathcal{P}^*) \simeq R \simeq S(\mathcal{P}^*) \geq \inf_{\mathcal{P}} \ S(\mathcal{P}).$$

Since R, $\sup_{\mathcal{P}} \ s(\mathcal{P})$, and $\inf_{\mathcal{P}} \ S(\mathcal{P})$ are observable, we must have

$$\sup_{\mathcal{P}} \ s(\mathcal{P}) \geq \inf_{\mathcal{P}} \ S(\mathcal{P}) \quad \text{and hence} \quad \sup_{\mathcal{P}} \ s(\mathcal{P}) = \inf_{\mathcal{P}} \ S(\mathcal{P}) = R.$$

This shows that f is Darboux integrable on $[a, b]$ and the Darboux integral has value R.

Assume that f is Darboux integrable and let D be the value of the Darboux integral. By definition of the supremum and infimum, there exist partitions $\mathcal{P}_1$ and $\mathcal{P}_2$ such that

$$s(\mathcal{P}_1) \simeq D \quad \text{and} \quad D \simeq S(\mathcal{P}_2).$$

By refining $\mathcal{P}_1 \cup \mathcal{P}_2$ if necessary, we can find a fine partition $\mathcal{P}^*$ such that

$$s(\mathcal{P}^*) \simeq D \simeq S(\mathcal{P}^*).$$

By the previous theorem ((1) implies (2)), this is true for all fine partitions $\mathcal{P}$; but

$$s(\mathcal{P}) \leq \sum(f; \mathcal{P}, \mathcal{T}) \leq S(\mathcal{P}), \quad \text{so} \quad \sum(f; \mathcal{P}, \mathcal{T}) \simeq D.$$

Hence f is Riemann integrable and the value of the Riemann integral is D. $\qquad \square$

9.3 Additional Exercises

Exercise 9.1
For the function $f(x) = x^2$ on $[0, 1]$ compute the Riemann sum $\sum(f; \mathcal{P}, \mathcal{T})$ where $\mathcal{P}$ is the even partition of $[0, 1]$ and $\mathcal{T}$ is the left tagging [right tagging, respectively].

Exercise 9.2
If $\mathcal{P}$ is a partition of $[a, b]$, we let $\|\mathcal{P}\| = \max\{dx_i : i = 0, \cdots, n-1\}$. Let f be a bounded function on $[a, b]$. Prove that the following statements are equivalent:

(1) $\int_a^b f(x) \cdot dx = R.$

(2) For every $\varepsilon > 0$ there exists $\delta > 0$ such that for every partition $\mathcal{P}$ with $\|\mathcal{P}\| < \delta$ and for every tagging $\mathcal{T}$ we have $|R - \sum(f; \mathcal{P}, \mathcal{T})| < \varepsilon.$

Exercise 9.3
Prove: If f is Riemann integrable on $[a, b]$ and $(\mathcal{P}_n, \mathcal{T}_n)$ is a sequence of tagged partitions such that $\lim_{n \to \infty} \|\mathcal{P}_n\| = 0$, then

$$\int_a^b f(x) \cdot dx = \lim_{n \to \infty} \sum(f; \mathcal{P}_n, \mathcal{T}_n).$$

Exercise 9.4
Let f be bounded on $[-a, a]$. Prove that

(1) If f is even (that is, $f(x) = f(-x)$ for all $x \in [0, a]$), then $\int_{-a}^{a} f(x) \cdot dx = 2 \cdot \int_0^a f(x) \cdot dx$.
(2) If f is odd (that is, $f(x) = -f(-x)$ for all $x \in [0, a]$), then $\int_{-a}^{a} f(x) \cdot dx = 0$.

Exercise 9.5
Define the function $g : [0, 1] \to \mathbb{R}$ as follows:

$$g(x) = \begin{cases} \frac{1}{q} & \text{if } x = \frac{p}{q} \in \mathbb{Q}, \text{ where } p, q \text{ are relatively prime} \\ 0 & \text{otherwise.} \end{cases}$$

Show that g is Riemann integrable and $\int_0^1 g(x) \cdot dx = 0$.

Exercise 9.6
We call a function f defined on $[a, b]$ piecewise continuous if there is a partition $\{x_0, x_1, \ldots, x_N\}$ of $[a, b]$ such that f is continuous on each open interval (x_i, x_{i+1}). Prove that a function that is bounded and piecewise continuous on $[a, b]$ is Riemann integrable on $[a, b]$.

Exercise 9.7
Let f be a bounded function on $[a, b]$. If for every $\varepsilon > 0$ there exist Riemann integrable functions g and h such that $g(x) \leq f(x) \leq h(x)$ holds for all $x \in [a, b]$ and $\int_a^b (h(x) - g(x)) \cdot dx < \varepsilon$, then f is Riemann integrable on $[a, b]$.

Exercise 9.8
Prove: If f is Riemann integrable on $[a, b]$, then f^2 is Riemann integrable on $[a, b]$.

Hint: Let $|f|$ be bounded by M; then show that

$$\left|\sum(f^2; \mathcal{P}_1, \mathcal{T}_1) - \sum(f^2; \mathcal{P}_2, \mathcal{T}_2)\right| \leq 2M \cdot \left|\sum(f; \mathcal{P}_1, \mathcal{T}_1) - \sum(f; \mathcal{P}_2, \mathcal{T}_2)\right|$$

and apply Theorem 129.

Exercise 9.9
Give an example of a function f such that f is not Riemann integrable, but $|f|$ is Riemann integrable.

Exercise 9.10
Prove: If f and g are Riemann integrable on $[a, b]$, then $f \cdot g$ is Riemann integrable on $[a, b]$.
Hint: Use $(f + g)^2 = f^2 + 2f \cdot g + g^2$ and a previous exercise.

Exercise 9.11
(Hard; see the proof of Theorem 138.)
Prove that a bounded function f is Riemann integrable on $[a, b]$ if and only if there is an observable real number R and a fine partition $\mathcal{P}$ of $[a, b]$ such that

$$\sum(f; \mathcal{P}, \mathcal{T}) \simeq R,$$

for all $\mathcal{T}$ such that $(\mathcal{P}, \mathcal{T})$ is a tagged partition of $[a, b]$.

10

Topology of Real Numbers

10.1 Open and Closed Sets

The topology of $\mathbb{R}$ is concerned with the properties of sets of real numbers. We describe and study open and closed sets, dense sets and compact sets. We recall that real numbers x and y are *neighbors* (relative to a given context) if $x \simeq y$ (relative to the context). If $x \simeq y$ and x is observable, then x is the unique *observable neighbor* of y.

Definition 71. *Let A be a subset of $\mathbb{R}$.*

> *(1) We say that A is **open** if all neighbors of every observable $a \in A$ belong to A.*

> *(2) We say that A is **closed** if all observable $a \in \mathbb{R}$ that have a neighbor in A belong to A.*

The parameter in the above definitions is A. It is clear from the definition that $\emptyset$ and $\mathbb{R}$ are both open and closed. Rephrasing the definition, a set A is open if whenever a is observable, $a \in A$ and $x \simeq a$, then $x \in A$. A set A is closed if whenever a is observable and $a \simeq x$ for some $x \in A$, then $a \in A$.

Stating the contrapositive, we have: A is open if whenever x is not in A and x has an observable neighbor, then this neighbor is not in A.

Similarly, A is closed if whenever x has an observable neighbor and this neighbor is not in A, then x is not in A.

Recall that the **complement** of a set A is the set

$$B = A^c = \mathbb{R} \setminus A.$$

By definition, $x \in A$ if and only if $x \notin A^c$.

Exercise 64 (Answer page 261)

> (1) Show that the complement of an open set is closed.

> (2) Show that the complement of a closed set is open.

Exercise 65 (Answer page 261)

    (1) Show that an open interval is open.

    (2) Show that a closed interval is closed.

    (3) Show that $(a, b]$ is neither open nor closed.

Consider a family $\mathcal{A} = \{A_i : i \in I\}$ of subsets of $\mathbb{R}$; I is an arbitrary index set. Strictly speaking, this family is a function $i \mapsto A_i$ with domain I. The index set I is observable relative to $\mathcal{A}$ (but for emphasis we sometimes specify $\mathcal{A}$ and I as the context for the study of $\mathcal{A}$). For each $i \in I$, the set A_i is observable relative to $\mathcal{A}$ and i, by the Closure Principle.

Given a family of sets $\mathcal{A} = \{A_i : i \in I\}$, we define the **union of $\mathcal{A}$**, written $\cup \mathcal{A}$, by

$$\cup \mathcal{A} = \bigcup_{i \in I} A_i = \{x \mid \text{there is } i \in I \text{ such that } x \in A_i\}.$$

Similarly, we define the **intersection of $\mathcal{A} \neq \emptyset$**, written $\cap \mathcal{A}$, by

$$\cap \mathcal{A} = \bigcap_{i \in I} A_i = \{x \mid \text{for all } i \in I \text{ we have } x \in A_i\}.$$

By the Closure Principle, $\cup \mathcal{A}$ and $\cap \mathcal{A}$ are observable relative to $\mathcal{A}$.

Theorem 140.

    (1) Any union of open sets is open.

    (2) Any finite intersection of open sets is open.

    (3) Any intersection of closed sets is closed.

    (4) Any finite union of closed sets is closed.

Proof.   (1) Let $\mathcal{A} = \{A_i : i \in I\}$ be a family of open sets. Assume that x has an observable neighbor $a \in \cup \mathcal{A}$. Then there is $i \in I$ such that $a \in A_i$. Since a is observable, some such i is observable by the Closure Principle. Then A_i is observable, and since A_i is open and contains a, we have $x \in A_i$. This implies that $x \in \cup \mathcal{A}$, so the union is open.

(2) Let $\mathcal{A} = \{A_i : i \in I\}$ be a family of open sets with I finite. Notice that each $i \in I$ is observable, since I is finite (Theorem 5). The Closure Principle implies that each A_i is observable. If x has an observable neighbor $a \in \cap \mathcal{A}$, then $a \in A_i$ for each $i \in I$, so that $x \in A_i$ for each $i \in I$, since each A_i is open. Thus $x \in \cap \mathcal{A}$, so the intersection is open.

(3) and (4) follow by complementation using

$$\left(\bigcup_{i\in I} A_i\right)^c = \bigcap_{i\in I} A_i^c \quad \text{and} \quad \left(\bigcap_{i\in I} A_i\right)^c = \bigcup_{i\in I} A_i^c.$$

□

Exercise 66 (Answer page 262)
Prove items (3) and (4) directly from the definition of closed set.

Exercise 67 (Answer page 262)
Show that arbitrary intersections of open sets are not open in general, and arbitrary unions of closed sets are not closed in general.

Exercise 68 (Answer page 262)
Let U be a subset of $\mathbb{R}$. We say that $A \subseteq U$ is **open in** U (respectively **closed in** U) if there exists $A' \subseteq \mathbb{R}$ open (respectively closed) such that

$$A = A' \cap U.$$

(1) Show that $\emptyset$ and U are open in U and closed in U.

(2) Let $A \subseteq U$. Show that A is open in U if and only if $U \cap A^c$ is closed in U.

(3) Show that any union of sets which are open in U is open in U.

(4) Show that any finite intersection of sets which are open in U is open in U.

(5) Show that any intersection of sets which are closed in U is closed in U.

(6) Show that any finite union of sets which are closed in U is closed in U.

Theorem 141 (Nested Intervals Theorem). *Let $\{[a_n, b_n] : n \in \mathbb{N}\}$ be a collection of nested closed intervals, that is,*

$$[a_{n+1}, b_{n+1}] \subseteq [a_n, b_n], \quad \text{for each } n \in \mathbb{N}.$$

Then there exists $c \in \bigcap_{n\in\mathbb{N}} [a_n, b_n]$.

Proof. The context is given by the collection. Let N be a positive ultralarge integer. Then $a_N \in [a_N, b_N] \subseteq [a_0, b_0]$, so a_N is not ultralarge. Let c be the observable neighbor of a_N. Then $c \in [a_n, b_n]$ for each n that is observable, since $a_N \in [a_n, b_n]$ and $[a_n, b_n]$ is closed and is observable. By the Closure Principle, this implies that $c \in [a_n, b_n]$ for all $n \in \mathbb{N}$. □

Let $f : \mathbb{R} \to \mathbb{R}$ be a function and A, B be subsets of $\mathbb{R}$. We define the **image of A (under f)**, written $f(A)$, to be

$$f(A) = \{f(a) : a \in A\}.$$

We define the **inverse image of B (under f)**, written $f^{-1}(B)$, to be

$$f^{-1}(B) = \{x \in \mathbb{R} : f(x) \in B\}.$$

The inverse image behaves well with respect to continuous functions.

Theorem 142. *Let $f : \mathbb{R} \to \mathbb{R}$ be a function. The following conditions are equivalent:*

(1) f is continuous.

(2) The inverse image of every open set is an open set.

Proof. (1) implies (2): Let B be an open set. The parameters are f and B. Assume x has an observable neighbor $a \in f^{-1}(B)$. Since a is observable and f is continuous, we have $f(x) \simeq f(a)$. But $f(a) \in B$ and B is open, so $f(x) \in B$. This shows that $x \in f^{-1}(B)$. Thus $f^{-1}(B)$ is open.

(2) implies (1): Let a be a real number. The parameters are f and a. Let $x \simeq a$ and let $c > 0$ be observable. Then $B = (f(a) - c, f(a) + c)$ is an open set which contains $f(a)$ and is observable. Hence $f^{-1}(B)$ is an observable open set containing a. Since a is the observable neighbor of x, we have $x \in f^{-1}(B)$, and so $f(x) \in B$. But this shows that $f(x) \simeq f(a)$, because the distance between $f(x)$ and $f(a)$ is less than any observable $c > 0$. $\qquad\square$

Exercise 69 (Answer page 262)
Let $f : \mathbb{R} \to \mathbb{R}$ be a function. Show that f is continuous if and only if the inverse image of every closed set is a closed set.

Theorem 143. *Let A be a closed set. Then every convergent sequence of elements of A has its limit in A.*

Proof. Let (x_n) be a convergent sequence of elements of A whose limit is c. We assume that A and (x_n) are observable. By definition, c is the observable neighbor of x_N, for any ultralarge N. Since A is closed and $x_N \in A$, we have $c \in A$. $\qquad\square$

It follows from Theorem 140 and Exercise 65 that any union of open intervals is an open set. These are actually the only open sets.

Theorem 144. *Let A be an open set and $a \in A$. Then there exists $\varepsilon > 0$ such that*

$$(a - \varepsilon, \, a + \varepsilon) \subseteq [a - \varepsilon, \, a + \varepsilon] \subseteq A.$$

Proof. Let $a \in A$ be given. We work relative to a context where A and a are observable. Let $\varepsilon > 0$ be ultrasmall. If $y \in [a - \varepsilon, \, a + \varepsilon]$, then the observable neighbor of y is $a \in A$, so $y \in A$ since A is open. This shows that $[a - \varepsilon, \, a + \varepsilon] \subseteq A$. □

By the Closure Principle, we can find an observable $\varepsilon > 0$ in the previous theorem.

Theorem 145. *Let A be an open set. Then A is a union of a system of mutually disjoint open intervals.*

Proof. For each $a \in A$, define $I(a)$ by

$$I(a) = \bigcup \{(c, d) : a \in (c, d) \subseteq A\}.$$

Then $I(a) \subseteq A$ is open, nonempty by the previous theorem, and observable relative to the context specified by A and a. Moreover,

$$A = \bigcup_{a \in A} I(a).$$

It is enough to show that each $I(a)$ is an open interval, and that $I(a) \cap I(b) = \emptyset$, if $I(a) \neq I(b)$.

Let $a \in A$. Let

$$U = \{d \in \mathbb{R} : a \in (c, d) \subseteq A \text{ for some } c < d\}$$

and let

$$L = \{c \in \mathbb{R} : a \in (c, d) \subseteq A \text{ for some } d > c\}.$$

By the previous theorem, neither U nor L are empty. Let $d' = \sup(U)$ (or $d' = +\infty$, if U is not bounded above) and $c' = \inf(L)$ (or $c' = -\infty$, if L is not bounded below). Notice that (c', d') is observable. We now show that

$$I(a) = (c', d').$$

Let $x \in I(a)$. Then $x \in (c, d)$ where $a \in (c, d) \subseteq A$. But by definition of c' and d' we have $(c, d) \subseteq (c', d')$, so $x \in (c', d')$. For the converse, suppose that $x \in (c', d')$. Then there is $d_1 \leq d'$ and c_1 such that $x, a \in (c_1, d_1) \subseteq A$. Similarly, there is $c_2 \geq c'$ and d_2 such that $x, a \in (c_2, d_2) \subseteq A$. But $a \in (c_1, d_1) \cap (c_2, d_2)$, so $(c_1, d_1) \cup (c_2, d_2) = (\min\{c_1, c_2\}, \max\{d_1, d_2\}) \subseteq A$ is an interval containing a and x. This shows that $x \in I(a)$.

Now if $I(a) \cap I(b) \neq \emptyset$, then their union is an open interval containing a, so $I(a) \cup I(b) \subseteq I(a)$, so $I(b) \subseteq I(a)$. Similarly, $I(a) \subseteq I(b)$, so we have $I(a) = I(b)$. $\square$

Exercise 70 (Answer page 263)
Let U be a subset of $\mathbb{R}$. This exercise provides an alternative way of defining sets open in U and closed in U.

(1) Suppose that $A \subseteq U$ has the property

if a is observable, $a \in A$, $x \in U$ and $a \simeq x$, then $x \in A$.

Show that for each $a \in A$ there exists $\varepsilon > 0$ such that

$$(a - \varepsilon,\ a + \varepsilon) \cap U \subseteq A.$$

(2) Deduce that $A \subseteq U$ is open in U if and only if whenever a is observable, $a \in A$, $x \in U$ and $x \simeq a$, then $x \in A$.

(3) Show that $A \subseteq U$ is closed in U if and only if whenever a is observable, $a \in U$, $a \simeq x$ and $x \in A$, then $a \in A$.

Before the next exercise, we need to generalize the notion of continuity from functions defined on an interval to functions whose domain is an arbitrary set.

Definition 72.

(1) Let $f : A \to \mathbb{R}$ be a function and $a \in A$. We say that f is **continuous at** *a if $f(x) \simeq f(a)$ whenever $x \simeq a$ and $x \in A$.*

(2) We say that $f : A \to \mathbb{R}$ is **continuous** *if f is continuous at each $a \in A$.*

Exercise 71 (Answer page 263)
Let $U \subseteq \mathbb{R}$ and let $f : U \to \mathbb{R}$ be a function. Show that f is continuous if and only if the inverse image of every open set is open in U (see Exercises 68 and 70).

Item (2) in the next theorem is often used as the definition of continuity.

Theorem 146. *Let $f : A \to \mathbb{R}$ be a function and let $a \in A$. The following conditions are equivalent:*

(1) f is continuous at a.

(2) For each $\varepsilon > 0$ there exists $\delta > 0$ such that $|f(x)-f(a)| < \varepsilon$ whenever $|x - a| < \delta$, $x \in A$.

Proof. (1) implies (2): Let $\varepsilon > 0$ be given. The parameters are f, a, A and ε. Let $\delta > 0$ be ultrasmall. If $|x - a| < \delta$, $x \in A$, then $x \simeq a$, so that $f(x) \simeq f(a)$ by continuity. This implies that $|f(x) - f(a)| < \varepsilon$, since $\varepsilon > 0$ is observable.

(2) implies (1): The parameters are f, A and a. Consider $x \simeq a$ with $x \in A$. Let $\varepsilon > 0$ be observable. By (2) and Closure there is some observable $\delta > 0$ such that $|f(x) - f(a)| < \varepsilon$ whenever $|x - a| < \delta$, $x \in A$. But $x \simeq a$, so necessarily $|x - a| < \delta$, which implies that $|f(x) - f(a)| < \varepsilon$. As this is true for every observable $\varepsilon > 0$, we conclude that $f(x) \simeq f(a)$. □

To extend the notion of limit to functions $f : A \to \mathbb{R}$, where A is not necessarily an interval, we introduce the next concept.

Definition 73. *Let A be a set. We say that a is a **limit point of** A if there is $x \in A$, $x \neq a$ such that $x \simeq a$.*

Definition 74. *Let $f : A \to \mathbb{R}$ be a function. Let a be a limit point of A and L be a real number. We say that*

$$\lim_{x \to a} f(x) = L$$

if $f(x) \simeq L$ whenever $x \simeq a$, $x \in A$ and $x \neq a$.

Note that the assumption that a be a limit point of A ensures that there exist some $x \in A$, $x \neq a$, with $x \simeq a$. The parameters of the previous definition are f, A, a and L. The usual argument (see Theorem 15) shows that L is uniquely determined and observable relative to f, A and a. This definition coincides with our definition of limits for functions defined on an interval, as well as with one-sided limits at the endpoints of intervals.

We now turn to the study of the boundary of a set. Intuitively, an observable point is on the boundary of a set A if it has neighbors both inside and outside of A.

Definition 75. *Let A be a subset of $\mathbb{R}$ and $a \in \mathbb{R}$. We say that a belongs to the **boundary of** A, written $a \in \delta A$, if there are $x_1 \in A$ and $x_2 \notin A$ such that*

$$a \simeq x_1 \qquad and \qquad a \simeq x_2.$$

The parameters are A and a. The definition is internal, so the boundary δA is a set which is observable whenever A is observable.

It is clear from the definition that

$$\delta A = \delta(A^c).$$

By definition of the boundary, if $a \notin \delta A$, then either all neighbors of a are in A, or all neighbors of a are in the complement of A. Notice also that the boundary of an interval consists of its endpoints.

Theorem 147. *Let A be a subset of $\mathbb{R}$. Then the boundary of A is closed.*

Proof. The parameter is A. Let c be observable and $c \simeq a$ for some $a \in \delta A$. Consider the extended context given by A and a and write $\overset{+}{\simeq}$ when working relative to the extended context. Since a is in the boundary of A, there are $x_1 \in A$ and $x_2 \notin A$ such that $a \overset{+}{\simeq} x_1$ and $a \overset{+}{\simeq} x_2$. Since $a \simeq c$, we have $c \simeq x_1$ and $c \simeq x_2$. This implies that $c \in \delta A$ and proves that δA is closed. $\square$

Definition 76. *Let A be a subset of $\mathbb{R}$.*

*(1) The **interior of** A, written A^o, is the set $A \setminus \delta A$.*

*(2) The **closure of** A, written $\overline{A}$, is the set $A \cup \delta A$.*

By the Closure Principle, the sets A^o and $\overline{A}$ are observable whenever A is.

Since the boundary of an interval consists of its endpoints, the interior of an interval is the open interval with the same endpoints, and the closure of an interval is the closed interval with the same endpoints.

Exercise 72 (Answer page 263)
Let A be a subset of $\mathbb{R}$. Prove the following properties.

(1) A^o is open.

(2) $\overline{A}$ is closed.

The use of the word *closure* here should not lead to confusion with the Closure Principle. The former describes a topological property of sets, the latter a general property of the observability concept.

Exercise 73 (Answer page 264)
Let A and B be subsets of $\mathbb{R}$. Prove the following properties.

(1) If $A \subseteq B$, then $\overline{A} \subseteq \overline{B}$.

(2) If $A \subseteq B$, then $A^o \subseteq B^o$.

Exercise 74 (Answer page 264)
Let A be a subset of $\mathbb{R}$. Show that

(1) A is open if and only if $A = A^o$.

(2) A is closed if and only if $A = \overline{A}$.

The previous two exercises show that the interior is the largest open subset of A and the closure is the smallest closed set containing A.

10.2 Dense Sets

Definition 77. *Let D be a subset of $\mathbb{R}$. We say that D is **dense** if for every $x \in \mathbb{R}$ there is $d \in D$ such that $d \simeq x$ (relative to D).*

For example, the set of rational numbers $\mathbb{Q}$ is dense in $\mathbb{R}$, since for each $x \in \mathbb{R}$ we can find a rational $x' \simeq x$. Simply choose $N \in \mathbb{N}$ ultralarge and truncate the decimal expansion of x after N digits.

We say that two sets **intersect** if their intersection is nonempty.

Theorem 148. *If D is dense, then D intersects every nonempty open set.*

Proof. Let A be a nonempty open set. We work relative to A and D. By Closure there is an observable $a \in A$. By density, there is $d \in D$ such that $d \simeq a$ relative to D and A, so $d \in A$, since A is open. ☐

Exercise 75 (Answer page 264)
Prove the converse of the previous theorem.

When the dense set D is itself open, then it intersects every open set rather substantially.

Exercise 76 (Answer page 264)
Let D be a dense open set and let (a, b) be an open interval. Show that there are $a_1 < b_1$ such that

$$[a_1, b_1] \subseteq D \cap (a, b).$$

By the Closure Principle, a_1 and b_1 can be chosen to be observable relative to D, a and b.

We now have all the ingredients to prove the Baire Category Theorem.

Theorem 149 (Baire Category Theorem). *The intersection of a count-able family of dense open sets is dense.*

Proof. Let $\{D_n : n \in \mathbb{N}\}$ be a family of dense open sets and let $D = \bigcap_{n \in \mathbb{N}} D_n$. Fix $x \in \mathbb{R}$ and work relative to a context where (D_n) and x are observable. Fix $a_0 = x$ and $b_0 > a_0$ such that $b_0 \simeq a_0$.

Since each D_n is dense and open, by Exercise 76, we can choose real numbers $a_n < b_n$ inductively such that

$$[a_{n+1}, b_{n+1}] \subseteq D_n \cap (a_n, b_n), \quad \text{for each } n \in \mathbb{N}.$$

By the Nested Intervals Theorem (Theorem 141), there is $d \in \bigcap_{n \in \mathbb{N}} [a_n, b_n]$. In particular, $d \in [a_{n+1}, b_{n+1}] \subseteq D_n$, for each $n \in \mathbb{N}$, so that $d \in \bigcap_{n \in \mathbb{N}} D_n = D$. Also, $d \in [a_0, b_0]$, so $d \simeq a_0 = x$ by the choice of b_0. $\square$

Exercise 77 (Answer page 265)
Show directly from the definition that the intersection of any finite family of dense open sets is dense.

The equivalent property in the next theorem is sometimes used as the definition of density.

Theorem 150. *Let D be a subset of $\mathbb{R}$. D is dense if and only if $\overline{D} = \mathbb{R}$.*

Proof. Suppose first that D is dense. We prove that every $x \in \mathbb{R}$ is in $\overline{D}$. We work relative to D and x. If $x \in D$, then there is nothing to prove. Otherwise, $x \notin D$ and by density there is $d \in D$ such that $d \simeq x$. This shows that $x \in \delta D \subseteq \overline{D}$.

Suppose now that $\overline{D} = \mathbb{R}$. Let x be in $\mathbb{R}$. If $x \in D$, there is nothing to prove; so we may assume that $x \in \delta D$. But by definition of the boundary, this means that there is $d \in D$ such that $x \simeq d$. This shows that D is dense. $\square$

We now establish two results allowing us to represent open sets as unions of a sequence of open intervals.

Theorem 151. *Any open set A can be written as $A = \bigcup_{n \in \mathbb{N}} (a_n, b_n)$, where the open intervals (a_n, b_n) are mutually disjoint.*

Proof. Theorem 145 shows that any open set can be written as a union of a family of mutually disjoint open intervals. Since $\mathbb{Q}$ is dense in $\mathbb{R}$, for each interval (a, b) in the family we can choose $q \in \mathbb{Q}$ with $q \in (a, b)$. Since the intervals are mutually disjoint, different intervals correspond to different rationals. This shows that the family is countable, since $\mathbb{Q}$ is countable $\square$

Theorem 152. *Any open interval can be written as a union of an increasing sequence of open intervals with rational endpoints.*

Proof. Let $I = (a, b)$ and fix m such that $\frac{1}{m} < \frac{b-a}{2}$. For each positive integer $n \geq m$, choose a_n and b_n rational so that $a + \frac{1}{n+1} < a_n < a + \frac{1}{n}$ and $b - \frac{1}{n} < b_n < b - \frac{1}{n+1}$. Then $(a, b) = \bigcup_{n \geq m} (a_n, b_n)$. $\qquad\square$

Definition 78. *A family $\mathcal{A}$ is an **open cover** of a set B if each $A \in \mathcal{A}$ is open and $B \subseteq \cup \mathcal{A}$.*

Exercise 78 (Answer page 265)
Show that every open cover of a set B contains a countable subcover.

10.3 Compact Sets

Definition 79. *Let K be a subset of $\mathbb{R}$. We say that K is **compact** if each $x \in K$ has an observable neighbor in K.*

Theorem 153. *Let K be a subset of $\mathbb{R}$. Then K is compact if and only if K is closed and bounded.*

Proof. Suppose that K is compact. Since each $x \in K$ has an observable neighbor, no $x \in K$ is ultralarge. Thus, if M is an ultralarge positive number, then $K \subseteq [-M, M]$. This shows that K is bounded. K is clearly closed.

Conversely, suppose that K is closed and bounded and let an observable M be such that $K \subseteq [-M, M]$. Then no $x \in K$ is ultralarge, so every $x \in K$ must have an observable neighbor, and this observable neighbor is in K, because K is closed. K is therefore compact. $\qquad\square$

Theorem 154. *Let K be compact and let $f : K \to \mathbb{R}$ be continuous. Then $f(K)$ is compact.*

Proof. Let $y \in f(K)$. Then there is $x \in K$ such that $f(x) = y$. Since K is compact, there is an observable $a \simeq x$ and $a \in K$. Since f is continuous at a and $x \simeq a$, we must have $f(x) \simeq f(a)$, that is, $y \simeq f(a)$. By the Closure Principle $f(a)$ is observable, and by definition $f(a) \in f(K)$, so $f(a)$ is the observable neighbor of y and it is in $f(K)$. This shows that $f(K)$ is compact. $\qquad\square$

Exercise 79 (Answer page 265)
Show that if K is compact and $f : K \to \mathbb{R}$ is continuous, then f is uniformly continuous on K.

A function $f : A \to \mathbb{R}$ **achieves its maximum** (respectively, **minimum**) if there is $a \in A$ such that $f(a) \geq f(x)$ for all $x \in A$ (respectively, $f(a) \leq f(x)$ for all $x \in A$).

Theorem 155 (Extreme Value Theorem). *Let K be a compact set and let $f : K \to \mathbb{R}$ be continuous. Then f achieves its maximum and its minimum.*

Proof. Since $f(K)$ is compact, it is enough to show that every compact set has a maximum and a minimum, that is, there are $a, b \in K$ such that $a \leq x \leq b$ for all $x \in K$.

Let K be compact. We show that K has a maximum, as the case for the minimum is similar. Since K is bounded, by completeness of $\mathbb{R}$ there is an observable b such that $b = \sup K$. By definition of the supremum, there is $x \in K$ such that $x \simeq b$. But then since K is compact, we have $b \in K$. □

We want to examine several properties equivalent to compactness. For this, we need a few additional definitions.

Definition 80. *Let (u_n) be a sequence and c a number. We say that c is a **cluster point** of (u_n) if there exists an ultralarge N such that $u_N \simeq c$.*

The context in the previous definition is given by (u_n) and c. Note that any bounded sequence has a cluster point. If (u_n) is bounded, then in particular there is an observable bound, hence the members of the sequence are not ultralarge. Let N be an ultralarge positive integer. Then there is an observable $c \simeq u_N$, and this c is a cluster point. Since the definition is internal, if a sequence has a cluster point, then it has an observable one, by Closure.

Item (2) in the next theorem is often used as a definition of cluster points.

Theorem 156. *Let (u_n) be a sequence and c be a real number. The following conditions are equivalent:*

> *(1) c is a cluster point of (u_n).*
>
> *(2) For each $\varepsilon > 0$ and for each positive integer m there exists an integer $n \geq m$ such that $|u_n - c| < \varepsilon$.*

Proof. (1) implies (2): Let $\varepsilon > 0$ and m be given. We work relative to $(u_n), c, \varepsilon$ and m. Since c is a cluster point, there is N ultralarge such that $u_N \simeq c$. Then $N > m$ as N is ultralarge, and $|u_N - c| < \varepsilon$, since $\varepsilon > 0$ is observable.

(2) implies (1): We work relative to (u_n) and c. Let $\varepsilon > 0$ be ultrasmall and M be ultralarge. By (2) there is $N > M$ such that $|u_N - c| < \varepsilon$. This implies that N is ultralarge and $u_N \simeq c$, so c is a cluster point $\qquad \square$

It is easy to deduce from the previous theorem that c is a cluster point of (u_n) if and only if some subsequence of (u_n) converges to c.

We can give useful alternative descriptions of the smallest and the largest cluster point. Consider a bounded sequence $(x_n)_{n \geq 0}$. Let n be a positive integer. Then $y_n = \inf\{x_k : k \geq n\}$ exists. The sequence (y_k) is increasing and bounded, so it converges by the Monotone Convergence Theorem for sequences. We call the limit of (y_n) the **lower limit** of (x_n) and write

$$\liminf x_n = \lim_{n \to \infty} y_n.$$

Similarly, we can consider the decreasing bounded sequence (z_n) defined by $z_n = \sup\{x_k : k \geq n\}$. It converges; we call the limit of (z_n) the **upper limit** of (x_n) and write

$$\limsup x_n = \lim_{n \to \infty} z_n.$$

Exercise 80 (Answer page 265)
Let (x_n) be a bounded sequence. Show that the lower limit (respectively, the upper limit) of (x_n) is the smallest (respectively, largest) cluster point of (x_n).

Definition 81. *A family $\mathcal{A} = \{A_i : i \in I\}$ has the **finite intersection property** if for every finite subfamily $\mathcal{A}' \subseteq \mathcal{A}$ we have $\cap \mathcal{A}' \neq \emptyset$.*

Theorem 157. *Let $K \subseteq \mathbb{R}$. The following conditions are equivalent:*

(1) K is compact.

(2) Every sequence (u_n) of elements of K has a cluster point in K.

(3) Every open cover of K has a finite subcover.

(4) Every family of closed subsets of K with the finite intersection property has nonempty intersection.

Proof. (1) implies (2): Let (u_n) be a sequence of elements of K. The parameters are K and (u_n). Let N be an ultralarge positive integer. Then

$u_N \in K$, so u_N has an observable neighbor $c \in K$, as K is compact. As $u_N \simeq c$, this c is a cluster point of (u_n) in K.

(2) implies (3): By Exercise 78, we may assume that the cover is countable. Let $\mathcal{A} = \{A_n : n \in \mathbb{N}\}$ be a countable open cover of K. The parameters are $\mathcal{A}$ and K. By taking A_0, $A_0 \cup A_1$, $A_0 \cup A_1 \cup A_2, \ldots$ if necessary, we may assume that the family is increasing, that is,

$$A_n \subseteq A_{n+1}, \quad \text{for each } n \in \mathbb{N}.$$

Suppose, for a contradiction, that no A_n covers K, so for each $n \in \mathbb{N}$, we can find

$$u_n \in K \setminus A_n.$$

By Closure, we may assume that the sequence (u_n) is observable. By (2), it has a cluster point $c \in K$, and by Closure, we may assume that c is observable. Since $\mathcal{A}$ is a cover, there exists $n \in \mathbb{N}$ such that $c \in A_n$, and by Closure we may assume that n and A_n are observable (as c is observable). Let N be ultralarge such that $u_N \simeq c$. Then $u_N \in A_n$, as A_n is open. But since $n \leq N$, we have $A_n \subseteq A_N$, so $u_N \in A_N$, contradicting the choice of $u_N \in K \setminus A_N$.

(3) implies (4): Let $\mathcal{F} = \{C_i : i \in I\}$ be a family of closed subsets of K such that $\cap \mathcal{F} = \emptyset$. Then

$$\mathbb{R} = \emptyset^c = (\cap \mathcal{F})^c = \left(\bigcap_{i \in I} C_i\right)^c = \bigcup_{i \in I} C_i^c.$$

But each C_i^c is open and their union covers $\mathbb{R}$ and hence K. By (3) we can find finitely many $i_1, \ldots i_n \in I$ such that

$$K \subseteq C_{i_1}^c \cup \cdots \cup C_{i_n}^c.$$

But this implies that $\emptyset = C_{i_1} \cap \cdots \cap C_{i_n}$, so that $\mathcal{F}$ does not have the finite intersection property.

(4) implies (1): The parameter is K. We prove the contrapositive. Suppose that K is not compact. Then there is $x \in K$ which is either ultralarge, or not ultralarge but whose observable neighbor $a \notin K$. We distinguish three cases.

Suppose that x is ultralarge and positive. Consider the family

$$\mathcal{F} = \{K \cap [n, +\infty) : n \in \mathbb{N}\}.$$

The observable family $\mathcal{F}$ consists of closed subsets of K and clearly has empty intersection. We show that $\mathcal{F}$ has the finite intersection property, thus contradicting (4). By Closure, it is enough to show that each observable finite subfamily $\mathcal{F}'$ of $\mathcal{F}$ has nonempty intersection. Let $\mathcal{F}'$ be such subfamily. Then

$$\cap \mathcal{F}' = K \cap [n_{max}, +\infty),$$

where n_{max} is the maximum index n among the members of $\mathcal{F}'$. Since $\mathcal{F}'$ is observable, n_{max} is also observable. This shows that $n_{max} < x$, since x is ultralarge and positive, and thus $x \in \cap \mathcal{F}'$ and $\cap \mathcal{F}' \neq \emptyset$.

If x is ultralarge and negative, argue similarly with

$$\mathcal{F} = \{K \cap (-\infty, -n] : n \in \mathbb{N}\}.$$

If x is not ultralarge and a is observable with $x \simeq a \notin K$, then argue with

$$\mathcal{F} = \{K \cap \left[a - \frac{1}{n}, a + \frac{1}{n}\right] : n \in \mathbb{N}, \ n \neq 0\}.$$

$\square$

Theorem 158 (Dini's Theorem). *Let (f_n) be a sequence of functions continuous on a compact set K and such that $f_n(x) \geq f_{n+1}(x)$ holds for all n and all $x \in K$. If (f_n) converges pointwise to f and f is continuous on K, then (f_n) converges uniformly to f on K.*

Proof. We first prove the theorem for the special case $f = 0$. Each f_n is continuous on K and hence it attains its maximal value $M_n = \max_{x \in K} f_n(x)$ at some point $x_n \in K$. The sequence (M_n) is decreasing and bounded below by 0, so it converges to some $M \geq 0$. It suffices to show that $M = \lim_{n \to \infty} M_n = 0$. We assume that $M > 0$ and deduce a contradiction.

We work relative to a context where $(f_n), K, (x_n)$ and M are observable. Let N be ultralarge. The sequence (x_n) of elements of K has an observable cluster point $c \in K$. The sequence $(f_n(c))$ converges to 0, hence $f_N(c) \simeq 0$. As c is a cluster point of (x_n), there exists N', ultralarge relative to the context extended by N, such that $x_{N'} \overset{+}{\simeq} c$. The observations that $N' > N$, the sequence $(f_n(x_{N'}))$ is decreasing, and f_N is continuous, imply that $f_{N'}(x_{N'}) \leq f_N(x_{N'}) \overset{+}{\simeq} f_N(c) \simeq 0$, contradicting $f_{N'}(x_{N'}) = M_{N'} \geq M$.

The general case follows by applying the special case to the sequence of functions $(f_n - f)$. $\square$

10.4 Additional Exercises

Exercise 10.1
Determine whether the following sets are open, closed, or neither: $\{0, 1, 2, \ldots, n\}$, $\mathbb{Z}$, $\mathbb{Q}$, $\{1/n : n \geq 1\}$, $\{0\} \cup \{1/n : n \geq 1\}$.

Exercise 10.2
Let $(u_n)_{n \geq 0}$ be a sequence and let $\lim_{n \to \infty} u_n = L$. Prove that the set $\{u_n : n \in \mathbb{N}\} \cup \{L\}$ is closed.

Exercise 10.3
Let $\{[a_n, b_n] : n \in \mathbb{N}\}$ be a collection of nested closed intervals such that $\lim_{n \to \infty}(b_n - a_n) = 0$. Prove that there exists a unique $c \in \bigcap_{n \in \mathbb{N}} [a_n, b_n]$.

Exercise 10.4
Give an example of a continuous function $f : \mathbb{R} \to \mathbb{R}$ and a closed set A such that $f(A)$ is not closed.

Exercise 10.5
Prove the converse to Theorem 143: Let A be a set. If every convergent sequence of elements of A has its limit in A, then A is closed.

Exercise 10.6
Let $f : A \to \mathbb{R}$ be a function and let $a \in A$. Prove that the following conditions are equivalent:

(1) f is continuous at a.

(2) If (x_n) is any sequence such that $x_n \in A$ for all n and $\lim_{n \to \infty} x_n = a$, then $\lim_{n \to \infty} f(x_n) = f(a)$.

Hint: If f is not continuous at a, then there is some $x \simeq a$, $x \in A$, such that $|f(x) - f(a)| \geq r$, for some observable $r > 0$. Hence for each observable n there is $x_n \in A$ such that $|f(x_n) - f(a)| \geq r$. By Closure, this is true for every n.

Exercise 10.7
Find all limit points of the set $\{1/m + 1/n : m, n \geq 1\}$.

Exercise 10.8
Prove that every bounded infinite set has a limit point.

Exercise 10.9
Prove that $\overline{A} = \bigcap \{F : F \supseteq A, F \text{ closed}\}$.
Similarly, $A^\circ = \bigcup \{G : G \subseteq A, G \text{ open}\}$.

Exercise 10.10
Prove that the following conditions are equivalent:

(1) a is a limit point of A.

(2) For every $\delta > 0$ there exists $x \in A$, $x \neq a$, such that $|x - a| < \delta$.

(3) There is a sequence (x_n), with $x_n \in A$ and $x_n \neq a$ for all n, such that $\lim_{n \to \infty} x_n = a$.

Exercise 10.11
Let $f : A \to \mathbb{R}$ be a function and let a be a limit point of the set A. Show that $\lim_{x \to a} f(x) = L$ if and only if for each $\varepsilon > 0$ there exists $\delta > 0$ such that $|f(x) - L| < \varepsilon$ whenever $0 < |x - a| < \delta$, $x \in A$.

Exercise 10.12
Prove that $\overline{A \cup B} = \overline{A} \cup \overline{B}$ and $(A \cap B)^o = A^o \cap B^o$, for any sets A and B. Give examples of sets A and B for which $\overline{A \cap B} \neq \overline{A} \cap \overline{B}$ and $(A \cup B)^o \neq A^o \cup B^o$.

Exercise 10.13
Show that $\mathbb{R} \setminus (\{0\} \cup \{1/n + 1/m : n, m \geq 1\})$ is a dense open set.

Exercise 10.14
Give an example of an open cover of the set $A = (0, 1]$ that contains no finite subcover.

Exercise 10.15
Prove that the intersection of a compact set F and a closed set A is compact.

Exercise 10.16
Prove that the union of two compact sets is compact.

Exercise 10.17
Prove that a bounded set A is compact if and only if $\limsup x_n \in A$ for every sequence (x_n) of elements of A.

Exercise 10.18
Let $f : A \to \mathbb{R}$ be a function. Prove that the following conditions are equivalent:

(1) f is uniformly continuous on A.

(2) For each $\varepsilon > 0$ there exists $\delta > 0$ such that $|f(x) - f(y)| < \varepsilon$ whenever $|x - y| < \delta$, $x, y \in A$.

Exercise 10.19
Prove: If f and g are uniformly continuous on A and $c \in \mathbb{R}$, then $f + g$ and $c \cdot f$ are uniformly continuous on A. If, in addition, f and g are bounded, then $f \cdot g$ is uniformly continuous.

Exercise 10.20
Let (f_n) be a sequence of functions, each defined on A. Prove that the following statements are equivalent:

(1) (f_n) converges uniformly to f on A.

(2) For every $\varepsilon > 0$ there exists N such that $|f_n(x) - f(x)| < \varepsilon$ whenever $n \geq N$ and $x \in A$.

Exercise 10.21
Give an example of a sequence of functions (f_n) such that each f_n is nonnegative and continuous on $[0,1]$ and (f_n) converges to 0 on $[0,1]$ pointwise, but not uniformly.

Exercise 10.22
Prove: If f is uniformly continuous on $[a,b)$ and bounded on $[a,b)$ [that is, $|f(x)| \leq M$ for all $x \in [a,b)$ and some $M > 0$], then $\lim_{x \to b^-} f(x)$ exists.

Exercise 10.23
Assume that every rational number appears as a term in the sequence (r_n); show that every real number is a cluster point of (r_n).

Exercise 10.24
Show that a is a cluster point of the sequence (u_n) if and only if it is the limit of some subsequence (u_{n_k}) of (u_n).

Exercise 10.25
Show that a sequence (u_n) converges if and only if it has a unique cluster point.

Exercise 10.26
Prove: If a Cauchy sequence (u_n) has a subsequence (u_{n_k}) that converges to a, then the sequence (u_n) converges to a.

Answers to Exercises

Answer to Exercise 1, page 8

(1) Suppose that x is such that $0 < |x| < |\varepsilon|$ and ε is ultrasmall. Let $c > 0$ be observable. Then by the assumption on ε, we have $|\varepsilon| < c$. Hence $0 < |x| < c$. But c is arbitrary, so x is ultrasmall.

(2) Suppose that x is such that $|M| < |x|$ and M is ultralarge. Let $c > 0$ be observable. Then by the assumption on M we have $|M| > c$. Hence $|x| > c$; but c is arbitrary, so x is ultralarge.

Answer to Exercise 2, page 9

We have $0 = 1 - 1$, $2 = 1 + 1$, $4 = 2 + 2$ and $17 = 4 \cdot 4 + 1$.

Answer to Exercise 3, page 12

Let a, b be observable real numbers and $h \simeq 0$.

(1) Without loss of generality assume $a > 0$. Then $|h| < \frac{a}{2}$, so in particular $-\frac{a}{2} < h$. Hence $\frac{a}{2} < a + h$ and $a + h \not\simeq 0$.

(2) Let $x = a + \varepsilon$ and $y = b + \delta$ for ultrasmall ε and δ. As $b - a > 0$ (and is observable) we have $(b + \delta) - (a + \varepsilon) > 0$ by part (1). Hence $x < y$.

(3) If $b < a$, then $y < x$, by (2).

(4) We can have $x < y$ and $a = b$ [for example, let $x = a = b = 0$ and let $y > 0$ be ultrasmall]. Also, $a \leq b$ and $y < x$ is possible [for example, let $x = a = b = 0$ and $y < 0$ ultrasmall].

Answer to Exercise 4, page 13

(1) By Rule 5, all counterexamples must be ultralarge. Let $x = N$ be ultralarge, and $y = N + \frac{1}{N}$, so $x \simeq y$ but $x^2 = N^2 \not\simeq N^2 + 2 + \frac{1}{N^2} = y^2$.

(2) By Rule 5, all counterexamples must be ultrasmall. Let h be ultrasmall, and let $x = h$ and $y = h^2$. Then $x \simeq 0$ and $y \simeq 0$, hence $x \simeq y$. So $\frac{1}{h}$ and $\frac{1}{h^2}$ are both ultralarge and $\frac{1}{h^2} - \frac{1}{h} = \frac{1-h}{h^2}$. By Rule 2, this is ultralarge, hence $\frac{1}{x} \not\simeq \frac{1}{y}$.

Answer to Exercise 5, page 13

(1) Take $\varepsilon = \delta$; then $\frac{\varepsilon}{\delta} = 1$.

(2) Take $\delta = \varepsilon^2$; then $\frac{\varepsilon}{\delta} = \frac{1}{\varepsilon}$ is ultralarge.

(3) Take $\varepsilon = \delta^2$; then $\dfrac{\varepsilon}{\delta} = \delta$ is ultrasmall.

Answer to Exercise 6, page 14

The assumptions imply that there exist observable M and ε such that $|x \cdot y| \leq M$ and $0 < \varepsilon \leq |y|$. Hence $|x| = |x \cdot y|/|y| \leq M/\varepsilon$, which is observable by Closure, and so x is not ultralarge. The conclusion follows by Rule 5(2).

Answer to Exercise 7, page 14

(1) As $\frac{1}{\varepsilon}$ is ultralarge, $1 + \frac{1}{\varepsilon}$ is ultralarge.

(2) We have $\frac{\sqrt{\delta}}{\delta} = \frac{1}{\sqrt{\delta}}$, which is ultralarge.

   (As $\delta < c^2$ for any observable $c > 0$, $\sqrt{\delta} < c$ for any observable $c > 0$, and $\sqrt{\delta} \simeq 0$. Hence $\frac{1}{\sqrt{\delta}}$ is ultralarge.)

(3) Maybe surprisingly, this is ultrasmall. To see this we multiply and divide by the conjugate:

$$\sqrt{H+1} - \sqrt{H-1}$$

$$= \frac{(\sqrt{H+1} - \sqrt{H-1})(\sqrt{H+1} + \sqrt{H-1})}{\sqrt{H+1} + \sqrt{H-1}}$$

$$= \frac{(H+1) - (H-1)}{\sqrt{H+1} + \sqrt{H-1}} = \frac{2}{\sqrt{H+1} + \sqrt{H-1}}.$$

H is assumed positive, its square root (plus or minus 1) is also a positive ultralarge. The sum of two positive ultralarge numbers is ultralarge, hence the quotient is ultrasmall.

(4) $\dfrac{H + K}{HK} = \dfrac{1}{K} + \dfrac{1}{H}$ is ultrasmall.

(5) $\dfrac{2 + \varepsilon}{5 + \delta} - \dfrac{2}{5} = \dfrac{10 + 5\varepsilon - 10 - 2\delta}{25 + 5\delta} = \dfrac{\overbrace{5\varepsilon - 2\delta}^{\simeq 0}}{\underbrace{25 + 5\delta}_{\simeq 25}}$ is ultrasmall or

zero.

(6) $\dfrac{\overbrace{\sqrt{1 + \varepsilon} - 2}^{\simeq -1}}{\underbrace{\sqrt{1 + \delta}}_{\simeq 1}} \simeq -1$, hence not ultralarge and not ultrasmall.

Answer to Exercise 8, page 14

Let h be ultrasmall; we consider the case $h > 0$, the other case being similar. Suppose that $\sqrt{1+h} \not\simeq 1$. Then, $\sqrt{1+h} - 1 \not\simeq 0$, hence there is an observable c such that $1 < c < \sqrt{1+h}$. Hence, since $x \mapsto x^2$ is increasing, we have $1 < c^2 < (\sqrt{1+h})^2 = 1 + h$. By Closure, we have that c^2 is observable. Hence $0 < c^2 - 1 < (1+h) - 1 = h$. But $c^2 - 1 > 0$ is observable by Closure, so h is not ultrasmall. A contradiction.

A simpler proof can be given using a familiar trick. Assume that h is ultrasmall. Let $\varepsilon = \sqrt{1+h} - 1$ and write

$$\varepsilon = \sqrt{1+h} - 1 = \frac{(\sqrt{1+h} - 1)(\sqrt{1+h} + 1)}{\sqrt{1+h} + 1} = \frac{h}{\sqrt{1+h} + 1}.$$

As $\sqrt{1+h} + 1 > 1$, it follows that $|\varepsilon| < |h|$, so $\varepsilon \simeq 0$.

Answer to Exercise 9, page 14

Assume that N is ultralarge. Let $\varepsilon = \sqrt[N]{N} - 1$ and note that $\varepsilon > 0$. By the Binomial Theorem,

$$N = (1+\varepsilon)^N = 1 + \binom{N}{1} \cdot \varepsilon + \binom{N}{2} \cdot \varepsilon^2 + \ldots + \binom{N}{N} \cdot \varepsilon^N > 1 + \frac{N(N-1)}{2} \cdot \varepsilon^2.$$

Hence $2N - 2 > N(N-1) \cdot \varepsilon^2$ and $\varepsilon^2 < 2/N$. As N is ultralarge, we conclude that $\varepsilon \simeq 0$.

Answer to Exercise 10, page 14

For $x, y \in \mathbb{R}$ define: $x \sim y$ if $x - y$ is not ultralarge.

For Rule 3. Clearly $x - x = 0$ is not ultralarge, and if $x - y$ is not ultralarge, then $y - x$ is not ultralarge. Now suppose that $x - y$ is not ultralarge and $z - y$ is not ultralarge. Then there are numbers a, b, not ultralarge, such that $y = x + a$ and $z = y + b$. Hence $z - x = a + b$, which is not ultralarge by Rule 1(1).

For Rule 4, let $x \sim x'$ and $y \sim y'$. Then there are a, b not ultralarge such that $x' = x + a$ and $y' = y + b$. Then

$$x' \pm y' = x \pm y + (a \pm b).$$

By Rule 1 (1), we have $a \pm b$ is not ultralarge, so $x' + y' \sim x + y$.

For the product, we assume further that x and y are not ultralarge. This implies that x' and y' are not ultralarge, since sums of numbers that are not ultralarge are not ultralarge. Hence, we have

$$x' \cdot y' = (x + a) \cdot (y + b) = x \cdot y + a \cdot y + a \cdot b + x \cdot b.$$

But, by assumption, a, x, y, b are not ultralarge, so the products $a \cdot y$, $a \cdot b$ and $x \cdot b$ are not ultralarge. Also the sum of numbers that are not ultralarge is not ultralarge, and thus $x' \cdot y' \sim x \cdot y$.

For the quotient, we further assume that y and y' are not ultrasmall. Then

$$\frac{x'}{y'} - \frac{x}{y} = \frac{x' \cdot y - x \cdot y'}{y' \cdot y}.$$

Since x, x', y, y' are not ultralarge, the numerator is not ultralarge. Further, since $y, y' \not\simeq 0$, then $y \cdot y' \not\simeq 0$ (since $|y| > c$ and $|y'| > d$, with observable $c, d > 0$, we have $|y \cdot y'| > c \cdot d > 0$, where $c \cdot d$ is also observable, by Closure). Hence, by the last observation, the quotient on the right hand side is not ultralarge, so $\frac{x'}{y'} \sim \frac{x}{y}$.

Answer to Exercise 8 (2) gives an example of $x \simeq y$ with $\frac{1}{x} \not\simeq \frac{1}{y}$.

Answer to Exercise 11, page 15

Let $a < b$ and x such that $a \le x \le b$. As x is bounded by a and b, x is not ultralarge. Let c be the observable neighbor of x. Then $a \le c \le b$, by Exercise 3(3).

Answer to Exercise 12, page 19

Let $N > 0$. If $a = 3 + \frac{2}{N}$ is observable, then $N = \frac{2}{a-3}$ is observable, by Closure.

Similarly, $b = \sqrt{N}$ implies $N = b^2$ and $c = \sqrt[N]{3}$ implies $N = \frac{\ln(3)}{\ln(c)}$, so N is observable if b or c is.

If $A = \{n \in \mathbb{N} : n \le N\}$, then $N = $ the greatest element of A, so it is observable provided A is observable, by Closure.

Answer to Exercise 13, page 21

For a counterexample to (1) take $m \in \mathbb{N}$ ultralarge and let $x_i = 2$; then $\sum_{i=1}^{m} x_i = 2m$ and $\Pi_{i=1}^{m} \xi_i = 2^m$ are both ultralarge.

For a counterexample to (2) take $m \in \mathbb{N}$ ultralarge and let $x_i = 1/m$ and $a_i = 0$. Then $x_i \simeq 0$ for all i, but $\sum_{i=1}^{m} x_i = 1 \not\simeq 0$.

Similarly, for ultralarge m, $x = 1 + 1/m \simeq 1$, but, by the Binomial Theorem, $x^m = (1+1/m)^m = 1 + m \cdot \frac{1}{m} + \ldots > 2$, so $x^m \not\simeq 1$, providing a counterexample to (3) and (4).

Answer to Exercise 14, page 21

This is the contrapositive of the comment on page 21. If f were bounded above, it would have an observable upper bound. Hence if it attains ultralarge positive values, then it has no upper bound.

Answer to Exercise 15, page 22

Let f be a function defined on I. The following statement makes no reference to observability: *There is $c \in I$ such that $f(c) = 0$*. By Closure, *there is an observable $c \in I$ such that $f(c) = 0$*, which is what we had to show.

Answer to Exercise 16, page 22

Let f be a function. The statement "There exist M and L such that $f(x) = L$ for all $x \geq M$" makes no reference to observability. By Closure it remains true for some observable M and L. But if M is observable, then for each ultralarge x we have $x > M$. Hence $f(x) = L$ for all ultralarge x.

Answer to Exercise 17, page 22

We proceed by contradiction. Assume there are no observable x satisfying the statement. Then, every observable x satisfies the negation of the statement. By the universal form of Closure, this implies that all x satisfy the negation of the statement. Hence, it is false that the statement is true for some x.

Answer to Exercise 18, page 25

The first claim is obvious: The object on the list $q_1, \ldots, q_\ell$, relative to which x is observable, appears also on the extended list $p, q_1, \ldots, q_\ell$. The second claim: If x is observable relative to $p, q_1, \ldots, q_\ell$, then either x is observable relative to some q_i, $1 \leq i \leq \ell$, or x is observable relative to p. In the second case, p is observable relative to some q_i, $1 \leq i \leq \ell$, and by transitivity, x is observable relative to the same q_i. In either case, x is observable relative to $q_1, \ldots, q_\ell$.

Answer to Exercise 19, page 27

By definition, x is ultrasmall if $|x| < r$, for all observable $r > 0$. But, according to Exercise 18, r is observable relative to $p, q_1, \ldots, q_\ell$ if and only if it is observable relative to $q_1, \ldots, q_\ell$.

Answer to Exercise 20, page 31

(1) The statement is internal because it makes no reference to observability, and therefore defines a function. There are no parameters, hence the function is standard. Reminder: the independent variable x is not a parameter of the definition.

(2) The reciprocal is a function, as its definition makes no reference to observability. The parameter a occurs in a part of the definition, hence the function is observable whenever a is observable.

(3) A function, for the same reason as above. Moreover, the fraction $\frac{b}{2b} = \frac{1}{2}$, and the function is equal to $x \mapsto \frac{1}{2}x$. Hence, this function is standard.

(4) The statement is not internal (the parameters are x and p, but ultrasmallness is relative to p), so it does not define a function.

Answer to Exercise 21, page 46

Assume that $f(a) < f(b)$; we prove first that if $a < x < b$, then $f(a) < f(x) < f(b)$. Assume to the contrary that $f(x) > f(b)$ [$f(x) = f(b)$ is impossible because f is one-to-one]. Fix d such that $f(x) > d > f(b) > f(a)$. By the Intermediate Value Theorem there exist c_1, c_2 such that $a < c_1 < x < c_2 < b$ and $f(c_1) = d = f(c_2)$. This is a contradiction with f being one-to-one. Similarly, the assumption $f(x) < f(a)$ leads to a contradiction.

Conversely, if $f(a) < d < f(b)$, then there is some $x \in (a, b)$ such that $d = f(x)$ (Intermediate Value Theorem). This proves that $f([a, b]) = [f(a), f(b)]$.

We now prove that f is strictly increasing. If not, then there are $x, y \in (a, b)$ such that $x < y$ and $f(x) > f(y) > f(a)$. By the Intermediate Value Theorem there is some $c \in (a, x)$ such that $f(c) = f(y)$. This contradicts f being one-to-one.

The case $f(a) > f(b)$ is similar; f is strictly decreasing in this case.

Answer to Exercise 22, page 46

Suppose that f is one-to-one and neither strictly increasing nor strictly decreasing on I. Then there exist $x_1 < x_2$ in I such that $f(x_1) > f(x_2)$, and also $x_3 < x_4$ in I such that $f(x_3) < f(x_4)$. Let $a = \min\{x_1, x_3\}$ and $b = \max\{x_1, x_3\}$ and apply Exercise 21.

Answer to Exercise 23, page 46

Assume that f is strictly increasing on I. For $y_1, y_2 \in J$, $y_1 < y_2$, let $x_1 = f^{-1}(y_1)$, $x_2 = f^{-1}(y_2)$. If $x_1 \geq x_2$, then $f(x_1) \geq f(x_2)$ and $y_1 \geq y_2$, a contradiction. Hence $f^{-1}(y_1) = x_1 < x_2 = f^{-1}(y_2)$ and f^{-1} is strictly increasing on J.

Answer to Exercise 24, page 51

The parameters are g, f and a.

Let $\lim_{x \to a} g(x) = L_g$ and $\lim_{x \to a} f(x) = L_f$. The limits L_f and L_g are observable. Let $x \simeq a$. The rest follows from Rule 5:

(1) $f(x) \pm g(x) \simeq L_f \pm L_g$.

(2) $f(x) \cdot g(x) \simeq L_f \cdot L_g$.

(3) We already know that L_f is not ultralarge. We also have that $L_g \neq 0$. As $g(x) \simeq L_g$, we have $g(x) \not\simeq 0$, hence $\frac{f(x)}{g(x)} \simeq \frac{L_f}{L_g}$.

(4) Here we consider the function $\lambda : x \mapsto \lambda$. The only parameter is λ.

Answer to Exercise 25, page 52

This is a direct consequence of Stability. The property "$f(x) \simeq \infty$ whenever $x \simeq a$" holds when "$\simeq$" is interpreted relative to the context f, a if and only if it also holds when the symbol is interpreted relative to any extended context.

This extends the remarks on limits. If a function takes ultralarge values in a deleted neighborhood of a, then in that neighborhood of a the function takes ultralarge values relative to any context, so in fact it is unbounded.

Answer to Exercise 26 on page 56

(1) Using the binomial formula on $(1 + b)^n \geq 1^n + \binom{n}{1}b = 1 + nb$.

(2) Let $a > 1$. The parameter is a. Write $a = 1 + b$. Then b is observable and $b > 0$. Let n be an ultralarge positive integer. Then $a^n = (1 + b)^n \geq 1 + nb \simeq \infty$.

(3) Let $0 < a < 1$. The parameter is a. Let $c = 1/a$. Then c is observable and $c > 1$. Let n be a positive ultralarge integer. Then $a^n = \left(\frac{1}{c}\right)^n = \frac{1}{c^n} \simeq 0$, since $c^n \simeq +\infty$.

Answer to Exercise 27 on page 57

Let q be an ultralarge positive rational number. Let $n \leq q$ be a positive ultralarge integer (for example the only one such that $q \in [n, n + 1)$). By monotonicity, we have $a^q \geq a^n = +\infty$.

We deduce that if $q \simeq -\infty$ then $a^q = 1/a^{-q} \simeq 0$.

Suppose that $q \simeq 0$ and without loss of generality, assume that $q > 0$. Then $\frac{1}{q}$ is ultralarge and $k \leq \frac{1}{q} < k + 1$ for some ultralarge $k \in \mathbb{N}$. We have $q \leq \frac{1}{k}$, hence $a^q \leq a^{1/k}$ by monotonicity, so it is sufficient to prove that $a^{1/k} \simeq 1$. If not, then there would be an observable b such that $a^{1/k} \geq b > 1$ and $a \geq b^k \simeq +\infty$, a contradiction.

Answer to Exercise 28 on page 59

To show: If $1 < a$ and $0 < c$, then $1 < a^c$.

The parameters are a and c. By density of the rationals and Closure, there is an observable rational c_1 such that $0 < c_1 < c$. Let $c' \simeq c$ be rational. By definition we have $a^c \simeq a^{c'}$. We also have $0 < c_1 < c'$.

Thus, by the monotonicity properties with rational exponents we have $1 = 1^{c_1} < a^{c_1} < a^{c'} \simeq a^c$. But a^{c_1} and a^c are observable, so $1 < a^{c_1} \leq a^c$, and so $1 < a^c$.

We deduce (1) by dividing both sides of $a_1^c < a_2^c$ by a_1^c (which is positive), putting $a = a_2/a_1 > 1$, and using the fact that $a_2^c/a_1^c = (a_2/a_1)^c = a^c$.

We deduce (2) by dividing both sides of $a^{c_1} < a^{c_2}$ by a^{c_1} (which is positive), putting $c = c_2 - c_1 > 0$ and using $a^{c_2}/a^{c_1} = a^{c_2-c_1} = a^c$.

Answer to Exercise 29, page 69

Assume that f is differentiable at a. The parameters are f and a. Then the tangent line to f at a is defined by

$$T : x \mapsto f'(a) \cdot (x - a) + f(a).$$

Let dx be ultrasmall. Let $x = a + dx$. Then

$$\frac{f(x) - T(x)}{x - a} = \frac{f(a + dx) - (f'(a) \cdot dx + f(a))}{dx}$$

$$= \frac{f(a + dx) - f(a)}{dx} - f'(a) \simeq 0,$$

so we have $\lim_{x \to a} \frac{f(x) - T(x)}{x - a} = 0$.

Conversely, suppose there is $\ell : x \mapsto m \cdot (x - a) + f(a)$ satisfying

$$\lim_{x \to a} \frac{f(x) - \ell(x)}{x - a} = 0.$$

Let dx be ultrasmall and put $x = a + dx$. Then

$$\frac{f(a + dx) - (m \cdot dx + f(a))}{dx} = \frac{f(a + dx) - f(a)}{dx} - m \simeq 0.$$

This implies that

$$\frac{f(a + dx) - f(a)}{dx} \simeq m.$$

But m is observable, so f is differentiable at a and $f'(a) = m$.

Answer to Exercise 30, page 70

The Increment Equation gives

$$f(x_2) - f(x_1) = (f(x_2) - f(a)) + (f(a) - f(x_1))$$
$$= f'(a)(x_2 - a) + \varepsilon_2 \cdot (x_2 - a) + f'(a)(a - x_1)$$
$$+ \varepsilon_1 \cdot (a - x_1)$$
$$= f'(a)(x_2 - x_1) + \varepsilon_1 \cdot (a - x_1) + \varepsilon_2 \cdot (x_2 - a),$$

with $\varepsilon_1, \varepsilon_2 \simeq 0$. It remains to show that $\varepsilon_1 \cdot (a - x_1) + \varepsilon_2 \cdot (x_2 - a) = \varepsilon \cdot (x_2 - x_1)$ for some $\varepsilon \simeq 0$. This is clear if $x_2 - x_1 = 0$. Otherwise let $\varepsilon = \frac{\varepsilon_1 \cdot (a - x_1) + \varepsilon_2 \cdot (x_2 - a)}{x_2 - x_1}$ and notice that $|\varepsilon| \le |\varepsilon_1| + |\varepsilon_2|$, since $a - x_1 \le x_2 - x_1$, $x_2 - a \le x_2 - x_1$, and $x_2 - x_1 > 0$. Thus $\varepsilon \simeq 0$.

Answer to Exercise 31, page 72

The function f is standard. To calculate $f'(2)$, let dx be ultrasmall.

$$
\begin{aligned}
(2 + dx)^3 - (2 + dx)^2 \quad &-\quad (2^3 - 2^2) = 12dx + 6dx^2 + dx^3 - 4dx - dx^2 \\
&= \quad (12 - 4)dx + \underbrace{(6dx + dx^2 - dx)}_{\simeq 0} dx \\
&= \quad 8 \cdot dx + \varepsilon \cdot dx, \qquad \text{with } \varepsilon \simeq 0.
\end{aligned}
$$

Hence $f'(2) = 8$.

To calculate $f'(3 + h)$, we assume that dx is ultrasmall relative to h. Then $(3 + h + dx)^3 - (3 + h + dx)^2 - ((3 + h)^3 - (3 + h)^2)$ yields

$$
3h^2 \cdot dx + 16h \cdot dx + 21dx + \underbrace{8dx^2 + 3h \cdot dx^2 + dx^3}_{=\varepsilon \cdot dx \text{ with } \varepsilon \simeq 0},
$$

hence the derivative is $21 + 16h + 3h^2$, which is the same as $3x^2 - 2x\big|_{x=3+h}$.

Answer to Exercise 32, page 72

Consider $f(x) = \sqrt{x}$. Let $x > 0$ be given. Let h be ultrasmall relative to x. Then

$$
\begin{aligned}
\frac{\sqrt{x + h} - \sqrt{x}}{h} &= \frac{(\sqrt{x + h} - \sqrt{x})(\sqrt{x + h} + \sqrt{x}))}{h \cdot (\sqrt{x + h} + \sqrt{x})} \\
&= \frac{x + h - x}{h(\sqrt{x + h} + \sqrt{x})} = \frac{1}{(\sqrt{x + h} + \sqrt{x})} \simeq \frac{1}{2\sqrt{x}}.
\end{aligned}
$$

We used the fact that $\sqrt{x + h} \simeq \sqrt{x}$ (continuity).

Answer to Exercise 33, page 77

Let h be a fixed ultrasmall number. Consider

$$
H : x \mapsto \begin{cases} 0 & \text{if } x \le -h; \\ \frac{1}{2h}(x + h) & \text{if } -h < x < h; \\ 1 & \text{if } x \ge h. \end{cases}
$$

The function depends on the parameter h.

For $x < -h$ and $x > h$ we have $H'(x) = 0$, because on the intervals $(-\infty, -h)$ and (h, ∞) the function is constant.

For $-h < x < h$ we must take an ultrasmall increment dx relative to x and h.

$$H(x + dx) - H(x) = \frac{1}{2h}(x + dx + h) - \frac{1}{2h}(x + h) = \frac{1}{2h} \cdot dx,$$

hence $H'(x) = \frac{1}{2h}$, which is ultralarge (relative to 1).

The function is not differentiable at $x = h$ and $x = -h$. Note that the area under $H'(x)$ is equal to 1.

In a context where h is not observable, H is indistinguishable from the discontinuous function

$$x \mapsto \begin{cases} 0 & \text{if } x < 0; \\ 1/2 & \text{if } x = 0; \\ 1 & \text{if } x > 0, \end{cases}$$

known as the Heaviside function. The "derivative" of the Heaviside function (as a distribution) is Dirac's δ "function," which is 0 at every (observable) x except at $x = 0$, and yet has area under its graph equal to 1. The function H above and its derivative H' can thus be considered as representations of the Heaviside function and the Dirac's δ, respectively.

In the classical approach, Dirac's δ is not a function. With ultrasmall numbers, it is possible to represent distributions as functions. Exercise 37 on page 87 gives the description of a yet better representation (differentiable everywhere).

Answer to Exercise 34, page 78

Proof. Let $a, b \in I$ with $a < b$. Suppose, for a contradiction, that $f(a) \geq f(b)$. The context is fixed by f, a, and b. Let $N \in \mathbb{N}$ be ultralarge. Let $dx = \frac{b-a}{N}$ and $x_i = a + i \cdot dx$, for $i = 0, 1, \ldots, N$. Since $f(x_0) = f(a) \geq f(a)$ and $f(x_N) = f(b) \leq f(a)$, there exists i such that $f(x_i) \geq f(a)$ and $f(x_{i+1}) \leq f(a)$.

Let c be the observable neighbor of x_i. Then $c \simeq x_i$, $c \simeq x_{i+1}$, and $c \in [a; b] \subseteq I$. Since $f'(c)$ exists, f is continuous at c and we have $x_i \simeq c \implies f(c) \simeq f(x_i) \geq f(a)$ and also $x_{i+1} \simeq c \implies f(c) \simeq f(x_{i+1}) \leq f(a)$. Thus $f(c) \simeq f(a)$ so $f(c) = f(a)$ since both sides are observable by closure.

Now $f'(c) > 0$, so f is increasing at c. We distinguish two cases: If $x_{i+1} \leq c$ then $x_i < c$ then $f(x_i) < f(c) = f(a)$, contradicting the choice of x_i. If $x_{i+1} > c$, then $f(x_{i+1}) > f(c) = f(a)$, contradicting the choice of x_{i+1}. $\qquad\square$

Answer to Exercise 35, page 85

Let $d\theta$ be ultrasmall. We have

$$\frac{\tan(d\theta)}{d\theta} = \frac{\sin(d\theta)}{d\theta} \cdot \frac{1}{\cos(d\theta)} \simeq 1 \cdot 1 = 1.$$

Hence

$$\lim_{\theta \to 0} \frac{\tan(\theta)}{\theta} = 1.$$

Answer to Exercise 36, page 86

For the first part, we consider the case $d\theta > 0$ (the case $d\theta < 0$ is similar). Then $\sin(d\theta) \leq \overline{BC} \leq d\theta$, hence

$$1 \simeq \frac{\sin(d\theta)}{d\theta} \leq \frac{\overline{BC}}{d\theta} \leq \frac{d\theta}{d\theta} = 1.$$

This implies that $\dfrac{\overline{BC}}{d\theta} \simeq 1$, so there is $\varepsilon \simeq 0$ such that $\dfrac{\overline{BC}}{d\theta} = 1 + \varepsilon$. Hence $\overline{BC} = d\theta + \varepsilon \cdot d\theta$, with $\varepsilon \simeq 0$. Now, by trigonometry,

$$\cos(\gamma) = \frac{\overline{AC}}{\overline{BC}} = \frac{\Delta \sin(\theta)}{d\theta(1 + \varepsilon)}.$$

But $\gamma = \theta + \frac{d\theta}{2} \simeq \theta$ (since the triangle OBC is isosceles at O). It follows that

$$\frac{\Delta \sin(\theta)}{d\theta} = \cos(\gamma) \cdot (1 + \varepsilon) \simeq \cos(\theta),$$

by continuity of cosine at θ. Since $\cos(\theta)$ is observable, we have $\sin'(\theta) = \cos(\theta)$.

Answer to Exercise 37, page 87

Let $H : x \mapsto \frac{1}{2} + \frac{1}{\pi} \cdot \arctan\left(\frac{x}{\varepsilon}\right)$. The parameter of H is ε.
Then

$$H'(x) = \frac{\varepsilon}{\pi(x^2 + \varepsilon^2)}.$$

The function looks like the Heaviside function when ε is not observable.

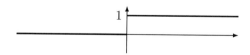

If the horizontal scale is expanded and the vertical scale is unchanged, continuity becomes clear.

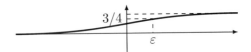

For numbers that are standard, $H'(x)$ is a horizontal line at $y = 0$ and has, at $x = 0$, a value ultralarge relative to 1. These features become discernible by zooming out on the vertical axis and zooming in on the horizontal axis. The area under this curve is 1.

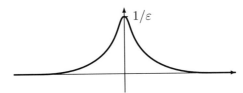

Answer to Exercise 38, page 87

(1) Assume that f is a continuous function such that $f(x) = \sin(1/x)$ for $x \neq 0$, and $f(0) = a$. Let N be an ultralarge positive integer. Then $\frac{2}{4N\pi+\pi}$ and $\frac{2}{4N\pi+3\pi}$ are ultrasmall. But $f\left(\frac{2}{4N\pi+\pi}\right)$ and $f\left(\frac{2}{4N\pi+3\pi}\right)$ cannot both be ultraclose to a, because $\sin(2N\pi + \pi/2) = 1$ and $\sin(2N\pi + 3\pi/2) = -1$. In other words, $\lim_{h\to 0} \sin\left(\frac{1}{h}\right)$ does not exist.

(2) Let g be defined by

$$g : x \mapsto \begin{cases} x \sin\left(\frac{1}{x}\right) & \text{if } x \neq 0; \\ 0 & \text{if } x = 0. \end{cases}$$

Let h be ultrasmall. Then

$$\frac{g(h) - g(0)}{h} = \frac{h \sin\left(\frac{1}{h}\right) - 0}{h} = \sin\left(\frac{1}{h}\right).$$

But, we saw in the previous item that $\lim_{h\to 0} \sin\left(\frac{1}{h}\right)$ does not exist.

(3) Finally, consider the function h given by

$$h : x \mapsto \begin{cases} x^2 \cdot \sin\left(\frac{1}{x}\right) & \text{if } x \neq 0; \\ 0 & \text{if } x = 0. \end{cases}$$

The derivative at $x \neq 0$ can be computed using the rules of differentiation:

$$h'(x) = 2x \sin(1/x) - \cos(1/x).$$

At $x = 0$ we use the definition:

$$\frac{h(x + dx) - h(x)}{dx} = \frac{(dx)^2 \cdot \sin(1/dx) - 0}{dx} = dx \cdot \sin(1/dx) \simeq 0$$

(because $-1 \leq \sin(1/dx) \leq 1$), hence $h'(0) = 0$ and we can write

$$h' : x \mapsto \begin{cases} 2x\sin(1/x) - \cos(1/x) & \text{if } x \neq 0; \\ 0 & \text{if } x = 0. \end{cases}$$

The derivative function h' is not continuous at 0. The $2x\sin(1/x)$ part is ultraclose to 0 when x is ultrasmall, but similarly as above, for any ultralarge integer N, $x_1 = \frac{1}{(2N+1)\cdot\pi}$ and $x_2 = \frac{1}{2N\cdot\pi}$ are ultrasmall, while $\cos(2N\pi) = 1$ and $\cos((2N+1)\pi) = -1$, so $h'(x_1) \simeq -1$ and $h'(x_2) \simeq 1$.

Answer to Exercise 39, page 89

Let f be twice differentiable at a and let $q(x) = b_0 + b_1 \cdot (x-a) + b_2 \cdot (x-a)^2$. The parameters are a, f, b_0, b_1, and b_2. We assume that for each $x \simeq a$, there is $\varepsilon \simeq 0$ such that

$$f(x) = q(x) + \varepsilon \cdot (x-a)^2.$$

For $x = a$, this implies that $f(a) = q(a) = b_0$.

Let $x \simeq a$, with $x \neq a$. We have

$$f(x) = f(a) + b_1 \cdot (x-a) + b_2 \cdot (x-a)^2 + \varepsilon \cdot (x-a)^2$$
$$= f(a) + b_1 \cdot (x-a) + \delta \cdot (x-a),$$

where $\delta = b_2 \cdot (x-a) + \varepsilon \cdot (x-a) \simeq 0$. By the Increment Equation, we have $b_1 = f'(a)$. Finally, by the Increment Equation of order two, there is $h \simeq 0$ such that

$$f(x) = f(a) + f'(a) \cdot (x-a) + \frac{f''(a)}{2} \cdot (x-a)^2 + h \cdot (x-a)^2.$$

By comparing the two expansions for $f(x)$ we get

$$b_2 + \varepsilon = \frac{f''(a)}{2} + h, \quad \text{that is,} \quad b_2 \simeq \frac{f''(a)}{2}.$$

Since b_2 and $\frac{f''(a)}{2}$ are observable, they must be equal.

Answer to Exercise 40, page 104

The constant function $x \mapsto c$ is continuous, so $\int_a^b c \cdot dx$ exists. Let N be ultralarge and $dx = (b-a)/N$. Then

$$\int_a^b c \cdot dx \simeq \sum_{i=0}^{N-1} c \cdot dx = c \cdot \sum_{i=0}^{N-1} dx = c \cdot (b-a).$$

Since the left-hand side and the right-hand side are observable, they are equal.

Answer to Exercise 41, page 104

Let f be continuous on $[a, b]$. The parameters are a, b and f. By the Extreme Value Theorem, there is $c \in [a, b]$ such that $f(c) \leq f(x)$ for all $x \in I$. Notice that $f(c)$ is observable. Let $N \in \mathbb{N}$ be ultralarge, $dx = (b - a)/N$ and $x_i = a + i \cdot dx$, for $i = 0, \dots, N - 1$. Then

$$\int_a^b f(x) \cdot dx \simeq \sum_{i=0}^{N-1} f(x_i) \cdot dx \geq \sum_{i=0}^{N-1} f(c) \cdot dx = f(c) \cdot (b - a).$$

Since both sides are observable, we have $\int_a^b f(x) \cdot dx \geq f(c) \cdot (b - a)$. If $f(x) > 0$ for all $x \in [a, b]$, then $f(c) > 0$, so $\int_a^b f(x) \cdot dx > 0$. The other cases are proved similarly.

Answer to Exercise 42, page 111

Let H be an antiderivative of $u \mapsto f(g(u)) \cdot g'(u)$, where g is a one-to-one correspondence whose inverse is differentiable. Then

$$(H \circ g^{-1})'(x) = H'(g^{-1}(x)) \cdot (g^{-1})'(x)$$
$$= f\left(g(g^{-1}(x))\right) \cdot g'(g^{-1}(x)) \cdot \frac{1}{g'(g^{-1}(x))} = f(x).$$

Answer to Exercise 43, page 114

Let $a \in I$. The parameters are I, f, g and a. Let $x \simeq a$, with $x \in I$. Without loss of generality, we may assume that $f(a) \leq g(a)$, so $h(a) = g(a)$. By continuity, we have $f(x) \simeq f(a)$ and $g(x) \simeq g(a)$. If $g(x) \geq f(x)$, then $h(x) = g(x) \simeq g(a) = h(a)$. Otherwise, $f(x) \geq g(x)$, and so

$$f(x) \geq g(x) \simeq g(a) \geq f(a) \quad \text{and} \quad f(x) \simeq f(a).$$

This implies that $f(x) \simeq g(a)$, and again $h(x) = f(x) \simeq h(a)$.

Answer to Exercise 44, page 119

Let x denote the distance from A to a point of B, so x ranges from 3 to 9. Let $N \in \mathbb{N}$ be ultralarge. The linear object B is sliced into N parts of length $\frac{6}{N} = dx$. The mass of each slice is $\frac{18}{N} = 3\,dx$. The force $\Delta F(x)$ between A and the slice of B from x to $x + dx$ has magnitude between $G \cdot \frac{6 \cdot 3\,dx}{(x+dx)^2}$ (this would be the force if the entire slice were located at its remote end) and $G \cdot \frac{6 \cdot 3\,dx}{x^2}$ (this would be the force if the entire slice were located at its near end). A simple calculation shows that

$1/x^2 - 1/(x+dx)^2 \simeq 0$; it follows that $\Delta F(x) = 18G \cdot \frac{dx}{x^2} + \varepsilon \cdot dx$, for $\varepsilon \simeq 0$. Hence

$$F = \sum_{i=0}^{N-1} \Delta F(x_i) \simeq \sum_{i=0}^{N-1} 18\,G \cdot \frac{dx}{x_i^2} \simeq 18\,G \cdot \int_3^9 \frac{dx}{x^2} = 4\,G.$$

Answer to Exercise 45, page 122

To compute the integral $\int_a^{a\cdot b} \frac{1}{x} \cdot dx$ (with $a, b > 0$), we use the substitution $u = \frac{x}{a}$. Then $x = a \cdot u$ and $dx = a \cdot du$. Moreover, $u = 1$ if $x = a$, and $u = b$ if $x = a \cdot b$. Hence

$$\int_a^{a\cdot b} \frac{1}{x} \cdot dx = \int_1^b \frac{1}{a \cdot u} \cdot a \cdot du = \int_a^b \frac{1}{u} \cdot du = \ln(b).$$

Therefore

$$\ln(a \cdot b) = \int_1^{a\cdot b} \frac{1}{x} \cdot dx = \int_1^a \frac{1}{x} \cdot dx + \int_a^{a\cdot b} \frac{1}{x} \cdot dx = \ln(a) + \ln(b).$$

Answer to Exercise 46, page 125

Let $z \in \mathbb{R}$. For $z = 0$ there is nothing to show, so let us assume that $z \neq 0$. We have $\left(1 + \frac{z}{x}\right)^x = \exp\left(x \cdot \ln(1 + \frac{z}{x})\right)$. Let x be ultralarge relative to z. Then z/x is ultrasmall, so by definition of the derivative of $\ln$ at 1 we have

$$x \cdot \ln(1 + \frac{z}{x}) = z \cdot \frac{\ln(1 + \frac{z}{x}) - \ln(1)}{\frac{z}{x}} \simeq z \cdot \ln'(1) = z.$$

But $\exp$ is continuous at z, so

$$\exp\left(x \cdot \ln(1 + \frac{z}{x})\right) \simeq \exp(z) = e^z.$$

Answer to Exercise 47, page 126

Let a be a real number.

(1) Let $h > 0$ be ultrasmall relative to a. Let $b = \frac{1 - e^{-h}}{h}$. Then $bh = 1 - e^{-h}$ is positive ultrasmall, since $e^{-h} < 1$ and $e^{-h} \simeq 1$. Hence $x = 1/bh$ is positive ultralarge. By the previous exercise we have $e^{-1} \simeq (1 - 1/x)^x = (e^{-h})^{1/bh} = e^{-1/b}$. This shows that $b \simeq 1$.

(2) Now let h be negative ultrasmall relative to a. Then $-h$ is positive ultrasmall, so by the first part $1 \simeq \frac{1 - e^{-(-h)}}{-h} = \frac{e^h - 1}{h}$. It follows that

$$\frac{e^{a+h} - e^a}{h} = e^a \cdot \frac{e^h - 1}{h} \simeq e^a.$$

Answer to Exercise 48, page 134

The context is given by α. We have $\int_0^b f_\alpha(x) \cdot dx = \frac{1}{\alpha} \cdot e^{\alpha x}\big|_0^b = \frac{e^{\alpha b}-1}{\alpha}$, when $\alpha \neq 0$, and $\int_0^b f_0(x) \cdot dx = b - 1$. Let $b \simeq +\infty$. Then $e^{\alpha b} \simeq +\infty$ if $\alpha > 0$ and $e^{\alpha b} \simeq 0$ if $\alpha < 0$. It follows that $\int_0^\infty f_\alpha(x) \cdot dx$ converges if and only if $\alpha < 0$, and in this case $\int_0^\infty f_\alpha(x) \cdot dx = 1/(-\alpha)$.

Answer to Exercise 49, page 136

The parameter is α. Let $h > 0$ be ultrasmall. Then $\int_h^1 \frac{1}{x} \cdot dx = -\ln(h)$ if $a = 1$, and $\int_h^1 \frac{1}{x^\alpha} \cdot dx = \frac{1-h^{1-\alpha}}{1-\alpha}$, if $\alpha \neq 1$. But $-\ln(h) \simeq +\infty$, and $h^{1-\alpha} \simeq 0$ if $\alpha < 1$ and $h^{1-\alpha} \simeq +\infty$ if $\alpha > 1$. It follows that $\int_0^1 \frac{1}{x^\alpha} \cdot dx = \frac{1}{1-\alpha}$ if $\alpha < 1$, and $\int_0^1 \frac{1}{x^\alpha} \cdot dx$ diverges otherwise.

Answer to Exercise 50, page 137

The parameters are a, b, f and g. We assume $-g(x) \leq f(x) \leq g(x)$, for each $x \in (a, b]$. Let $h > 0$ be ultrasmall. By Exercise 4.9, $-\int_{a+h}^b g(x) \cdot dx \leq \int_{a+h}^b f(x) \cdot dx \leq \int_{a+h}^b g(x) \cdot dx$. Since $\int_a^b g(x) \cdot dx$ converges, the numbers $\pm \int_{a+h}^b g(x) \cdot dx$ are not ultralarge, and so $\int_a^b f(x) \cdot dx$ is not ultralarge. Now suppose that $h, k > 0$ are ultrasmall, say $h < k$. By Exercise 4.9 again, we have $-\int_{a+h}^{a+k} g(x) \cdot dx \leq \int_{a+h}^{a+k} f(x) \cdot dx \leq \int_{a+h}^{a+k} g(x) \cdot dx$. But the left-hand side and the right-hand side are ultraclose to 0 since $\int_a^b g(x) \cdot dx$ converges, so $\int_{a+h}^{a+k} f(x) \cdot dx \simeq 0$ and therefore $\int_a^b f(x) \cdot dx$ converges.

Answer to Exercise 51, page 144

We have $0 < \frac{1}{n} + \frac{1}{m} \leq 2$ for all $n, m \geq 1$. As $2 = \frac{1}{1} + \frac{1}{1} \in A$, 2 is the least upper bound of A, so $\sup A = 2$. For ultralarge n, m we have $x = \frac{1}{n} + \frac{1}{m} \simeq 0$, so $\inf A = 0$. Note that $0 \notin A$.

Answer to Exercise 52, page 147

The statement $\mathcal{P}(n)$ given by "$\lim_{x \to 0^+} x^n = 0$" is internal, so we can use the Principle of Mathematical Induction. For $n = 1$, it is clear. Assume that $\mathcal{P}(n)$ is true. Then by our rules on limits of products

$$\lim_{x \to 0^+} x^{n+1} = \lim_{x \to 0^+} x^n \cdot x = \left(\lim_{x \to 0^+} x^n \right) \cdot \left(\lim_{x \to 0^+} x \right) = 0 \cdot 0 = 0.$$

This shows that $\mathcal{P}(n + 1)$ is true, and by induction, $\mathcal{P}(n)$ is true for all $n \geq 1$.

The proof for the other limit is similar.

Answer to Exercise 53, page 147

The statement $\mathcal{P}(n)$ given by "$x \mapsto x^n$ is differentiable everywhere and $(x^n)' = nx^{n-1}$" is internal. We use the Principle of Mathematical Induction. For $n = 0$, it is clear since the function $x \mapsto x^0$ is the constant function $x \mapsto 1$, so it is differentiable everywhere and $(x^0)' = 0 = 0 \cdot x^{-1}$. Now let $n \geq 0$ and assume that $\mathcal{P}(n)$ is true. Since $x^{n+1} = x^n \cdot x$, the function $x \mapsto x^{n+1}$ is differentiable everywhere by the product rule and

$$(x^{n+1})' = (x^n \cdot x)' = (x^n)' \cdot x + x^n \cdot x' = nx^{n-1} \cdot x + x^n \cdot 1 = (n+1)x^n.$$

This shows that $\mathcal{P}(n+1)$ is true, so by the Principle of Mathematical Induction $\mathcal{P}(n)$ is true for all $n \in \mathbb{N}$.

Answer to Exercise 54, page 158

We sketch the proof in the $0/0$ case. Assume that $\lim\limits_{x \to \infty} f(x) = 0$, $\lim\limits_{x \to \infty} g(x) = 0$, and $\lim\limits_{x \to \infty} \dfrac{f'(x)}{g'(x)} = L$. Let x be ultralarge relative to the context f, g and let y be ultralarge relative to the extended context f, g, x. Then $x < y$ and, by Cauchy's Theorem (Theorem 39), there is $c \in (x, y)$ such that

$$\Big(f(x) - f(y)\Big) \cdot g'(c) = \Big(g(x) - g(y)\Big) \cdot f'(c).$$

The rest of the proof closely follows the $\lim\limits_{x \to a}$ case. The ∞/∞ case can be handled similarly.

Answer to Exercise 55, page 158

Consider the limit

$$\lim_{x \to \infty} \left(1 + \frac{z}{x}\right)^x.$$

By applying ln and rewriting the product we obtain

$$\lim_{x \to \infty} x \, \ln\left(1 + \frac{z}{x}\right) = \lim_{x \to \infty} \frac{\ln\left(1 + \frac{z}{x}\right)}{\frac{1}{x}},$$

which has the form $0/0$. Hence, by L'Hôpital's Rule, we have

$$\lim_{x \to \infty} \frac{\ln\left(1 + \frac{z}{x}\right)}{\frac{1}{x}} = \lim_{x \to \infty} \frac{-\frac{z}{x^2}}{\left(1 + \frac{z}{x}\right) \cdot \left(-\frac{1}{x^2}\right)} = z.$$

By applying exp and using the continuity of exp at z we get

$$\lim_{x \to \infty} \left(1 + \frac{z}{x}\right)^x = \exp(z) = e^z.$$

Answer to Exercise 56, page 160

Let f be n-times differentiable at a. Let $p(x) = \sum_{k=0}^{n} c_k \cdot (x-a)^k$ and assume that for all $x \simeq a$ there is $\varepsilon \simeq 0$ such that $f(x) = p(x) + \varepsilon \cdot (x-a)^n$.

The parameters are f, n and $c_0, \dots, c_n$. We show that $c_k = \frac{f^k(a)}{k!}$ by induction on $k = 0, \dots, n$.

For $k = 0$ it is clear since $f(a) = p(a) = c_0$.

Assume that it is true up to $k < n$. Let $x \simeq a$, with $x \neq a$. Then, by the Taylor formula, there is $\delta \simeq 0$ such that

$$f(x) = \sum_{i=0}^{k} \frac{f^{(i)}(a)}{i!}(x-a)^i + \sum_{i=k+1}^{n} \frac{f^{(i)}(a)}{i!}(x-a)^i + \delta(x-a)^n,$$

and by assumption there is $\varepsilon \simeq 0$ such that

$$f(x) = \sum_{i=0}^{k} \frac{f^{(i)}(a)}{i!}(x-a)^i + \sum_{i=k+1}^{n} c_i(x-a)^i + \varepsilon(x-a)^n.$$

The right hand sides are therefore equal, and after some simplifications and a division by $(x-a)^{k+1}$ we obtain that

$$\frac{f^{(k+1)}(a)}{(k+1)!} + Q = c_{k+1} + P,$$

where

$$Q = \sum_{i=k+2}^{n} \frac{f^{(i)}(a)}{i!}(x-a)^{i-(k+1)} + \delta(x-a)^{n-(k+1)} \simeq 0$$

since $x - a \simeq 0$, $\delta \simeq 0$ and all the coefficients are observable, and

$$P = \sum_{i=k+2}^{n} c_i(x-a)^{i-(k+1)} + \varepsilon(x-a)^{n-(k+1)} \simeq 0$$

since $x - a \simeq 0$, $\varepsilon \simeq 0$ and all the coefficients are observable. Hence

$$\frac{f^{(k+1)}(a)}{(k+1)!} \simeq c_{k+1}, \quad \text{and so} \quad \frac{f^{(k+1)}(a)}{(k+1)!} = c_{k+1},$$

since they both are observable. This finishes the induction.

Answer to Exercise 57, page 164

Let M be such that $u_n = u'_n$ for all $n \geq M$. The parameters are u, u' and M. Assume $\lim_{n \to \infty} u_n = L$. Hence for any ultralarge N we have

$u_N \simeq L$. As $N > M$, we have $u'_N = u_N \simeq L$, hence $u'_N \simeq L$. It follows that $\lim_{n\to\infty} u'_n = L$.

Answer to Exercise 58, page 164

Let $(u_n)_{n\geq k}$ converge to L. The context is specified by the sequence; L and k are observable. Let $s_n = \frac{u_k+\cdots+u_n}{n}$. Then $(s_n)_{n\geq k}$ is observable. Let N be ultralarge. We must show that

$$s_N = \frac{u_k + \cdots + u_N}{N} \simeq L.$$

We use Theorem 89. Let M and N be such that M is ultralarge and that N is still ultralarge relative to (u_n) and M. We use $\overset{+}{\simeq}$ when we work relative to the context extended by M. Since M is ultralarge and (u_n) converges to L, for each $i \geq M$ there is $\varepsilon_i \simeq 0$ such that $u_i = L + \varepsilon_i$. We thus have

$$\frac{\sum_{i=k}^N u_i}{N} = \underbrace{\frac{\sum_{i=k}^M u_i}{N}}_{\overset{+}{\simeq}0} + \frac{\sum_{i=M+1}^N(L+\varepsilon_i)}{N} \overset{+}{\simeq} \frac{(N-M)\cdot L}{N} + \sum_{i=M+1}^N \varepsilon_i \cdot \frac{1}{N}$$

$$= L - \underbrace{\frac{M}{N}\cdot L}_{\overset{+}{\simeq}0} + \underbrace{\sum_{i=M+1}^N \varepsilon_i \cdot \frac{1}{N}}_{\simeq 0} \simeq L.$$

Answer to Exercise 59, page 166

Let $F : [a,\infty) \to \mathbb{R}$ be defined by $F(x) = \int_a^x f(t)\cdot dt$. The parameters are thus a and f. By the Definition Principle, F is an observable function. The only thing we need to show is that if $\int_b^c f(x) \cdot dx \simeq 0$, for every b, c positive ultralarge, then $F(c)$ is not ultralarge for any positive ultralarge c. Fix b positive ultralarge. By assumption, $F(c) \simeq F(b)$ for every $c \geq b$ (since b, c are ultralarge). This shows that

$$F(b) - 1 \leq F(c) \leq F(b) + 1, \quad \text{for every } c \geq b.$$

So, there are real numbers M_1, M_2 and b such that

$$M_1 \leq F(c) \leq M_2, \quad \text{for every } c \geq b.$$

This statement does not mention observability, so by Closure, there are observable b, M_1, and M_2 with this property. But this implies that $F(c) = \int_a^c f(x) \cdot dx$ is not ultralarge for any positive ultralarge c.

Answer to Exercise 60, page 170

Consider the series $\sum_{n\geq k} u_n$ and the positive integer $m \geq k$. The parameters are (u_n), k and m. Let N be a positive integer such that $k \leq m \leq N$. Then

$$\sum_{n=k}^{N} u_n = \underbrace{u_k + u_{k+1} + \cdots + u_{m-1}}_{=s} + \sum_{n=m}^{N} u_n.$$

The finite sequence $u_k, \ldots, u_{m-1}$ is defined from the parameters, and hence it is observable, by Closure. For the same reason, its sum s is also observable. Hence $\lim_{N\to\infty} \sum_{n=m}^{N} u_n$ exists if and only if $\lim_{N\to\infty} \sum_{n=k}^{N} u_n$ exists, because

$$\lim_{N\to\infty} \sum_{n=k}^{N} u_n = s + \lim_{N\to\infty} \sum_{n=m}^{N} u_n \text{ and } \lim_{N\to\infty} \sum_{n=m}^{N} u_n = -s + \lim_{N\to\infty} \sum_{n=k}^{N} u_n.$$

Answer to Exercise 61, page 183

Fix $a \in A$. By Stability, if N, M are ultralarge relative to (f_n), a, then for all $x \in A$, $f_N(x) \simeq f_M(x)$ relative to (f_n), a holds. In particular, $f_N(a) \simeq f_M(a)$ relative to (f_n), a holds. This shows that the sequence $(f_n(a))$ is Cauchy and therefore converges. Define $f(a) = \lim_{n\to\infty} f_n(a)$ for all $a \in A$. The sequence (f_n) converges pointwise to f on A; it only remains to show that the convergence is uniform.

Let N be ultralarge relative to (f_n). Take some M ultralarge relative to (f_n), a. Then $f_M(a) \simeq f(a)$ relative to (f_n), a, by pointwise convergence at a, and $f_N(a) \simeq f_M(a)$ relative to (f_n) by the assumption. We conclude that $f_N(a) \simeq f(a)$ relative to (f_n).

Answer to Exercise 62, page 197

Let $t \in I$. The parameters are F, y, I, J, t. Let $t' \simeq t$. Since $y(t)$ is observable and $y(t') \simeq y(t)$ by continuity of y, we have $f(t') = F(t', y(t')) \simeq F(t, y(t)) = f(t)$ by continuity of F at $\langle t, y(t)\rangle$. This shows that f is continuous on I.

Answer to Exercise 63, page 199

Assume $a < b$. Let $|f(x)| \leq M$ for all $x \in [a, b]$, where M is observable. From $z_{N-1} = a + (N-1) \cdot dz \leq b \leq z_N = a + N \cdot dz$ it follows that

$$dx = \frac{b-a}{N} \leq dz \leq \frac{b-a}{N-1}.$$

$$\text{Hence } 0 \leq dz - dx \leq \frac{b-a}{N-1} - \frac{b-a}{N} = \frac{b-a}{N(N-1)}$$

and $z_i - x_i = (a + i \cdot dz) - (a + i \cdot dx) = i \cdot (dz - dx) \leq \dfrac{i \cdot (b - a)}{N(N-1)} \leq \dfrac{b-a}{N}$,

for $0 \leq i \leq N - 1$. As dz is ultrasmall, N has to be ultralarge, so in particular $z_i \simeq x_i$. Hence, by continuity of f, $\varepsilon_i = |f(z_i) - f(x_i)| \simeq 0$, for $0 \leq i \leq N - 1$. We now have

$$\left| \sum_{i=0}^{N-1} f(z_i) \cdot dz - \sum_{i=0}^{N-1} f(x_i) \cdot dx \right| \leq \sum_{i=0}^{N-1} |f(z_i) \cdot dz - f(x_i) \cdot dx|.$$

From $|f(z_i) \cdot dz - f(x_i) \cdot dx| = |f(z_i) \cdot (dz - dx) + (f(z_i) - f(x_i)) \cdot dx| \leq |f(z_i)| \cdot (dz - dx) + |f(z_i) - f(x_i)| \cdot dx \leq M \cdot \frac{b-a}{N(N-1)} + \varepsilon_i \cdot dx$ we obtain by summation that

$$\left| \sum_{i=0}^{N-1} f(z_i) \cdot dz - \sum_{i=0}^{N-1} f(x_i) \cdot dx \right| \leq M \cdot \frac{b-a}{N-1} + \sum_{i=0}^{N-1} \varepsilon_i \cdot dx.$$

Both terms are $\simeq 0$, and the claim is proved.

Answer to Exercise 64, page 221

Let A be a set and let $B = A^c$. We show that A is open if and only if B is closed. The parameters are A and B. Let c be the observable neighbor of x (if it exists). We have A is open if and only if $c \in A$ implies $x \in A$ (whenever x has an observable neighbor) if and only if $x \notin A$ implies $c \notin A$ (whenever x has an observable neighbor) if and only if $x \in B$ implies $c \in B$ (whenever x has an observable neighbor) if and only if B is closed.

Answer to Exercise 65, page 222

(1) Let $A = (a, b)$. The parameters are a and b. Let x be not ultralarge and c be its observable neighbor. If $c \in A$, then $a < c < b$. Since $x \simeq c$ and a, b are observable, we must have $a < x < b$, so $x \in A$. The cases $(-\infty, b)$ and (a, ∞) are similar.

(2) Let $A = [a, b]$. Let x be not ultralarge and c be its observable neighbor and suppose that $x \in A$. Then $a \leq x \leq b$. Since $x \simeq c$ and c is observable, we must have $a \leq c \leq b$, so $c \in A$. The cases $(-\infty, b]$ and $[a, \infty)$ are similar.

(3) Let $A = (a, b]$. Let $x > b$ such that $x \simeq b$. Then b is the observable neighbor of x, $b \in A$, but $x \notin A$, so A is not open. Now let $x > a$ such that $x \simeq a$. Then $x \in A$, a is the observable neighbor of x, but $a \notin A$, so A is not closed.

Answer to Exercise 66, page 223

(3) The parameter is $\mathcal{A}$; then $\cap \mathcal{A}$ is observable. Let x be not ultralarge such that $x \in \bigcap_{i \in I} A_i$ and let c be the observable neighbor of x. Then $x \in A_i$ for each $i \in I$; hence, for each observable i, we have $c \in A_i$ (since A_i is closed). Thus, by Closure, $c \in A_i$ is true for all $i \in I$. This implies that $c \in \bigcap_{i \in I} A_i$ and proves that $\bigcap_{i \in I} A_i$ is closed.

(4) The parameter is $\mathcal{A}$; then $\cup \mathcal{A}$ is observable. Let x be not ultralarge such that $x \in \bigcup_{i \in I} A_i$, with I finite, and let c be the observable neighbor of x. Then $x \in A_i$ for some $i \in I$ and, since I is finite, this i is observable and so A_i is observable, by Closure. Hence $c \in A_i$, since A_i is closed, so $c \in \bigcup_{i \in I} A_i$. We conclude that $\bigcup_{i \in I} A_i$ is closed.

Answer to Exercise 67, page 223

The intersection of the family of open sets $\{(0,\ 1+1/n) : n \in \mathbb{N}\}$ is $(0,1]$, which is not open.

The union of the family of closed sets $\{[1/n,\ 1] : n \in \mathbb{N}\}$ is $(0,1]$, which is not closed.

Answer to Exercise 68, page 223

(1) is clear since $\emptyset = \emptyset \cap U$ and $U = \mathbb{R} \cap U$.

(2) If A is open in U, then there is $A' \subseteq \mathbb{R}$ open such that $A = A' \cap U$. By Exercise 64 we have $(A')^c$ is closed. But $A^c \cap U = (A')^c \cap U$, so $A^c \cap U$ is closed in A.

(3) Let $A_i \subseteq U$ be open in U for $i \in I$. For each $i \in I$ we choose one open set A'_i such that $A_i = A'_i \cap U$. Then $\bigcup_{i \in I} A_i = \bigcup_{i \in I}(A'_i \cap U) = (\bigcup_{i \in I} A'_i) \cap U$. But $\bigcup_{i \in I} A'_i$ is open by Theorem 140, hence $\bigcup_{i \in I} A_i$ is open in U. The appeal to the Axiom of Choice can be eliminated by replacing A'_i in the above argument with $B_i = \bigcup \{A' \subseteq \mathbb{R} \text{ open} : A_i = A' \cap U\}$; note that B_i is open and $A_i = B_i \cap U$.

(4) – (6) are similar to (3).

Answer to Exercise 69, page 224

Let $f : \mathbb{R} \to \mathbb{R}$ be a continuous function and $B \subseteq \mathbb{R}$ a closed set. Let $x \in f^{-1}(B)$ be not ultralarge. Let a be observable such that $a \simeq x$. Since f is continuous, we have $f(a) \simeq f(x)$. But $f(a)$ is observable, so $f(a)$ is the neighbor of $f(x)$. Now $f(x) \in B$ and B is closed, so $f(a) \in B$ and $a \in f^{-1}(B)$. This shows that $f^{-1}(B)$ is closed.

Another proof can be given using Exercise 64.

Answer to Exercise 70, page 226

(1) Let $a \in A$. The parameters are A, U and a. Let $\varepsilon > 0$ be ultrasmall. Let $x \in (a - \varepsilon, \, a + \varepsilon) \cap U$. Then $a \simeq x$, so $x \in A$ by the given property of A.

(2) Suppose that $A \subseteq U$ is open in U. Let A' be open and observable such that $A = A' \cap U$. The parameters are U, A and A'. Let $x \in U$ be such that its observable neighbor $a \in A$. Then $a \in A'$, so $x \in A'$ since A' is open. But $x \in U$, so $x \in A' \cap U = A$.

For the converse, we use (1). Let $A' = \bigcup_{a \in A, \varepsilon} (a - \varepsilon, \, a + \varepsilon)$, where ε is as in (1). Then A' is a union of open sets, so it is open. Furthermore, $A = A' \cap U$ by (1). This shows that A is open in U.

(3) follows by complementation.

Answer to Exercise 71, page 226

Assume that $f : U \to \mathbb{R}$ is continuous. Let $B \subseteq \mathbb{R}$ be open. The parameters are f, U and B. We use Exercise 70 to show that $f^{-1}(B)$ is open in U. Let $x \in U$ be such that its observable neighbor $a \in f^{-1}(B)$. Since a is observable and f is continuous, we have $f(x) \simeq f(a)$. But $f(a) \in B$ and B is open, so $f(x) \in B$. This shows that $x \in f^{-1}(B)$. Thus $f^{-1}(B)$ is open in U.

For the converse, let $a \in U$ be a real number. The parameters are f, U and a. Let $x \simeq a$, $x \in U$. Let $c > 0$ be observable. Then $B = (f(a) - c, \, f(a) + c)$ is observable, it is an open set, and it contains $f(a)$. Hence $f^{-1}(B)$ is open in U and contains a. Since $a \simeq x$ and $x \in U$, we have $x \in f^{-1}(B)$, and so $f(x) \in B = (f(a) - c, \, f(a) + c)$. But this shows that $f(x) \simeq f(a)$, since the distance between $f(x)$ and $f(a)$ is less than any observable $c > 0$.

Answer to Exercise 72, page 228

(1) The parameter is A. Assume that x has an observable neighbor $a \in A^o$. We must show that $x \in A^o$. We have $a \in A \backslash \delta A$, which implies by definition that all $y \simeq a$ are in A (since $a \in A$). Thus $x \in A$. Now $x \notin \delta A$, for otherwise there is $y \notin A$ such that $y \overset{+}{\simeq} x$ (relative to the context extended by x), which implies that $y \simeq a$, so $a \in \delta A$, a contradiction. In all, $x \in A \backslash \delta A = A^o$, which shows that A^o is open.

(2) The parameter is A. Let x be not ultralarge such that $x \in \overline{A}$ and let a be its observable neighbor. We must show that $a \in \overline{A}$. If $x \in \delta A$, then $a \in \delta A$ since δA is closed. Suppose that $x \in A$. If $a \notin A$, then $x_1 = x \in A$ and $x_2 = a \notin A$ both satisfy $a \simeq x_1$ and $a \simeq x_2$. This shows that $a \in \delta A$. In any case, $a \in \overline{A}$, which proves that $\overline{A}$ is closed.

Answer to Exercise 73, page 228

Suppose that $A \subseteq B$.

(1) Let $x \in \overline{A}$, so $x \in A \cup \delta A$. The parameters are A, B and x. If $x \in B$, then we are done. Otherwise $x \notin B$ and therefore $x \notin A$, so $x \in \delta A$. Thus, there is $x_1 \in A \subseteq B$ such that $x_1 \simeq x$. Let $x_2 = x \notin B$. Hence we have $x \in \delta B$, so $x \in \overline{B}$.

(2) Let $x \in A^o$. The parameters are A, B and x. We have $x \in A \setminus \delta A$, hence $x \in B$, and we prove that $x \notin \delta B$. Let $x_1 = x \in A$. Suppose that $x \in \delta B$; then there is $x_2 \notin B$ such that $x \simeq x_2$. But then $x \simeq x_1$ and $x \simeq x_2$ with $x_1 \in A$ and $x_2 \notin A$, so $x \in \delta A$, a contradiction.

Answer to Exercise 74, page 229

(1) Suppose that A is open. We want to show that A is disjoint from its boundary. Suppose, for a contradiction, that there is $x_1 \in \delta A \cap A$. The parameters are A and x_1. Since $x_1 \in \delta A$, there is $x_2 \notin A$ such $x_2 \simeq x_1$. But $x_2 \simeq x_1 \in A$ (and x_1 is observable), so $x_2 \in A$ since A is open. This is a contradiction.

(2) Suppose that A is closed. We want to show that A contains its boundary. Suppose, for a contradiction, that there is $x_2 \in \delta A \setminus A$. The parameters are A and x_2. Since $x_2 \in \delta A$, there is $x_1 \in A$ such that $x_1 \simeq x_2$. But A is closed and x_2 is observable, so $x_2 \in A$, a contradiction.

Answer to Exercise 75, page 229

Suppose that D intersects every nonempty open set. Let x be given. The parameters are x and D. Let $\varepsilon > 0$ be ultrasmall. Then $(x-\varepsilon,\ x+\varepsilon)$ is open, so there is $d \in D \cap (x - \varepsilon,\ x + \varepsilon)$, by assumption. But $d \simeq x$. This shows that D is dense.

Answer to Exercise 76, page 229

Let D be a dense open set and let (a, b) be an open interval. The parameters are a, b and D. Since D is dense, it must intersect the open set (a, b), so let $a_1 \in D \cap (a, b)$. By the Closure Principle, we may assume that a_1 is observable. Let $\varepsilon > 0$ be ultrasmall. Let $b_1 = a_1 + \varepsilon$. Then each $y \in [a_1, b_1]$ satisfies $y \simeq a_1 \in D \cap (a, b)$ and $D \cap (a, b)$ is an open set, so we must have $y \in D \cap (a, b)$. This shows that $[a_1, b_1] \subseteq D \cap (a, b)$. By Closure, there exist observable a_1, b_1 with this property.

Answer to Exercise 77, page 230

It is enough to show that the intersection of two dense open sets is dense and open (and then proceed by induction). Let D_1, D_2 be two dense open sets. The parameters are D_1 and D_2. Let $a \in \mathbb{R}$. Let $d_1 \simeq a$ be in D_1 and let $d_2 \in D_2$ be such that $d_2 \stackrel{+}{\simeq} d_1$, where $\stackrel{+}{\simeq}$ is relative to the context extended by d_1. Then d_1 is observable relative to the extended context, $d_2 \stackrel{+}{\simeq} d_1 \in D_1$ and D_1 is open, so we must have $d_2 \in D_1$. This implies that $d_2 \in D_1 \cap D_2$ and $d_2 \stackrel{+}{\simeq} d_1 \simeq a$, so $d_2 \simeq a$. Thus $D_1 \cap D_2$ is dense. The fact that $D_1 \cap D_2$ is open follows from Theorem 140.

Answer to Exercise 78, page 231

Let $\mathcal{A} = \{A_i : i \in I\}$ be an open cover of a set B. Each A_i can be written as a countable union of open intervals with rational endpoints. Since there are only countably many intervals with rational endpoints, only countably many of these suffice to cover B, say (a_n, b_n), for $n \in \mathbb{N}$. We now extract a countable subcover of $\mathcal{A}$: For each $n \in \mathbb{N}$, choose an $i(n) \in I$ such that $(a_n, b_n) \subseteq A_{i(n)}$. Then $B \subseteq \bigcup_{n \in \mathbb{N}} (a_n, b_n) \subseteq \bigcup_{n \in \mathbb{N}} A_{i(n)}$.

Answer to Exercise 79, page 232

The parameters are f and K. Let $x \simeq y$ such that $x, y \in K$. Since K is compact, the observable neighbor a of x is in K. Since f is continuous at a and $x \simeq a$ and $y \simeq a$, we have $f(x) \simeq f(a)$ and $f(y) \simeq f(a)$. This shows that $f(x) \simeq f(y)$, so f is uniformly continuous.

Answer to Exercise 80, page 233

We prove the case of the lower limit. The case of the upper limit can be deduced immediately since $\limsup x_n = -\liminf -x_n$.

Let c be a cluster point of (x_n). The context is specified by (x_n) and c. By definition, we can find a positive ultralarge integer N such that $c \simeq x_N$. Let $d = \lim_{n \to \infty} y_n$. The sequence (y_n) is observable, so $d \simeq y_N$. But $y_N \leq x_N$ because $y_N = \inf\{x_n : n \geq N\}$. It follows that $d \leq c$, that is, $\liminf x_n \leq c$.

Appendix: Foundations and Relative Set Theory

The principles on which this book is founded employ concepts from logic, like "statement" and "parameter," that we have never defined. In this respect, our book is no different from the usual textbooks of elementary analysis (advanced calculus) that do not attempt to rigorously define, for example, what is meant by "statement" in the Principle of Mathematical Induction. Along with the authors of the traditional textbooks, we believe that readers will develop an intuitive understanding of these concepts, sufficient to use them correctly, as they work through the mathematics. An excessive emphasis on rigor at this level would obscure the underlying mathematical ideas.

However, readers with more exposure to advanced mathematics may well desire to see a more formal presentation of the logical and axiomatic framework. It is also necessary to have such framework available in case any doubts about correctness of an argument should arise. Finally, there is the important question whether the nontraditional ideas and tools introduced here are consistent with traditional mathematics.

This Appendix is intended to be a bridge to research literature, where these issues are addressed at length. It describes the axiomatic foundations of analysis with ultrasmall numbers in a more formal manner. It is written at a somewhat more advanced level. Although we hope that any reader who has got this far can benefit from it, some familiarity with mathematical logic (e.g., Enderton [3]) and axiomatic set theory (e.g., Hrbacek and Jech [12]) would be helpful.

Language, Logic, and Set Theory

Every mathematical theory has to have some primitive concepts—concepts that are not explicitly defined in terms of other, simpler notions. The work of mathematicians in the 20th century showed that a single primitive concept, set-theoretic membership $\in$, suffices; all other notions of traditional mathematics can be defined from it.

Thus the mathematical language initially has a single primitive **binary predicate symbol** $\in$. In addition, there is a symbol $=$ for **equality**, and a potentially infinite list of symbols for variables over sets. Mathematical **statements**, also known as well-formed formulas, are generated from these symbols by the following rules.

(1) If x and y are variables, then $(x \in y)$ and $(x = y)$ are statements. We read them "x is an element of y" and "x is equal to y," respectively.

(2) If $\mathcal{P}$ and $\mathcal{Q}$ are statements, then $(\neg\mathcal{P})$, $(\mathcal{P} \wedge \mathcal{Q})$, $(\mathcal{P} \vee \mathcal{Q})$, $(\mathcal{P} \to \mathcal{Q})$ and $(\mathcal{P} \leftrightarrow \mathcal{Q})$ are statements. We read them "not $\mathcal{P}$," "$\mathcal{P}$ and $\mathcal{Q}$," "$\mathcal{P}$ or $\mathcal{Q}$," "if $\mathcal{P}$, then $\mathcal{Q}$" and "$\mathcal{P}$ if and only if $\mathcal{Q}$," respectively. The symbols $\neg, \wedge, \vee, \to$ and $\leftrightarrow$ are called **logical connectives**.

(3) If $\mathcal{P}$ is a statement and x is a variable, then $(\exists x \mathcal{P})$ and $(\forall x \mathcal{P})$ are statements. We read them "there exists x such that $\mathcal{P}$" and "for all x $\mathcal{P}$," respectively. The symbols $\exists$ and $\forall$ are called the **existential** and **universal quantifier**, respectively.

(4) All statements are obtained by successive application of rules (1) through (3).

For emphasis, we call statements that are generated by the rules (1) through (4) the $\in$-*statements*. When actually writing a statement, we often add or remove parentheses to improve legibility. The language described above is quite rudimentary; nevertheless, many decades of experience show that all mathematical assertions can *in principle* be expressed in this language. The words "in principle" are essential; it would be excruciatingly cumbersome to state Fermat's Last Theorem in this language. An important part of the development of mathematics is that its language is always being enriched by new symbols denoting new concepts, defined in terms of those introduced earlier.

Before we describe this process, we have to establish a distinction between **free variables** (also known as **parameters**) and **bound variables** of a statement. Roughly speaking, bound variables are the variables quantified by an existential or universal quantifier. The precise definition is given by the following rules.

(1) Variables x and y occur free in the statements $(x \in y)$ and $(x = y)$; these statements have no bound variables.

(2) A variable occurs free (respectively bound) in $(\neg \mathcal{P})$, $(\mathcal{P} \wedge \mathcal{Q})$, $(\mathcal{P} \vee \mathcal{Q})$, $(\mathcal{P} \rightarrow \mathcal{Q})$ or $(\mathcal{P} \leftrightarrow \mathcal{Q})$ if it occurs free (respectively bound) in $\mathcal{P}$ or $\mathcal{Q}$.

(3) A variable occurs free (respectively bound) in $(\exists x \mathcal{P})$ or $(\forall x \mathcal{P})$ if it occurs free in $\mathcal{P}$ and it is different from x (respectively if either it is x or it occurs bound in $\mathcal{P}$).

A variable can occur both free and bound in a statement, but we assume that such usage is avoided.

We now return to the matter of definitions of new concepts. We write $\mathcal{P}(x_1, \ldots, x_k)$ to indicate that the free variables of the statement $\mathcal{P}$ are among $x_1, \ldots, x_k$. We can enrich the language by a new **predicate symbol**, say P, and define a k-ary predicate $Px_1, \ldots, x_k$ by

$$\forall x_1 \ldots \forall x_k \ (Px_1, \ldots, x_k \leftrightarrow \mathcal{P}(x_1, \ldots, x_k)).$$

That is, $Px_1, \ldots, x_k$ serves as shorthand for the possibly very complicated **defining statement** $\mathcal{P}(x_1, \ldots, x_k)$.

Here is an example. Let $\mathcal{P}(x, y)$ be the statement $\forall z \ (z \in x \rightarrow z \in y)$. We introduce a new symbol $\subseteq$ and define

$$\forall x \forall y \ (\subseteq x, y \leftrightarrow \forall z (z \in x \rightarrow z \in y)).$$

For binary predicates like $\subseteq$ it is customary to write $x \subseteq y$ in place of $\subseteq x, y$. $\subseteq$ is of course just the set-theoretic inclusion; we read it as "x is a subset of y." It is a concept defined in terms of $\in$ by the definition given above.

Besides predicates, it is handy to be able to define **operations**. Let $\mathcal{P}(x_1, \ldots, x_k, y)$ be a statement such that

(i) $\forall x_1 \ldots \forall x_k \ \exists y \ \mathcal{P}(x_1, \ldots, x_k, y)$, and

(ii) $\forall x_1 \ldots \forall x_k \ \forall y_1 \ \forall y_2 \ [\mathcal{P}(x_1, \ldots, x_k, y_1) \wedge \mathcal{P}(x_1, \ldots, x_k, y_2) \rightarrow y_1 = y_2]$.

Condition (i) asserts that there exists a y such that $\mathcal{P}(x_1, \ldots, x_k, y)$ holds, and condition (ii) asserts that this y is uniquely determined. We abbreviate the conjunction of (i) and (ii) as

$$\forall x_1 \ldots \forall x_k \ \exists ! y \ \mathcal{P}(x_1, \ldots, x_k, y).$$

In this situation we can give a name to this unique y; the name should indicate the dependence of y on $x_1, \ldots, x_k$. Formally, we introduce a new **operation symbol**, say F, and define the k-ary operation

$$F(x_1, \ldots, x_k)$$

by postulating

$$\forall x_1 \ldots \forall x_k \; \mathcal{P}(x_1, \ldots, x_k, F(x_1, \ldots, x_k)).$$

Example. Let $\mathcal{P}(x, y)$ be the statement

$$\forall z(z \in y \leftrightarrow z \in x \lor z = x).$$

It can be proved (see the axioms below) that for every x there is a unique y such that $\mathcal{P}(x, y)$ holds. This justifies introducing a new symbol S and defining the unary operation $S(x)$ by

$$\forall x \; \mathcal{P}(x, S(x)),$$

that is,

$$\forall x \forall z \; (z \in S(x) \leftrightarrow z \in x \lor z = x).$$

The operation $S(x)$ is called the **successor** of x; in terms of the more familiar operations $\cup$ and $\{\cdot\}$, $S(x) = x \cup \{x\}$. The operations $\cup$ and $\{\cdot\}$ are defined as follows:

$$\forall x \forall y \forall z \; (z \in x \cup y \leftrightarrow z \in x \lor z \in y)$$

and

$$\forall x \forall z \; (z \in \{x\} \leftrightarrow z = x).$$

An important special case is when $\mathcal{P}$ has only one free variable. If

$$\exists! y \; \mathcal{P}(y),$$

we can introduce a new **constant symbol** C and define it by $\mathcal{P}(C)$. For example, it can be shown that

$$\exists! y \; \forall z(z \in y \leftrightarrow \neg z = z).$$

This y has no elements; we introduce a new constant symbol $\emptyset$ and define it by $(\forall z)(z \in \emptyset \leftrightarrow \neg z = z)$. Of course, $\emptyset$ is the empty set. We mention in passing that specific natural numbers can be defined by repeated application of the successor operation to the empty set: $0 = \emptyset$, $1 = S(0)$, $2 = S(1)$, $3 = S(2)$, and so on.

Definitions of new predicates, operations and constants keep enriching the language of mathematics as needed to make it easy to talk about any objects of interest. In principle, every statement of the enriched language can be converted into an equivalent $\in$-statement by replacing all defined concepts step-by-step by their definitions. We call statements in the language enriched by such definitions *extended $\in$-statements*.

The main purpose of mathematics is to prove statements about mathematical objects of interest. Even before just defining a new operation or constant, one has to prove the required uniqueness. We need to say something about proofs.

In every mathematical theory there have to be some statements that are accepted without proof; they are called **axioms** (also principles, postulates). The generally accepted axioms for set theory have been formulated in the early 20th century; they are known as **Zermelo-Fraenkel set theory**, or **ZFC** for short (**C** stands for the Axiom of Choice). We do not list all of the axioms of **ZFC** here; the reader is referred to [12] or other textbooks on axiomatic set theory. We state the few of them that are particularly relevant to the discussion in this Appendix.

- **The Axiom of Extensionality:**

$$(\forall x)(\forall y)[x = y \leftrightarrow (\forall z)(z \in x \leftrightarrow z \in y)].$$

Two sets are equal if and only if they have the same elements.

- **The Axiom of Union:**

$$(\exists w)(\forall z)\,(z \in w \leftrightarrow z \in x \lor z \in y).$$

Given any sets x and y, there is a set w whose elements are precisely the elements of x and the elements of y. By the Axiom of Extensionality, the set w is uniquely determined. These two axioms together establish the conditions (i) and (ii) needed to define the operation of union $x \cup y$.

There are similar axioms for other important set theoretic operations.

- **Axioms of Separation:**

Let $\mathcal{P}(x, x_1, \ldots, x_k)$ be an $\in$-statement.

$$(\forall A)(\exists B)(\forall x)(x \in B \leftrightarrow x \in A \land \mathcal{P}(x, x_1, \ldots, x_k)).$$

Given a set A, there is a set B whose elements are precisely those elements of A for which the statement $\mathcal{P}$ is true. Again, for given $x_1, \ldots, x_k$, this set is unique, by the Axiom of Extensionality; it is usually written

$$\{x \in A : \mathcal{P}(x, x_1, \ldots, x_k)\}.$$

- **Axioms of Replacement:**

Let $\mathcal{P}(x, y, x_1, \ldots, x_k)$ be an $\in$-statement. If

$$(\forall x \in A)(\exists! y)\mathcal{P}(x, y, x_1, \ldots, x_k),$$

then there is a function f with domain A such that

$$(\forall x \in A)\mathcal{P}(x, f(x), x_1, \ldots, x_k).$$

It remains to say something about how statements are deduced from the axioms. In this book, we use informal arguments. Formal rules of deduction are the subject of mathematical logic. Logicians accept certain statements without proof, as logically true; they are the tautologies, like

$$(\forall x)(\forall y)(x \in y \ \lor \ \neg \ x \in y).$$

They give deduction rules that lead from true statements to other true statements. For example, the rule of Modus Ponens: If $\mathcal{P}$ is true and $(\mathcal{P} \to \mathcal{Q})$ is true, then $\mathcal{Q}$ is true. We refer the reader to [3] for more on mathematical logic.

Relative Set Theory

Relative concepts, such as "ultrasmall" or "neighbor," cannot be expressed in the language of Zermelo-Fraenkel set theory. In order to be able to do so, one has to introduce a new primitive concept, observability. Formally, we enlarge the $\in$-language by a new primitive unary predicate symbol **S**. We replace rule (1) on page 268 by

(1') If x and y are variables, then $(x \in y)$, $(x = y)$ and $(\mathbf{S}(x))$ are statements. We read $\mathbf{S}(x)$ as "x is observable" (or "x is standard").

The $\in$-**S**-statements are obtained by successive application of the rules (1'), (2) and (3).

The discussion of free and bound variables applies to $\in$-**S**-statements. The rules of logic remain the same. The axioms of **ZFC** are taken over unchanged. In particular, the implicit assumption that the statement $\mathcal{P}$ in the axioms of Separation and Replacement is an $\in$-statement remains in force! These axioms do not have to hold when $\mathcal{P}$ is an $\in$-**S**-statement; for example, there is no set B such that

$$(\forall x)(x \in B \ \leftrightarrow \ x \in \mathbb{N} \land \mathbf{S}(x)).$$

The nonstandard set theory on which this book is based is known as **BST** (*Bounded Set Theory*). Before stating its axioms we introduce important notation. Let $\mathcal{P}$ be any $\in$-statement. Then $\mathcal{P}^{\mathbf{S}}$ denotes the *relativization of* $\mathcal{P}$ *to* **S**, the statement obtained from $\mathcal{P}$ by restricting all quantifiers to **S**. In more detail, this means replacing each occurrence of the existential quantifier $\exists$ in $\mathcal{P}$ by $\exists^{\mathbf{S}}$, where $(\exists^{\mathbf{S}})\ldots$ is shorthand for $(\exists x)(\mathbf{S}(x)\wedge\ldots)$, and replacing each occurrence of the universal quantifier $\forall$ by $\forall^{\mathbf{S}}$, where $(\forall^{\mathbf{S}})\ldots$ is shorthand for $(\forall x)(\mathbf{S}(x)\rightarrow\ldots)$.

The notation $\overline{x}$ is used as shorthand for a list of variables $x_1,\ldots,x_k$.

The axioms of **BST** are:

- **ZFC in S:**
$$\mathcal{P}^{\mathbf{S}}, \text{ where } \mathcal{P} \text{ is any axiom of } \mathbf{ZFC}.$$

- **Boundedness:**
$$(\forall x)(\exists^{\mathbf{S}} y)(x \in y).$$

- **Transfer:**
$$(\forall^{\mathbf{S}} x_1)\ldots(\forall^{\mathbf{S}} x_k)\,(\mathcal{P}^{\mathbf{S}}(x_1,\ldots,x_k) \leftrightarrow \mathcal{P}(x_1,\ldots,x_k))$$

  where $\mathcal{P}(x_1,\ldots,x_k)$ is any statement in the $\in$-language.

- **Standardization:**
$$(\forall\overline{x})(\forall x)(\exists^{\mathbf{S}} y)(\forall^{\mathbf{S}} z)(z \in y \leftrightarrow z \in x \,\wedge\, \mathcal{P}(z,x,\overline{x};\mathbf{S}))$$

  where $\mathcal{P}(z,x,\overline{x};\mathbf{S})$ is any statement in the $\in$-**S**-language.

- **Bounded Idealization:**
$$(\forall\overline{x})(\forall^{\mathbf{S}} A)[(\forall^{\mathbf{S}} a \in P^{\mathbf{fin}} A)(\exists y)(\forall x \in a)\,\mathcal{P}(x,y,A,\overline{x})$$
$$\leftrightarrow (\exists y)(\forall^{\mathbf{S}} x \in A)\mathcal{P}(x,y,A,\overline{x})]$$

  where $\mathcal{P}(x,y,A,\overline{x})$ is any statement in the $\in$-language; $P^{\mathbf{fin}} A$ is the set of all finite subsets of A

We make some comments on the meaning of those axioms and derive from them the principles used in this book.

The first on the list of axioms is the assertion that the universe of observable sets satisfies the axioms of **ZFC**. If we identify observable sets with the familiar standard sets, as we do in Sections 1.2–1.4, the reasons for this assertion become obvious.

The axioms of **ZFC** can be expressed by statements $\mathcal{P}$ with no parameters. One then gets immediately from Transfer that $\mathcal{P}^{\mathbf{S}} \leftrightarrow \mathcal{P}$, and hence that $\mathcal{P}$ holds. So the universe of all sets (nonstandard ones included) also satisfies all the axioms of **ZFC**. This is a rigorous form of the assumption we make throughout the book.

Boundedness asserts that every new, ideal object added to the standard universe is an element of some standard set.

The Closure Principle is an easy consequence of the Transfer axiom. Indeed, assume that $\mathcal{P}(x_1, \ldots, x_k)$ is a statement in the $\in$-language and $\mathbf{S}(x_1), \ldots, \mathbf{S}(x_k)$ and $(\exists x)\ \mathcal{P}(x, x_1, \ldots, x_k)$ hold. By Transfer then also $(\exists^{\mathbf{S}} x)\ \mathcal{P}^{\mathbf{S}}(x, x_1, \ldots, x_k)$ holds. Fix x such that $\mathbf{S}(x)$ and $\mathcal{P}^{\mathbf{S}}(x, x_1, \ldots, x_k)$ holds. By Transfer once more, the statement $\mathcal{P}(x, x_1, \ldots, x_k)$ holds as well, and therefore we conclude that $(\exists^{\mathbf{S}} x)\ \mathcal{P}(x, x_1, \ldots, x_k)$.

We now discuss Standardization. As pointed out above, for an arbitrary $\in$-$\mathbf{S}$-statement $\mathcal{P}$, there does not necessarily exist a set B such that

$$(\forall x)(x \in B \ \leftrightarrow\ x \in A \ \wedge\ \mathcal{P}(x)).$$

(For simplicity, in this discussion we suppress spelling out the additional parameters $x_1, \ldots, x_n$ and the explicit mention of $\mathbf{S}$ in $\mathcal{P}$.) Standardization provides an approximation to this possibly nonexistent set. It asserts that, for every A, there is an observable set B such that

$$(\forall^{\mathbf{S}} x)[(x \in B \ \leftrightarrow\ x \in A \ \wedge\ \mathcal{P}(x))].$$

Theorem 159. *The observable set B whose existence is given by Standardization is uniquely determined.*

Proof. Let B' be observable and such that $(\forall^{\mathbf{S}} x)[(x \in B' \ \leftrightarrow\ x \in A \ \wedge\ \mathcal{P}(x))]$. Then $(\forall^{\mathbf{S}} x)(x \in B \ \leftrightarrow\ x \in B')$; hence, $(B = B')^{\mathbf{S}}$ by Extensionality in $\mathbf{S}$, and $B = B'$ by Transfer. $\square$

An important special case is when $\mathcal{P}(x)$ is $x = x$ (or some other statement that is always true). Then $(\forall^{\mathbf{S}} x)(x \in B \ \leftrightarrow\ x \in A)$. The unique observable set B that has exactly the same *observable* elements as the set A is called the **observable neighbor** of A.

The Observable Neighbor Principle is a consequence of Transfer and Standardization.

Theorem 160. *The Observable Neighbor Principle follows from the axioms of $\mathbf{BST}$.*

Proof. In Chapter 5 we deduce completeness of $\mathbb{R}$ from the Observable Neighbor Principle. For this proof, we have to assume that completeness of $\mathbb{R}$ has been established without relying on the Observable Neighbor Principle, to avoid circularity. Textbooks of set theory (for example, [12]) show how this can be done.

Assume $x \in \mathbb{R}$ is not ultralarge; say $|x| \leq r$ for some observable r. By Standardization, there is an observable set B such that

$$(\forall^{\mathbf{S}} z)(z \in B \leftrightarrow z \in \mathbb{R} \wedge z \leq x).$$

Note that $B \neq \emptyset$ (since $-r \in B$) and B is bounded above by r. By completeness of $\mathbb{R}$, B has a supremum b, and by Closure, b is observable. We now show that $x \simeq b$, so b is the observable neighbor of x.

If not, then $|b - x| > s > 0$ for some observable s. This means that either $x > b + s$ or $x < b - s$. In the first case, $b + s \in B$, contradicting $b = \sup B$. In the second case, $b - s$ is an upper bound on B, again contradicting $b = \sup B$. ∎

The following consequence of Standardization is used in the proof of Theorem 125.

Theorem 161. *Let $f : I \to \mathbb{R}$ be a function with $|f(t)| \leq M$ for all $t \in I$, where I and M, but not necessarily f, are observable. There is an observable function $F : I \to \mathbb{R}$ such that $F(t) \simeq f(t)$ for all observable $t \in I$.*

Proof. From the Standardization Principle we get an observable set F such that, for all observable t, y,

$$\langle t, y \rangle \in F \leftrightarrow \langle t, y \rangle \in I \times \mathbb{R} \wedge y \simeq f(t).$$

As "$\langle t, y \rangle \in F \to \langle t, y \rangle \in I \times \mathbb{R}$" holds for all observable t, y, it holds for all t, y, by Closure; in other words, $F \subseteq I \times \mathbb{R}$. We have to prove that F is a function defined on I. For every fixed observable $t \in I$ there is y for which $\langle t, y \rangle \in F$, namely the observable neighbor of $f(t)$ (it exists because $f(t)$ is not ultralarge). By Closure, for every $t \in I$ there is y for which $\langle t, y \rangle \in F$. Similarly, if t is observable, then, for all observable y_1, y_2, $\langle t, y_1 \rangle \in F \wedge \langle t, y_2 \rangle \in F \to y_1 = y_2$ (because $y_1 \simeq f(t) \simeq y_2$). By Closure, this is true for all y_1, y_2. By Closure again, for every $t \in I$ and for all y_1, y_2, $\langle t, y_1 \rangle \in F \wedge \langle t, y_2 \rangle \in F \to y_1 = y_2$. We conclude that for every $t \in I$ there is a unique y for which $\langle t, y \rangle \in F$, so F is a function defined on I. ∎

We deduce one more consequence of Standardization.

Theorem 162 (Principle of Finite Choice). *Let $\mathcal{P}(x, y, \overline{x}; \mathbf{S})$ be any $\in$-$\mathbf{S}$-statement. If A is observable and finite and*

$$(\forall x \in A)(\exists y)\, \mathcal{P}(x, y, \overline{x}; \mathbf{S}),$$

then there is a function f with domain A such that

$$(\forall x \in A)\, \mathcal{P}(x, f(x), \overline{x}; \mathbf{S}).$$

Proof. Let $A = \{a_1, \ldots, a_n\}$, where n is observable. By Standardization, there is an observable set $B \subseteq \{1, \ldots, n\}$ such that, for all observable m, $m \in B$ if and only if there exists a function f_m with domain $\{1, \ldots, m\}$ such that $\mathcal{P}(a_i, f_m(a_i), \overline{x})$ holds for all $i = 1, \ldots, m$.

We prove that $n \in B$. If not, the set $\{1, \ldots, n\} \setminus B$ has a least element $\overline{n}$. If $\overline{n} = 1$, we take some y_1 such that $\mathcal{P}(a_1, y_1, x_1, \ldots, x_k; \mathbf{S})$ holds and define $f_1(a_1) = y_1$. If $\overline{n} > 1$, we fix some $f_{\overline{n}-1}$ and some $y_{\overline{n}}$ such that $\mathcal{P}(a_{\overline{n}}, y_{\overline{n}}, x_1, \ldots, x_k; \mathbf{S})$ holds, and define $f_{\overline{n}}(a_i) = f_{\overline{n}-1}(a_i)$ for $i < \overline{n}$, In either case $(\forall i \leq \overline{n})\mathcal{P}(a_i, f_{\overline{n}}(a_i), \overline{x}; \mathbf{S})$, so $\overline{n} \in B$, a contradiction.

The function $f = f_n$ has the required properties. $\qquad\square$

We now turn to Idealization. In this book we only use one simple consequence of Bounded Idealization, namely the existence of natural numbers that are not observable.

Theorem 163. *There is $n \in \mathbb{N}$ such that n is not observable.*

Proof. Let $A = \mathbb{N}$. If $a = \{n_1, \ldots, n_\ell\} \subseteq \mathbb{N}$ is observable and finite, then there is $y \in \mathbb{N}$ such that $y \neq x$ for any $x \in a$ (choose for example $y = \max\{n_1, \ldots, n_\ell\} + 1$ if $a \neq \emptyset$, or $y = 1$ if $a = \emptyset$). Take $(y \in \mathbb{N} \wedge y \neq x)$ as $\mathcal{P}(x, y)$. Idealization enables us to conclude that there exists $y \in \mathbb{N}$ such that $y \neq x$ holds for all observable $x \in \mathbb{N}$. $\qquad\square$

Corollary (Existence Principle). *There exist ultrasmall real numbers.*

The next theorem shows that Idealization implies existence of many other ideal elements.

Theorem 164 (Saturation). *Let $\mathcal{F}$ be a nonempty observable collection of sets with the finite intersection property. Then there exists y such that $y \in X$ for all observable $X \in \mathcal{F}$.*

Proof. If $a = \{X_1, \ldots, X_\ell\} \subseteq \mathcal{F}$ is finite and observable, then the finite intersection property gives that $\bigcap_{i=1}^{\ell} X_i \neq \emptyset$, that is, there exists y such that $y \in X$ for all $X \in a$. We now take $\mathcal{P}(X, y)$ to be $(y \in X)$, let $A = \mathcal{F}$ and use Idealization to conclude that there exists y such that $y \in X$ holds for all observable $X \in \mathcal{F}$. $\qquad\square$

Theorem 165. *There is an ultrasmall number h such that $h \neq x_n$ for any observable infinite sequence (x_n) and any $n \in \mathbb{N}$.*

Proof. Let $\mathcal{F} = \{(-r, r) : r > 0\} \cup \{\mathbb{R} \setminus C : C \text{ is at most countable}\}$. The observable collection $\mathcal{F}$ has the finite intersection property. Any h that belongs to all observable elements of $\mathcal{F}$ satisfies the claim of the theorem. $\qquad\square$

Example. Let $K \subseteq \mathbb{R}$ be a compact set. We can use Saturation to show that any nonempty collection of closed subsets of K with the finite intersection property has nonempty intersection.

Let $\mathcal{F} = \{C_i : i \in I\}$ be a nonempty family of closed subsets of K with the finite intersection property. By Saturation, there exists y such that $y \in C_i$ for all observable $i \in I$; in particular, $y \in K$. Since K is compact, the observable neighbor of y exists and belongs to K; call it c. If $i \in I$ is observable, then C_i is observable; since C_i is closed, necessarily $c \in C_i$. Hence $c \in C_i$ for each observable $i \in I$. By the Closure Principle, this is true for all $i \in I$. This shows that c belongs to the intersection of $\mathcal{F}$, so the intersection is nonempty.

We conclude our discussion of **BST** with a statement of a very important theoretical result about it. For the proof, which is outside the scope of this book, see Kanovei and Reeken [14], Theorem 3.2.3.

Reduction

Suppose that $\mathcal{P}(x_1, \ldots, x_k; \mathbf{S})$ is a statement in the $\in$-$\mathbf{S}$-language. Then there exists a statement $\mathcal{Q}(x_1, \ldots, x_k)$ in the $\in$-language such that the following is a theorem of **BST**:

$$(\forall^{\mathbf{S}} x_1) \ldots (\forall^{\mathbf{S}} x_k)(\mathcal{P}(x_1, \ldots, x_k; \mathbf{S}) \leftrightarrow \mathcal{Q}(x_1, \ldots, x_k) \leftrightarrow \mathcal{Q}^{\mathbf{S}}(x_1, \ldots, x_k)).$$

For an elementary exposition of calculus the theory **BST** has a big drawback. While most, if not all, results in this book can be proved in it *for observable objects*, it does not have the means to apply these proofs uniformly to *all* objects, or even to give "infinitesimal" definitions of the basic concepts of calculus (continuity, limit, derivative) for all functions and all arguments. We solve this problem by relativizing the notion of observability. For any object p we have a notion of *observability relative to p*, which has all the properties postulated by **BST**.

To formalize this idea, we enlarge the $\in$-language by a new primitive binary predicate symbol $\sqsubseteq$. The rule (1) for generation of $\in$-$\sqsubseteq$-statements allows $x \sqsubseteq y$, which we read as "x is observable relative to y."

The basic axiom of the *Relative Bounded Set Theory* **RBST** is **Relativization:**

(1) $(\forall p)(p \sqsubseteq p)$;

(2) $(\forall p)(\forall q)(\forall r)(p \sqsubseteq q \ \wedge \ q \sqsubseteq r \rightarrow p \sqsubseteq r)$;

(3) $(\forall p)(\forall q)(p \sqsubseteq q \ \vee \ q \sqsubseteq p)$;

(4) $(\forall p)(0 \sqsubseteq p)$;

(5) $(\forall p)(\exists q)(p \sqsubseteq q \ \wedge \ \neg q \sqsubseteq p)$.

Items (1) through (3) express reflexivity, transitivity, and comparability of the observability relation $\sqsubseteq$. These imply the Relative Observability Principle. (4) postulates that 0 is observable relative to every p. (5) asserts that for every p there is q that is not observable relative to p.

For the statements of the remaining axioms we use the notation $\mathbf{S}_p(q)$ in place of $q \sqsubseteq p$. Intuitively, $\mathbf{S}_p$ is the universe of objects observable relative to p, and occasionally we write $q \in \mathbf{S}_p$ for $\mathbf{S}_p(q)$.

RBST *postulates that the axioms of* **BST**, *to wit,* **ZFC** *in* **S**, *Boundedness, Transfer, Standardization and Bounded Idealization, hold with* **S** *replaced by* $\mathbf{S}_p$, *for all* p.
For example,

• **Relative Transfer:**

$$(\forall p)(\forall^{\mathbf{S}_p} x_1) \ldots (\forall^{\mathbf{S}_p} x_k)\ (\mathcal{P}^{\mathbf{S}_p}(x_1, \ldots, x_k) \leftrightarrow \mathcal{P}(x_1, \ldots, x_k))$$

where $\mathcal{P}(x_1, \ldots, x_k)$ is any statement in the $\in$-language.

• **Relative Standardization:**

$$(\forall p)(\forall \overline{x})(\forall x)(\exists^{\mathbf{S}_p} y)(\forall^{\mathbf{S}_p} z)(z \in y \leftrightarrow z \in x \ \wedge\ \mathcal{P}(z, x, \overline{x}; \mathbf{S}_p))$$

where $\mathcal{P}(z, x, \overline{x}; \mathbf{S})$ is any statement in the $\in$-**S**-language.

The formal statement of the Stability Principle is as follows.

Stability: Let $\mathcal{P}(x_1, \ldots, x_k; \mathbf{S})$ be a statement in the $\in$-**S**-language. For any $p \sqsubseteq q$,

$$(\forall^{\mathbf{S}_p} x_1) \ldots (\forall^{\mathbf{S}_p} x_k)\ (\mathcal{P}(x_1, \ldots, x_k; \mathbf{S}_p) \ \leftrightarrow\ \mathcal{P}(x_1, \ldots, x_k; \mathbf{S}_q)).$$

Theorem 166. *Stability Principle follows from the axioms of* **RBST**.

Proof. Since all the axioms of **BST** hold with **S** replaced by any $\mathbf{S}_p$, the same is true about Reduction. Assuming $p \sqsubseteq q$ and $\mathbf{S}_p(x_1), \ldots, \mathbf{S}_p(x_k)$, we have also $\mathbf{S}_q(x_1), \ldots, \mathbf{S}_q(x_k)$, and Reduction applies to both $\mathbf{S}_p$ and $\mathbf{S}_q$ to give

$$\mathcal{P}(x_1, \ldots, x_k; \mathbf{S}_p) \ \leftrightarrow\ \mathcal{Q}(x_1, \ldots, x_k) \leftrightarrow \mathcal{P}(x_1, \ldots, x_k; \mathbf{S}_q).$$

$\square$

We need to relate the above formal statement of Stability to the one that is given in Section 1.5 and used throughout the book. First, our informal version of Stability Principle talks about contexts, that is, lists of parameters. The next theorem shows that a list of parameters can be treated as a single parameter.

Theorem 167. *q is observable relative to $p_1, \ldots, p_k$ if and only if q is observable relative to p, for $p = \langle p_1, \ldots, p_k \rangle$.*

Proof. The length k of the list is an explicitly given natural number (such as 1, 2, 3, 17,...) and so it is standard, as is every $i \leq k$. Hence every p_i ($1 \leq i \leq k$) is observable relative to p (it is uniquely defined from p and i, as the i-th term in the k-tuple p). Therefore every q observable relative to some p_i is observable relative to p.

For the converse it suffices to notice that there is a p_i such that all p_j for $1 \leq j \leq k$ are observable relative to p_i (by linearity of $\sqsubseteq$). As p is uniquely defined from $p_1, \ldots, p_k$ (it is the k-tuple with these terms, in this order), it is observable relative to p_i, by Closure. Hence every q that is observable relative to p is observable relative to some p_i. $\square$

We recall that a statement is called *internal* if the context of all relative concepts that occur in it is given by the parameters of the statement. The relative concepts "ultrasmall," "ultralarge," "ultraclose" and "observable neighbor" have simple definitions in terms of observability. If we replace them in an internal statement by their definitions, we obtain a statement where observability is the only relative concept. So formally:

Internal statements are statements of the form

$$\mathcal{P}(x_1, \ldots, x_k; \mathbf{S}_{\langle x_1, \ldots, x_k \rangle}).$$

The informal version of Stability from Chapter 1 can now be formulated as follows:

$$\mathcal{P}(x_1, \ldots, x_k; \mathbf{S}_{\langle x_1, \ldots, x_k \rangle}) \leftrightarrow \mathcal{P}(x_1, \ldots, x_k, \mathbf{S}_{\langle x_1, \ldots, x_k, y_1, \ldots, y_\ell \rangle}).$$

It is an immediate consequence of Stability as stated before Theorem 166.

Reduction in **RBST** asserts that for every statement $\mathcal{P}$ in the $\in$-**S**-language there exists a statement $\mathcal{Q}$ in the $\in$-language such that **RBST** proves

$$(\forall p)(\forall^{\mathbf{S}_p} x_1) \ldots (\forall^{\mathbf{S}_p} x_k)(\mathcal{P}(x_1, \ldots, x_k; \mathbf{S}_p) \leftrightarrow \mathcal{Q}(x_1, \ldots, x_k)).$$

Let $p = \langle x_1, \ldots, x_k \rangle$ in the above to conclude that

$$\mathcal{P}(x_1, \ldots, x_k; \mathbf{S}_{\langle x_1, \ldots, x_k \rangle}) \leftrightarrow \mathcal{Q}(x_1, \ldots, x_k).$$

In sum:

Every internal statement is equivalent to a statement in the $\in$-language.

This is a result of great importance. In this book we define a number of internal concepts: limit, derivative and definite integral, to name just a few. For some of them, we explicitly show their equivalence to traditional definitions, that is, definitions in the extended $\in$-language. The above result implies that in principle the same can be done for every internal concept. In fact, the proof of Reduction provides an algorithm for doing just that.

This result also easily justifies the remaining practices used in this book. First, informally, internal statements are allowed to refer to previously defined internal concepts. The internal statement $\mathcal{R}(z_1, \ldots, z_n)$ defining an internal predicate $R(z_1, \ldots, z_n)$ is equivalent to some statement $\mathcal{Q}(z_1, \ldots, z_n)$ in the $\in$-language. Any statement $\mathcal{P}(x_1, \ldots, x_k; \mathbf{S}_{\langle x_1, \ldots, x_k \rangle})$ in the $\in$-**S**-R-language is equivalent to the statement in the $\in$-**S**-language obtained by replacing each occurrence of $R(z_1, \ldots, z_n)$ by $\mathcal{Q}(z_1, \ldots, z_n)$, and the resulting statement is internal.

Second, since every internal statement is equivalent to a statement in the $\in$-language, the Closure Principle for internal statements follows from the Closure Principle for statements in the $\in$-language.

The Definition Principle for sets and functions reduces to the axioms of Separation and Replacement, respectively, upon replacing the internal defining statements by equivalent $\in$-statements. We state this observation formally as a theorem.

Theorem 168. *The Definition Principle follows from the axioms of* **RBST**. *In more detail:*

*(1) Let $\mathcal{P}(x, x_1, \ldots, x_k; \mathbf{S})$ be a statement in the $\in$-**S**-language. For every set A there is a set B, observable relative to $A, x_1, \ldots, x_k$, such that*

$$(\forall x)(x \in B \leftrightarrow x \in A \wedge \mathcal{P}(x, x_1, \ldots, x_k, \mathbf{S}_{\langle x, x_1, \ldots, x_k \rangle})).$$

*(2) Let $\mathcal{P}(x, y, x_1, \ldots, x_k; \mathbf{S})$ be a statement in the $\in$-**S**-language. If*

$$(\forall x \in A)(\exists! y \in B)\mathcal{P}(x, y, x_1, \ldots, x_k; \mathbf{S}_{\langle x, y, x_1, \ldots, x_k \rangle}),$$

then there is a function $f : A \to B$, observable relative to $A, B, x_1, \ldots, x_k$, and such that

$$(\forall x \in A)\mathcal{P}(x, f(x), A, B, x_1, \ldots, x_k, \mathbf{S}_{\langle x, y, x_1, \ldots, x_k \rangle}).$$

The following theorem extends the applicability of the Definition Principle. We use special cases of it to show that our definitions of limit (Theorem 15), integral (Exercise 4.8), a^b (Section 2.4) and supremum (Exercise 5.11) are equivalent to internal statements.

Theorem 169. *Let* $\mathcal{P}(x_1, \ldots, x_k, y; \mathbf{S}_{\langle x_1, \ldots, x_k \rangle})$ *be a statement such that for all* $y, z \in \mathbf{S}_{\langle x_1, \ldots, x_k \rangle}$

$$\mathcal{P}(x_1, \ldots, x_k, y; \mathbf{S}_{\langle x_1, \ldots, x_k \rangle}) \wedge \mathcal{P}(x_1, \ldots, x_k, z; \mathbf{S}_{\langle x_1, \ldots, x_k \rangle}) \rightarrow y = z$$

(that is, there is at most one $y \in \mathbf{S}_{\langle x_1, \ldots, x_k \rangle}$ *for which*

$$\mathcal{P}(x_1, \ldots, x_k, y; \mathbf{S}_{\langle x_1, \ldots, x_k \rangle})$$

holds), then the following statements are equivalent.

(1)
$$y \in \mathbf{S}_{\langle x_1, \ldots, x_k \rangle} \wedge \mathcal{P}(x_1, \ldots, x_k, y; \mathbf{S}_{\langle x_1, \ldots, x_k \rangle}).$$

(2)
$$\mathcal{P}(x_1, \ldots, x_k, y; \mathbf{S}_{\langle x_1, \ldots, x_k, y \rangle}).$$

Note that (2) is an internal statement.

Proof. (1) implies (2): If $y \in \mathbf{S}_{\langle x_1, \ldots, x_k \rangle}$, then

$$\mathcal{P}(x_1, \ldots, x_k, y; \mathbf{S}_{\langle x_1, \ldots, x_k \rangle}) \rightarrow \mathcal{P}(x_1, \ldots, x_k, y; \mathbf{S}_{\langle x_1, \ldots, x_k, y \rangle})$$

by Stability.

(2) implies (1): Assume that $\mathcal{P}(x_1, \ldots, x_k, y; \mathbf{S}_{\langle x_1, \ldots, x_k, y \rangle})$ holds. Then

$$(\exists z \in \mathbf{S}_{\langle x_1, \ldots, x_k, y \rangle}) \, \mathcal{P}(x_1, \ldots, x_k, z; \mathbf{S}_{\langle x_1, \ldots, x_k, y \rangle})$$

holds (take $z = y$). By Stability,

$$(\exists z \in \mathbf{S}_{\langle x_1, \ldots, x_k \rangle}) \, \mathcal{P}(x_1, \ldots, x_k, z; \mathbf{S}_{\langle x_1, \ldots, x_k \rangle})$$

holds. We fix z_0 such that $z_0 \in \mathbf{S}_{\langle x_1, \ldots, x_k \rangle} \wedge \mathcal{P}(x_1, \ldots, x_k, z_0; \mathbf{S}_{\langle x_1, \ldots, x_k \rangle})$. It suffices to show that $y = z_0$.

By Stability, $\mathcal{P}(x_1, \ldots, x_k, z_0; \mathbf{S}_{\langle x_1, \ldots, x_k, y \rangle})$. So we have both

$$\mathcal{P}(x_1, \ldots, x_k, y; \mathbf{S}_{\langle x_1, \ldots, x_k, y \rangle}) \text{ and } \mathcal{P}(x_1, \ldots, x_k, z_0; \mathbf{S}_{\langle x_1, \ldots, x_k, y \rangle}).$$

Stability applied to the uniqueness assumption yields that for $y, z_0 \in \mathbf{S}_{\langle x_1, \ldots, x_k, y \rangle}$

$$\mathcal{P}(x_1, \ldots, x_k, y; \mathbf{S}_{\langle x_1, \ldots, x_k, y \rangle}) \wedge \mathcal{P}(x_1, \ldots, x_k, z_0; \mathbf{S}_{\langle x_1, \ldots, x_k, y \rangle}) \rightarrow y = z_0,$$

and we conclude that $y = z_0$. $\square$

Consistency, History and Philosophy

Every time an extension of the traditional set theory **ZFC** is put forward, the foremost question on one's mind has to be whether it is consistent; that is, free from contradictions. A very famous theorem of Kurt Gödel shows that no consistent theory can prove its own consistency (if it is strong enough to at least prove the results of elementary arithmetic). Gödel's theorem applies in particular to **ZFC**, the workhorse of modern mathematics, and it applies to **RBST** as well: Neither **ZFC** nor **RBST** can prove its own consistency. We believe in the consistency of **ZFC** because set theorists have developed a convincing intuitive picture of a universe of sets where the axioms of **ZFC** hold (the so-called cumulative hierarchy of sets), and because of its proven track record (over a century of intensive development with no hint of a contradiction). Relative set theory does not have such a track record—it is much more recent—and an intuitive picture of its workings has to be acquired by way of some exposure to it. However, the following fundamental theoretical result shows that it is no less safe than **ZFC**.

Theorem 170. *If **ZFC** is consistent, then **RBST** is consistent. In fact, **RBST** is a conservative extension of **ZFC**.*

The first sentence assures us that if, as is generally believed, there are no contradictions in **ZFC**, then there are no contradictions in **RBST** either. The second sentence means that any statement in the ∈-language of **ZFC** that can be proved in **RBST** is provable already in **ZFC**. However, the proof using **RBST** may be simpler, shorter, and/or more transparent than a proof in **ZFC**, as—we hope—this book illustrates. This is the principal reason for working in relative set theory.

A proof of Theorem 170 is outside the scope of this book. It consists of constructing an interpretation ("model") for **RBST** inside **ZFC**. The constructions use some more advanced tools from set theory and logic. The models are complicated and do not offer much help in understanding of, or working with, relative set theory. Here we provide a brief guide to the history of the subject and references to the literature where such proofs can be found.

The axiomatic approach to nonstandard mathematics was developed in the mid-seventies, independently and at about the same time, by the first author, Edward Nelson and Petr Vopěnka. [1]

---

[1] The first author's paper [7] was accepted by *Fundamenta Mathematicae* in May 1975 and appeared in the first issue of the journal in 1978. Nelson's paper [18] was

The first author in [7] proposed several theories which allow both *internal sets* (what we call simply *sets* in this book) and *external sets*. The internal sets in all these theories satisfy axioms equivalent to **BST**, and a proof that **BST** is a conservative extension of **ZFC** is implicit in this paper.

Nelson [18] introduced a theory of internal sets only, which he called **IST**. This theory differs from **BST** by allowing also objects that are not elements of any standard set (*unbounded sets*). As a consequence, Reduction in **IST** holds only for "bounded" statements, but not for all statements (see [14], Exercise 3.5.8 for a counterexample). Nelson gave a proof that **IST** is a conservative extension of **ZFC** (see also [14], Proposition 4.4.1 and Corollary 4.4.2). From this the same claim for **BST** can be deduced by observing that the bounded sets of **IST** satisfy the axioms of **BST** (see [14], Theorem 3.4.5).

The theory **BST** was explicitly formulated by Kanovei in 1990. The monograph of Kanovei and Reeken [14] is the standard reference for the axiomatic approach to nonstandard analysis. In particular, they prove (Theorem 4.1.10 (i)) that **BST** has an interpretation in **ZFC** in which the universe of standard sets is isomorphic to the universe of all sets of **ZFC**. In [11] such interpretations are called *realizations*. This is a stronger result than mere conservativity; it implies for example that every model of **ZFC** can be extended to a model of **BST** in which the sets of the original model are exactly the standard sets. As a theory, **BST** has many other pleasing metamathematical properties; for example, it is a *maximal* conservative extension of **ZFC**, in the sense that if $\mathcal{P}$ is a statement in the $\in$-**S**-language and **BST** $+ \mathcal{P}$ is a conservative extension of **ZFC**, then **BST** proves $\mathcal{P}$ [11]. On this evidence, we consider **BST** to be the "right" way to axiomatize the extension of the mathematical universe of **ZFC** by ideal objects. (See also Kanovei–Reeken [14] for an extensive discussion of the advantages of **BST**.)

The idea that observability (or standardness, as it is usually called in more technical literature) should be made relative was suggested by Guy Wallet in the mid-1980s, and elaborated into an axiomatic system called **RIST** (Relative Internal Set Theory) by Yves Péraire [21]. [2] This theory relativizes Nelson's **IST**. For reasons outlined above we prefer to relativize **BST**; hence **RBST**.

Péraire [21] gives a detailed proof that **RIST** is a conservative extension of **ZFC**. It is not too hard to verify that the bounded sets of **RIST** satisfy the axioms of **RBST**. Hence **RBST** is a conservative extension

---

submitted to *Bull. Amer. Math. Soc.* in November 1976 and published in November 1977. The earliest publication presenting Vopěnka's Alternative Set Theory seems to be Sochor [26] from 1976.

[2]A different approach to relative set theory was proposed by Evgeni Gordon [5, 6].

of **ZFC**. A self-contained proof along these lines can be found on the website www.ultrasmall.org. This is perhaps the most accessible venue for the reader who wants to see a proof of Theorem 170.

Another possibility is to use the construction of a realization of **BST** in **ZFC** from Kanovei–Reeken [14], Section 4.3, as a starting point. As noted above, this construction implies that every model **M** of **ZFC** can be extended to a model $\overline{\mathbf{M}}$ of **BST** in which the sets of **M** are exactly the standard sets. Starting with a model $\mathbf{M}_0$ and proceeding inductively, we define $\mathbf{M}_{n+1} = \overline{\mathbf{M}_n}$ and $\mathbf{N} = \bigcup_{n \in \mathbb{N}} \mathbf{M_n}$. It is not hard to prove that **N** determines a model of **RBST**. With more care, one can obtain a stronger result, namely a realization of **RBST** in **ZFC**, with implications similar to those this fact has for **BST**. However, Kanovei-Reeken construction is of necessity much more complicated than the one in Péraire [21].

Unlike **BST**, **RBST** is *not* a maximal conservative extension of **ZFC** (for statements in the $\in$-$\sqsubseteq$-language); other, potentially useful axioms can be added to it. In [9], the first author developed a powerful extension of **RBST** called **FRBST** [actually, two slightly different versions of it] and proved that it is a conservative extension of **ZFC**. Theorem 170 follows immediately from this result. Unfortunately, this proof is technically very complicated.

The ultimate extension of **RBST** is the theory **GRIST** introduced in [11]. **GRIST** is a maximal conservative extension of **ZFC** for statements in the $\in$-$\sqsubseteq$-language and has a realization in **ZFC**.

In this book we take the view that standard sets (i.e., the sets observable relative to every context) are to be identified with the sets of traditional mathematics, but that these sets (at least when infinite) have also unobservable, ideal elements. This is the view taken in the first author's [8], but it is not the only possible one. The philosophy of Nelson's **IST** is to identify the universe of **IST** with the universe of sets we are accustomed to. The standardness predicate is to be viewed as a linguistic device that singles out certain objects for special attention. In this view no new elements are being added to the usual universe. This is an elegant idea, and our book was originally written from this "internal viewpoint" (see [13]). Our experience showed, however, that most working mathematicians find the idea that some predicates do not satisfy the Principle of Mathematical Induction incompatible with their view of the "usual" natural numbers. The main point we want to make here is that the choice between these two views is purely a matter of personal philosophy—the differences are in how to interpret the concepts of **RBST** intuitively. The mathematics—axioms, definitions, theorems, proofs—is exactly the same in either view.

One thing that both of these views have in common is avoidance of external sets. Certain statements describe "collections" that cannot be sets; the most basic example is $\{x \in \mathbb{N} : \mathbf{S}(x)\}$. This describes the "bare" collection of standard natural numbers only, separated from the ideal elements of $\mathbb{N}$, and neither view has a place for such objects. Relative set theory can in fact be extended to allow such "external" objects, but the price to pay is a substantial increase in complexity. For every statement $\mathcal{P}(x)$ in the $\in$-$\mathbf{S}$ or $\in$-$\sqsubseteq$-language (possibly with parameters) one can consistently postulate the existence of a **class** X such that, for all sets x, $x \in X \leftrightarrow \mathcal{P}(x)$. Subclasses of sets are called *external sets* (or *semisets*). It is only necessary to keep in mind that external sets need not be sets or have the usual properties of sets. For example, the external set $^{\circ}\mathbb{N} = \{x \in \mathbb{N} : \mathbf{S}(x)\}$ is a nonempty subclass of $\mathbb{N}$ and it is bounded above (by every ultralarge natural number), yet it does not have a greatest element. Some interesting external sets are the *monad* of a, $\mathbf{m}(a) = \{x \in \mathbb{R} : x \simeq a\}$, and the *galaxy* of a, $\mathbf{g}(a) = \{x \in \mathbb{R} : x \sim a\}$.

In general, one has to strictly distinguish between the set $\mathbb{N}$ of all natural numbers (standard or not) and the external set $^{\circ}\mathbb{N}$ of standard numbers only. Similarly, there is the set of real numbers $\mathbb{R}$ (containing among other things the ultrasmall numbers) and the external set $^{\circ}\mathbb{R} = \{x \in \mathbb{R} : \mathbf{S}(x)\}$ that contains only the standard real numbers. Every familiar set comes in two versions, the "full" (internal) and the "bare" (external). This is the price one pays for mixing the traditional view with our view (or the internal view). For elementary calculus and much else the external sets are not needed and the complications they introduce are best avoided—as we do scrupulously throughout this book.

In more advanced parts of nonstandard analysis external sets are sometimes essential; we mention in this context only Loeb's nonstandard measure theory and the neutrices and external numbers of van den Berg. If one desires to use external sets extensively, it is perhaps natural to adopt the point of view that the external sets (such as $^{\circ}\mathbb{N}$, $^{\circ}\mathbb{R}$,...) rather than the internal sets ($\mathbb{N}$, $\mathbb{R}$,...) should be identified with the usual sets of traditional mathematics. In such "external view" one would reserve the notation $\mathbb{N}$, $\mathbb{R}$,... for what we here call $^{\circ}\mathbb{N}$, $^{\circ}\mathbb{R}$,..., and use say $^{*}\mathbb{N}$, $^{*}\mathbb{R}$,... for the sets we here call $\mathbb{N}$, $\mathbb{R}$,... In general, one has an embedding $*$ of the traditional universe into an extended universe, in which every standard set X has a counterpart $^{*}X$ (usually called the *standard copy* of X). The axiomatic presentation of nonstandard analysis from the external point of view can be found in Kanovei and Reeken [14], especially Chapters 1 and 2. This view and notation are quite close to the usual model–theoretic framework based on superstructures. We refer the reader to Vakil [28] and Goldblatt [4] for further study of nonstandard analysis in a model–theoretic framework. Some additional discussion of external sets in the setting of **RBST** can be found on the website www.ultrasmall.org.

Bibliography

[1] Robert Bartle and Donald Sherbert. *Introduction to Real Analysis, Third Edition.* John Wiley and Sons, Inc, 2001.

[2] George Berkeley. The analyst: a discourse addressed to an infidel mathematician. http://www.maths.tcd.ie/pub/HistMath/People/Berkeley/Analyst/, 1734.

[3] Herbert B. Enderton. *Mathematical Introduction to Logic.* Harcourt/Academic Press, 2001.

[4] Robert Goldblatt. *Lectures on the Hyperreals.* Springer, 1998.

[5] Evgeni Gordon. Relatively nonstandard elements in the theory of internal sets of E. Nelson. *Siberian Mathematical Journal (Russian)*, 30:89–95, 1989.

[6] Evgeni Gordon. *Nonstandard Methods in Commutative Harmonic Analysis.* American Mathematical Society, Providence, Rhode Island, 1997.

[7] Karel Hrbacek. Axiomatic foundations for nonstandard analysis. *Fundamenta Mathematicae*, 98:1–19, 1978.

[8] Karel Hrbacek. Nonstandard set theory. *American Mathematical Monthly*, 86:659–677, 1979.

[9] Karel Hrbacek. Internally iterated ultrapowers. In A. Enayat and R. Kossak, editors, *Nonstandard Models of Arithmetic and Set Theory*, pages 87–120. American Mathematical Society, Providence, RI, 2004.

[10] Karel Hrbacek. Stratified analysis? In V. Neves and I. van den Berg, editors, *Nonstandard Methods and Applications in Mathematics*, pages 47–63. Springer, 2007.

[11] Karel Hrbacek. Relative set theory: Internal view. *Journal of Logic and Analysis*, 1(8):1–108, 2009.

[12] Karel Hrbacek and Thomas Jech. *Introduction to Set Theory. Third Edition, Revised and Expanded.* Marcel Dekker, Inc., 1999.

[13] Karel Hrbacek, Olivier Lessmann, and Richard O'Donovan. Analysis with ultrasmall numbers. *American Mathematical Monthly*, 117(9), November 2010.

[14] Vladimir Kanovei and Michael Reeken. *Nonstandard Analysis, Axiomatically.* Springer-Verlag, Berlin, 2004.

[15] H. Jerome Keisler. *Foundations of Infinitesimal Calculus.* University of Wisconsin, 2007.

[16] H. Jerome Keisler. *Elementary Calculus, An Infinitesimal Approach.* University of Wisconsin, 2013.

[17] John Kimber and Richard O'Donovan. Nonstandard analysis at pre-university level, magnitude analysis. In M. Di Nasso N. J. Cutland and D. A. Ross, editors, *Nonstandard Methods and Applications in Mathematics*, Lecture Notes in Logic 25, pages 235–248. Association for Symbolic Logic, 2006.

[18] Edward Nelson. Internal set theory: a new approach to nonstandard analysis. *Bull. Amer. Math. Soc.*, 83:1165–1198, 1977.

[19] Richard O'Donovan. Pre-university analysis. In I. van den Berg V. Neves, editor, *The Strength of Nonstandard Analysis*, pages 395–401. Springer-Verlag, 2007.

[20] Richard O'Donovan. Teaching analysis with ultrasmall numbers. *Mathematics Teaching-Research Journal*, 3(3):1–22, August 2009.

[21] Yves Péraire. Théorie relative des ensembles internes. *Osaka Journal of Mathematics*, 29:267–297, 1992.

[22] Yves Péraire. Formules absolues dans la théorie relative des ensembles internes. *Rivista di matematica pura ed applicata*, 19:27–55, 1996.

[23] Alain Robert. *Analyse non standard.* Presses polytechniques romandes, 1985.

[24] Abraham Robinson. *Non-standard Analysis: Studies in Logic and the Foundations of Mathematics.* North Holland, 1966.

[25] Kenneth A. Ross. *Elementary Analysis: The Theory of Calculus.* Springer-Verlag, Berlin, 1980.

[26] Antonín Sochor. The alternative set theory. In *Set Theory and Hierarchy Theory*, Lecture Notes in Math. 537, pages 259–273. Springer-Verlag, Berlin, 1976.

[27] Keith D. Stroyan. *Calculus, The Language of Change*. Academic Press, 1993.

[28] Nader Vakil. *Real Analysis through Modern Infinitesimals*. Cambridge University Press, 2011.

[29] Petr Vopěnka. *Mathematics in the Alternative Set Theory*. Teubner-Texte zur Mathematik, 1979.

Index